Beyond Sustainability

Second Edition

Also by Tim Delaney and Tim Madigan
and from McFarland

Friendship and Happiness: And the Connection Between the Two (2017)

Sportsmanship: Multidisciplinary Perspectives (2016)

The Sociology of Sports: An Introduction, 2d ed. (2015)

Beyond Sustainability

A Thriving Environment

Second Editon

TIM DELANEY *and*
TIM MADIGAN

McFarland & Company, Inc., Publishers
Jefferson, North Carolina

ISBN (print) 978-1-4766-8236-5
ISBN (ebook) 978-1-4766-4429-5

Library of Congress and British Library
cataloguing data are available

Library of Congress Control Number 2021027801

Printed in the United States of America

McFarland & Company, Inc., Publishers
Box 611, Jefferson, North Carolina 28640
www.mcfarlandpub.com

Table of Contents

The Doomsday Clock reads
100 seconds to midnight

Preface

The authors of this book come from two disciplines—sociology and philosophy—which, while having certain similarities, usually address issues in a very different way. Nonetheless, both sociologists and philosophers have concerns about our environment's ability to not only sustain itself, but also reach a point where it can actually thrive. And this helps to explain why we feel that it is especially important to take an interdisciplinary approach to the study of the environment.

We are children of the "Environmental Seventies" and as such have always been well aware of the attacks on the environment and attempts to help it thrive via social movements and the passage of a great deal of legislation designed to protect it on behalf of humanity. At this time, both Democrats and Republicans agreed that the environment was important enough to protect. But then came the 1980s and all of that changed. The past four decades have been characterized by the growing power and influence of the fossil fuel industry, countered by large groups of people who are still fighting the good fight. The authors have done their share of promoting environmental causes throughout this time, but during the past decade they have become increasingly dedicated to contributing to the conversation of making a positive change through education.

In the spirit of the socio-philosophical approach to the study of the environment, we attended and gave presentations at the Seventh International Conference on Environmental, Cultural, Economic and Social Sustainability, held January 5–7, 2011, at the University of Waikato in Hamilton, New Zealand. Hundreds of dedicated individuals from throughout the world gathered together to share their perspectives and offer possible strategies for furthering the cause of sustainability. It was a rich cultural experience, enhanced by the natural beauty of the surroundings. We took the time to explore much of the North Island and among other things became enamored with the indigenous culture of the Maori (a Polynesian people that arrived in New Zealand around 950 CE) and the spirit of *manaakitanga*—a deep-rooted concept in Maori culture that implies guardianship—over their land (whenua), treasures (taonga), people (tangata), and visitors (manuhiri). Manuhiri are expected to abide by the Maori sentiment of "take nothing but photographs, leave nothing but footprints." Such a motto helps to sustain the natural environment.

This was an amazing experience for us, and we decided to become more involved with the pro-environment movement by attending other such conferences but also by conducting a great deal of research. Our research led us to publishing the first edition of *Beyond Sustainability: A Thriving Environment* in 2014 and now the second edition. We have each also developed environmental courses at our respective college campuses that we teach regularly (as online courses to help save the environment) and have become a part of the sustainability programs at our colleges (see Chapter 10 for a description).

We have always been struck by the idea that a focus on "sustainability" was misguided. Instead, we have proposed that the goal for the environment should be "thrivability." After all, if the environment is already compromised, as it certainly is, why would anyone promote *sustaining* a comprised environment? If one is in debt, sustaining the current situation is not the goal; the hope would be to get out of debt and thrive. Thus, in *Beyond Sustainability* we emphasize that the environment needs to be repaired to the point that it can thrive.

In this, the second edition of *Beyond Sustainability*, we have made a very large number of updates both in terms of more recent data and examples and also with the introduction of new topics and concepts, all the while emphasizing more clearly the need to lessen our dependency on fossil fuels in order to halt the significant and negative impact humans have inflicted upon the environment. In many ways, then, this is like a new book.

In Chapter 1 we examine the differences between sustainability and thrivability and discuss such topics as sociological and philosophical environmentalism, the environment and its many ecosystems, mass extinctions and their causes, and social movements.

In Chapter 2, we examine the concept of "carrying capacity"; our reliance on fossil fuels; climate change with an emphasis on carbon dioxide and global warming, climate change and the ozone, and climate change and the greenhouse effect; the effects of climate change including melting glaciers, ice caps and the thawing permafrost; ocean acidification; storms and severe weather. We also place an emphasis on the need to increase our use of renewable energy sources.

In Chapter 3 we look at the effects of overpopulation on the environment and discuss the "Five Horrorists" concept (an updated version of the "Four Horsemen") that includes a description of "enviromares"—environmental nightmares (various forms of human caused pollution).

Chapter 4 begins with a close examination of the extraction of fossil fuels; the creation and mass abuse of plastics; food waste; harmful agricultural practices; deforestation; marine debris; electronic waste (e-waste); and medical waste. These activities contribute, either directly or indirectly, to the sixth mass extinction.

In Chapter 5, we look at the role of nature (e.g., volcanic eruptions, lightning strikes, wildfires, storms, invasive species, vapors from sulfur springs, and outer-worldly forces) as the potential cause of the sixth mass extinction and examine the skepticism (e.g., the role of politics, Big Business, religion, anti-science rhetoric, the conservative and radical right media, and the finite pool of worry) held by some toward the role of humans in the sixth mass extinction.

Chapter 6 provides a philosophical and ethical look at sustainability and thrivability and addresses such issues as whether or not it is humanity's responsibility to try and protect the environment.

In Chapter 7 we examine ways to help the environment thrive including a discussion on such topics as the meaning of "going green"; measures designed to protect and save natural resources; the development of new technology; and the idea that the environment itself should have the right to thrive.

Chapter 8 examines the meaning of happiness and describe the many different ways of achieving happiness. We conclude with a discussion on the concept of Gross National Happiness (GNH).

Chapter 9 highlights the need for humans to change their behavior if we hope to prolong the sixth mass extinction as long as possible; provides two case study examples that demonstrates that humans can change if properly motivated; explores the concept of the "Change to Green"; and describes the environmental movement toward thrivability.

Chapter 10 discusses ways in which we need to educate people on environmental sustainability and thrivability. The chapter concludes with a checklist of ways people can help the environment thrive in the home, the yard, at school or work, while on vacation, in the car, and green RVing.

Each chapter ends with a popular culture section. The purpose of these is to demonstrate that one of the most effective ways to influence change, as well as to educate people about the ways in which the environment is being compromised, is through a variety of popular culture mediums. Didactic lecturing seldom causes individuals to change, and quite often has the opposite effect of raising resentment, whereas popular works can often have an immediate impact on one's behavior. Thus, popular culture has direct application for changing people's actions.

It is also our hope that this book will help to raise the awareness of others about the dire issues connected to our compromised environment. From a personal standpoint we also hope to pay off some of the carbon footprints we generated by traversing the planet in our pursuit of knowledge and cultural awareness of a multitude of diverse cultures. Ultimately, it is our goal to educate others on the need to change our behaviors immediately if we hope to save humanity from itself.

We wrapped up the revisions of this second edition of *Beyond Sustainability* while the November 3, 2020, U.S. elections for president and several other governmental positions were held. These elections were critical for a number of socio-economic reasons including the future of the environment. The reelection of Donald Trump would have led to a continued assault on the environment while the election of Joe Biden would signal a rebirth in environmentalism. As we demonstrate throughout this text, Trump took innumerable steps to gut not only the eight years of pro-environmental legislation ushered in by former president Barack Obama and vice-president Joe Biden, but he also sided with the fossil fuel industry repeatedly much to the chagrin of those who would prefer such basics as clean air, water, and sustainable land and those who would promote renewable forms of energy and a lessening of our dependency on fossil fuels. Under Trump, the environment would not only fail to thrive, but it would also no longer sustain itself. Conversely, Biden promised, among other things, to create millions of good-paying green jobs (many of them union), invest in a green infrastructure, create a clean energy future, establish a plan to reach his goal of net-zero emissions by 2050, and recommit the United States to international climate agreements (i.e., the Paris Agreement). Biden won the election (in what people referred to as a "landslide" victory), giving hope to all environmentalists across the globe that, perhaps, the environment could be saved.

It should also be noted that this new edition of *Beyond Sustainability* was written during the COVID-19 pandemic and as such there are many references to this disease, primarily in direct relationship to the environment. A few politicians, including Boris Johnson, UK prime minister; Jair Bolsonaro, Brazil's president; and U.S. president Donald Trump were referenced with regard to the coronavirus and all three of these men would minimize its threat and contract the disease. Johnson contracted the disease in early April; Bolsonaro in July; and Trump in October. All three survived. By the start of November 2020, other world leaders would also contract the disease including Juan

Orlando Hernandez (Honduras); Alexander Lukashenko (Belarus); Prince Albert II (Monaco); Jeanine Anez (Bolivia); Luis Abinader (Dominican Republic); and top officials from many other nations including Iran, India, Israel, South Africa, South Sudan, Gambia, and Guinea-Bissau. Globally, more than 110 million people had contracted COVID-19 and more than 2.4 million had died from COVID-19 by the start of November. In the United States, more than 28 million Americans had contracted the disease and nearly a half million deaths were recorded by mid-February 2021. As alarming as these statistics are, they represented data just as the second global wave of COVID-19 was beginning.

CHAPTER 1

Environmental Thrivability, the Ecosystem and Mass Extinctions

When you hear the expression "The sky is falling!" what image comes to mind? Perhaps it is the folktale from "Chicken Little" in which a chicken thinks the sky is falling when an acorn falls on his head, with the corresponding moral of not succumbing to the paranoia and mass hysteria of others? Or maybe you think of the version of "Chicken Little" wherein one chick learns to heed the warning that a danger, in the form of a fox, is imminent. The chick that heeded the warning lived while all the others perished (eaten by the fox).

Cornelia Griggs, a general surgeon affiliated with Massachusetts General Hospital at the time of the early American stages of the COVID-19 global pandemic, warned in mid-March 2020, "The sky is falling. I'm not afraid to say it. A few weeks from now you may call me an alarmist; and I can live with that. Actually, I will keel over with happiness if I'm proven wrong" (Griggs 2020). COVID-19, also known as the novel Coronavirus, was first reported in Wuhan, China, and quickly spread worldwide. As explained by the Cleveland Clinic (2020), "Coronaviruses are a family of viruses that can cause respiratory illness in humans. They get their name, 'corona,' from the many crown-like spikes on the surface of the virus. Severe acute respiratory syndrome (SARS), Middle East respiratory syndrome (MERS) and the common cold are examples of Coronaviruses that cause illness in humans." Initially, a number of people including such politicians as U.S. president Donald Trump, UK prime minister Boris Johnson and Brazilian president Jair Bolsonaro did not heed the warnings of medical experts and were slow to respond. It's as if they cried, "Fake fox." Predictably, these three nations would have among the highest number of infected persons. Other global and local leaders (i.e., Jacinda Ardern of New Zealand; Lee Hsien Loong, Singapore; and Moon Jae-in, South Korea) reacted more quickly, fearing that the proverbial fox was already in the henhouse. (See Chapter 3 for a further discussion on COVID-19 specifically and pandemics in general.)

The "sky is falling" expression has been applied to all sorts of social activities, including concern about the environment. For many years pro-environment extremist advocates have proclaimed the idea that we are all doomed unless we drastically change our behaviors. And, while some inroads have been achieved by pro-environmentalists in their attempt to better protect the environment, their voices have, historically, often gone unheard. Why is this? Because they are perceived as people who have been hit on the head with an acorn and have now developed a heightened sense of paranoia? For people who doubt that the proverbial sky is falling, environmentalists and their proclamation that humans are harming the environment are merely the product of ecohype. *Ecohype* refers to the constant proclamations among certain environmentalists that the environment is doomed (the sky *is* falling) unless humans drastically change their behaviors.

As the evidence of a compromised environment becomes too overwhelming and convincing for most people to doubt, however, a pro-environment movement has grown. This social movement is generally expressed, in some variation, as environmental sustainability. For years now, all sorts of people have used the terms "environment" and "sustainability" together. Proclaiming an interest in environmental sustainability has become almost ubiquitous. Most college campuses, places of business, and governmental agencies express a growing concern about taking sustainability to heart, and an increasingly large number of task forces are devoted to various projects related to this.

Unfortunately, in some cases, promoting environmental sustainability or environmentally friendly products has merely served as a gimmick by corporations hoping to sell products promoted as "green" or politicians desperately trying to project some sort of cultural meaningfulness by expressing "green" concerns and yet holding back their votes on actual green initiatives. A number of colleges and universities, perhaps to appease environmentally concerned students and faculty members, have implemented policies that, essentially, incorporate the same desired ends of corporations and politicians. There also exists, however, academic persons who are very concerned about the environment for the same reason that many individuals and truly pro-environmental groups care about the environment; that is, to save the environment so that both humans and all other life forms may thrive.

It is necessary to explain the two terms "environment" and "sustainability." The *environment* refers to the totality of social and physical conditions that affect nature (land, water, air, plants, and animals) and humanity and their influence on the growth, development, and survival of organisms found in a given surrounding (e.g., a limited proximity or the earth as a whole). Miller and Spoolman (2018) concur with our perspective and define the environment as everything that surrounds us including "all the living things (such as plants and animals) and the non-living things (such as air, water, and sunlight) with which you interact" (p. 5). *Sustainability* refers to the ability of the environment to hold, endure, or bear the weight of a wide variety of social and natural forces that may compromise its functionality. Miller and Spoolman (2018) define sustainability as "the capacity of the earth's natural systems that support life and human social systems to survive or adapt to changing environmental conditions indefinitely" (p. 4).

As individuals with a deep concern about the environment, the two authors of this text used the concept of "sustainability" to express our perspectives on how to address such matters. After conducting research and teaching courses on the environment, however, each of us came to realize just how compromised the earth's environment really is. As a result, *sustaining* the environment seems too insignificant of a goal. And that is why we are now promoting the ideal of "thrivability." *Thrivability* refers to a cycle of actions which reinvest energy for future use and stretch resources further; it transcends sustainability by creating an upward spiral of greater possibilities and increasing energy. Each cycle would build a world of ever-increasing possibilities. Thrivability emerges from the persistent intention to create more value than one consumes. Thrivability incorporates all that is implied with sustainability but expands beyond its limitations. By no means will we ignore the relevant aspects of sustainability; instead, we try to emphasize the importance of thrivability as an ultimate goal of environmentalism.

Although the authors come from different academic backgrounds (sociology and philosophy, respectively) they nonetheless share a single voice of concern about the environment. But each of their respective fields has its own unique perspective on environmentalism.

Environmental Sociology

There exists in the field of sociology a concept known as the "sociological imagination." The term coined by C. Wright Mills highlights the importance of the social environment's influence on human behavior (Delaney 2012). As Mills explained, "The sociological imagination enables its possessor to understand the larger historical scene in terms of its meanings for the inner life and the external career of a variety of individuals" (1959: 5). The sociological imagination concept is a staple within the field and has been applied to explain numerous scenarios. Many decades after Mills first used this term, Lawrence Buell created the concept of the "environmental imagination." According to Buell (1995), any text that purports to provide a sociological perspective on the environment must address the four key aspects of the environmental imagination:

1. The nonhuman environment is present not merely as a framing device but as a presence that begins to suggest that human history is implicated in natural history.
2. The human interest is understood not to be the only legitimate interest.
3. Human accountability to the environment is part of the text's ethical orientation.
4. Some sense of the environment as a process rather than as a constant or a given is at least implicit in the text.

Buell and other sociologists that utilize an environmental imagination will be pleased with the contents of this text. In addition to addressing a sociological imagination, it addresses the four major concerns.

Sociologists look at the environment as the totality of social and physical conditions that affects nature (land, water, air, plants, and animals), and humanity and its influence on the growth, development, and survival of organisms. "Sociologists examine how people waste or conserve natural resources. They track public opinions on issues such as global warming and identify what categories of people support one side or the other of environmental issues. But the most important contribution sociologists make is in demonstrating how our society's technology, cultural patterns, and specific political and economic arrangements affect the natural environment" (Macionis 2010: 420). From this perspective, sociologists examine the environment from both the micro level, e.g., individual and small group behaviors, and the macro, e.g., governmental policies, business practices, and human ecology.

Environmental sociology in the United States has existed in one form or another since the mid–1930s with the formation of the Rural Sociological Society (RSS). The RSS was officially established on December 29, 1937, and had, among its goals a commitment to understanding the human impact on the physical environment (Rural Sociological Society 2010).

Today, environmental sociology examines natural and social forces that have an impact on the environment. The sociological perspective examines the many adverse effects that nature has on the environment, including volcanic eruptions, lightning strikes, wildfires, storms and invasive species. The sociological perspective is most concerned with the role of humans on the environment. Among the issues concerning environmental sociology are environmental social movements, overpopulation, human dependency on fossil fuels, the use of plastics, medical waste, e-waste, agricultural mismanagement and hydraulic fracturing.

As sociology is the authority on the study of human behavior in the context of human interaction within local and global communities and environments, Bell and Ashwood (2016) define environmental sociology as "the study of community in the largest possible sense" (p. 2). Bell and Ashwood (2016) describe how people, other animals, land, water and air are all closely interconnected and that together, they form a kind of solidarity, what we have come to call ecology.

Environmental Philosophy

Philosophers have traditionally attempted to address what the great thinker Immanuel Kant (1724–1804) called the Four Great Questions (or as Germans would say, "Die Kantischen Fragen"): "What can I know?" (epistemology); "What should I do?" (ethics); "What may I hope for?" (metaphysics); and "What is Man?" (anthropology). One can add two other major questions: "What is beauty" (aesthetics), and "How should I think?" (logic). These great questions were often asked about humans in isolation, without reference to the environment(s) in which they lived, or to the other living beings with which they interacted. Kant himself, for instance, felt that ethics only pertained to the actions of human beings, and had no particular relevance to other animals.

A sea-change in thinking has occurred in philosophy since at least the early 1960s, in part due to the impact of books like Rachel Carson's 1962 classic work *Silent Spring*. Environmental philosophers now ask questions. What can I know about the environment in which humanity exists? What should I do to both sustain and contribute to this environment? What may I hope for regarding the future of this environment? What is "humanity" in relationship to the ecosystem? What role does nature play in developing our sense of beauty? What are the most logical ways of addressing the role of humans in the environment?

This new emphasis on humans-in the-world rather than on humans-apart-from the-world has revitalized the entire field of philosophy. In particular, the realization that the existence of the human species itself is fragile and interconnected with forces previously little-understood has given a new urgency to all of the above questions. In the words of the philosopher Anthony Weston, "Environmental philosophy emerged to insist upon the next step: recognizing the richness and depth of the *more*-than-human world: the way in which physical embodiment, for example, does not end at the skin … and the ways in which we are and must always be profoundly *animal*—a distinctive animal, perhaps, but so are *all* animals—as well as kin even to stone and stars. Very large issues come up, from the cultural history of our alienation from nature to the resources for recovering a sense of connection *now*" (Weston 1999: 2–3).

Perhaps what makes the human animal distinctive is that it alone, as far as we know, even asks such questions, and it alone feels a sense of responsibility for preserving and improving upon the ecosystem to the best of its abilities. The International Association for Environmental Philosophy states that it "embraces a broad understanding of environmental philosophy, including not only environmental ethics, but also environmental aesthetics, ontology, and theology, philosophy of science, ecofeminism, and philosophy of technology" (IAEP 2020).

Environmental philosophy, therefore, attempts to unify the various fields of philosophy and provide a holistic approach to understanding the role of humans in the world.

Ecosystems and Ecology

Protecting the earth's wide variety of ecosystems is the primary goal of environmentalists and environmental social movements. But what exactly is an ecosystem? An ecosystem refers to the ecological network of interconnected and interdependent living organisms (plants, animals and microbes) in union with the nonliving aspects found in their immediate community, including air, water, minerals and soil. Although it could be argued that the entire planet is an ecosystem, the term is generally applied to limited spaces. In this regard, the concepts of "ecosystem" and "environment" are similar but not interchangeable. Charles Harper (2012) explains that an ecosystem refers to specific parts of the greater environment: "An ecosystem means the community of things that live and interact in parts of the geophysical environment.... [They] are composed of structural units that form a progressively more inclusive hierarchy" (p. 3). Miller and Spoolman (2018) define an ecosystem as "a set of organisms within a defined area of land or volume of water that interact with one another and their environment of nonliving matter and energy" (p. 5).

The term "ecosystem" was coined by the British botanist Arthur Tansley in his 1935 article "The Use and Abuse of Vegetational Concepts and Terms" (1935). He stressed that one should not focus solely upon organisms, but rather upon the interactions between organisms and their environments. He writes: "The more fundamental conception is, at is seems to me, the whole *system* (in the sense of physics), including not only the organisms-complex, but also the whole complex of physical factors forming what we call the environment of the biome—the habitat factors in the widest sense. It is the systems so formed which, from the point of view of the ecologist, are the basic units of nature on the face of the earth" (1935: 286). It was Tansley's goal to make ecology a legitimate area of scientific study, akin to biology, chemistry or physics, so as to differentiate it from more metaphysical concepts that looked upon organisms as purpose-driven or essentially unconnected with their environments. As Joseph R. Des Jardins points out, "The ecosystem concept preserves the key ecological idea that ecological wholes are a fundamental part of nature. The individualism that characterizes much of zoology and botany cannot tell the entire story of the biological sciences. Ecology contributes by observing and explaining the integration, connections, and dependencies within and among ecological wholes" (1997: 160).

The discussion of ecosystems goes hand-in-hand with the concept of ecology, as *ecology* is a branch of science that concerns itself with the interrelationship of organisms and their immediate environment (ecosystems). Perhaps the most influential of the early ecologists is Aldo Leopold. According to the Aldo Leopold Foundation (2020), Leopold is considered by many to be the father of wildlife ecology and the United States' wilderness system. He is most famous for his 1949 publication *A Sand County Almanac and Sketches Here and There*, and the oft-quoted line, "A thing is right when it tends to preserve the integrity, stability, and human beauty of the biotic community. It is wrong when it tends otherwise."

Ever since the works of Tansley and Leopold first appeared, the role of ecologists has been to examine the structures and functions of ecosystems, in particular the various ways in which life forms interact and rely upon each other. It is this network of relationships which we have since learned are so fragile and yet so resilient. Clearly, our understanding of ecosystems is still being mapped out and developed, which makes it all the more important to be cautious about policies which seem to degrade the environment or cause massive disruptions to its functionings.

The network that comprises any ecosystem is linked through a system of exchanges of energy flows, chemicals and nutrients that are interconnected and bind the various components together. The ecosystem is fueled by energy drawn from the sun that operates through a process known as photosynthesis. Photosynthesis assists the cycles and flows of carbon, nitrogen, oxygen, phosphorus, and water. The photosynthesis process allows energy and carbon to enter the ecosystem. Plants and animals play an important role in facilitating the movement of matter and energy through the system. For example, dead organic matter decomposes and releases carbon back into the atmosphere, which assists the nutrient cycle by converting the nutrients stored in the dead biomass back into a form of energy that can help fuel other plants and microbes. Most mineral nutrients are recycled within ecosystems.

Every ecosystem housed on our planet is potentially affected by many external and internal variables that either help to sustain life or compromise it. Among the most critical variables to affect an ecosystem is climate. Our planet possesses a very diverse biosphere, one that is capable of creating a wide range of climatic conditions that affects living organisms. "The biosphere is the biological component of the earth systems, which also include the lithosphere, hydrosphere, atmosphere and other 'spheres' (e.g., cryosphere, anthrosphere, etc.)" (Ellis 2009). The biosphere is the part of the earth and its atmosphere in which living organisms exist, or areas that are capable of supporting life. The biosphere is a relatively thin 12-mile zone of life that extends from the depths of the oceans and to the tops of the tallest mountains (Miller 2011). Living organisms, the dead organic matter produced by them, and ecosystems compose the biosphere's totality.

The diversity of the biosphere is the result of climate, and climate, in turn, determines the biome of an ecosystem. The biome refers to a large community of plants and animals that occupies specific areas. Climate dictates whether some regions will result in grasslands, tundras, deserts, rainforests and so on. There are a number of subcategories such as freshwater and marine biomes.

Among the other variables that may affect an ecosystem are internal periodic disturbances. Changes in climatic conditions—such as an unusual dry spell in marshland areas or an unusual amount of rainfall in an arid region or forces of nature such as an earthquake or hurricane—may negatively influence an environment. There are numerous instances where invasive species and human interference (to be discussed in subsequent chapters) have had devastating impacts on localized ecosystems.

The Evolving Ecosystem

Scientists have long been aware of the fact that biological species evolve through a process of natural selection and rare genetic mutations (Harper 2012). Ecosystems also evolve in the same manner and have done so since long before humans arrived on the planet. The biosphere itself has evolved from a single-celled organism that originated 3.5 billion years ago under atmospheric conditions that resemble those of neighboring planets Mars and Venus (Ellis 2009). The development of multi-cell organisms from single-cell organisms further led to the creation of more complex organisms, populations, species, biomes and, eventually, the biosphere.

Our current exploration of Mars is spirited by this evolutional realization of how life is formed, and it is hoped that as we learn more about what happened on Mars, we will learn more about the direction in which our own planet is headed. Among the things we know for sure about Mars is that it once had plenty of water, more than enough to sustain life. We also know that Mars (and Venus) have atmospheres composed primarily of carbon dioxide. On Earth, plants have released enough oxygen to create the oxygen-rich atmosphere we know and enjoy today. There is plenty of oxygen to (O_2) to breathe and sufficient oxygen in the stratospheric ozone (O_3) to protect us from the harmful effects of UV radiation, all of which has made life possible on Earth.

The influence of natural selection on the evolving ecosystem is reflected in the reality that competition among species exists in order to secure critical needs. These needs are expressed in terms of available energy (nutrients and food) found in the environment (Harper 2012). Each species needs energy to survive; hence, the battle of the species. Those species which successfully secure their energy needs are selected, by nature, to survive. In some instances, genetic mutations take hold within individual members of a species. If these mutations lead to further development of the species, the species gets stronger. If the mutation was harmful to the individual and was passed on to the population of a species, then that species risks dissolution. The survival of one species generally comes at the cost of one, or several other, species. In some cases, the cost may mean extinction. This aspect of the evolving ecosystem is referred to as the process of "ecological succession"—species that replace one another in gradual changes (Harper 2012).

There is an alternative to "winner" and "loser" categories in the battle for survival of the species. This is coevolution, or the reciprocal natural selection that forms relationships between different species, called symbiosis. Symbiosis can be mutually beneficial (mutualism) or (as parasitism) only beneficial to one species but not mutually beneficial, as when fungi or microorganisms infect humans and other species (Harper 2012: 6).

Humans often act as parasites and take from the environment what they please and, in many cases, more than what they need. Humans create monocultures wherein one species dominates and grows while others disappear. For example, as humans have moved to suburbs, they infringe upon wildlife areas. Where species such as foxes and deer once roamed free in grassy ecosystems designed to help sustain their lives, humans have moved in and altered the environment. Humans have cleared woods and brush and drained marshlands and replaced them with lawns maintained by chemical fertilizers, all the while chasing wild animals away from their habitants.

One has to wonder, will we reach the point where human intrusion alters the biosphere so dramatically that we risk the very viability of the planet? Such a concern is just one of many that will be addressed throughout this book. In the short term, we will conclude our discussion of the ecosystem by reviewing the "Seven Pillars of Ecosystem Management" put forth by Robert T. Lackey.

Seven Pillars of Ecosystem Management

If there is a need to save ecosystems, how do we begin such a task? In 1998, Lackey, then an associate director for science at the Environmental Protection Agency's Western Ecology Division, National Health and Environment Effects Research Laboratory, put forth the idea of seven core principles, or pillars, of ecosystem management. Although ecosystem management itself is subject to a number of interpretations, Lackey uses Overbay's (1992) definition: "The careful and skillful use of ecological, economic, social, and managerial principles in managing ecosystems to produce, restore, or sustain ecosystem integrity and desired conditions, uses, products, values, and services over the long term." Lackey states that there are two fundamentally different world views of ecosystems. The first is the *biocentric*, which considers maintenance of ecological health or integrity as the goal. All other aspects, including human needs (tangible or intangible), are of secondary consideration. The second view is *anthropocentric*, an approach that considers the needs of humans within any attempt to manage the environment. The seven-pillar approach reflects the anthropocentric approach.

A brief review of Lackey's (1998) seven pillars of ecosystem management approach is provided below:

1. *Ecosystem management reflects a stage in the continuing evolution of social values and priorities; it is neither a beginning nor an end.* This pillar acknowledges the evolving ecosystem model and as a result is mindful that changes in the ecosystem may be a part of an evolutionary process, and therefore any manipulations to the ecosystem have to be weighed against the potential of disturbing bigger issues and concerns related to the biosphere.

2. *Ecosystem management is place-based and the boundaries of the place of concern must be clearly and formally defined.* With pillar one in mind, attempts to protect the health of a particular ecosystem need to be weighed against the overall health of the totality of the environment. Assuring the health of ecosystems is directly tied to concerns of protecting the health of humans.

3. *Ecosystem management should maintain ecosystems in the appropriate condition to achieve desired social benefits; the desired social benefits are defined by society, not scientists.* If one is truly taking an anthropocentric approach to managing the environment, the needs of the public need to supersede those of scientists. Such a pillar may lead to debates between scientists who claim to be working in the best interests of people even when people do not want certain infringements. For example, if scientists provide evidence that hydro-fracking is dangerous to an ecosystem but people in a particular community want hydro-fracking because it will generate jobs, who should make the decision whether hydro-fracking can take place? The third pillar would state that the community should. The biocentric and ecological approach to ecosystem management would be suspect of this pillar.

4. *Ecosystem management can take advantage of the ability of ecosystems to respond to a variety of stressors, natural and man-made, but there is a limit in the ability of all ecosystems to accommodate stressors and maintain a desired state.* Ecosystems benefit from biological diversity. The anthropocentric approach, because of its concerns for

humans first, puts forth the idea that human societies need to weigh out how diverse species co-exist in given ecosystems. The biocentric and ecological approach would be very suspect of this pillar too.

5. *Ecosystem management may or may not result in emphasis on biological diversity as a desired social benefit. This fifth pillar follows the logic of the fourth pillar, and it is essentially claiming that human needs are more important than other biological species' needs.* Lackey claims that the "sustainability" movement has roots in this pillar's idealism. That is, the environment needs to be saved only to assure the sustainability of the human species; if that costs other species their extinction, so be it. Lackey defends this approach by referring back to the idea of the first pillar; that is, ecosystems are in a constant state of flux. There is no "natural" condition of any ecosystem because it is in flux. Seeking ways to create a balance with nature becomes passé because there is no homeostatic state for nature.

6. *The term sustainability, if used at all in ecosystem management, should be clearly defined—specifically, the time frame of concern, the benefits and costs of concern, and the relative priority of the benefits and costs.* The fifth pillar, which downplayed the importance of sustainability in the first place, is here being put to a deeper question. Of particular emphasis is the need for the creation of a measuring device to determine whether the goals of sustainability supporters are met through sustainability efforts.

7. *Scientific information is important for effective ecosystem management, but it is only one element in a decision-making process that is fundamentally one of public or private choice.* The anti-science bias of the ecosystem management approach is emphasized once again.

It seems a little odd that a member of the Environmental Protection Agency (EPA) (Lackey) would downplay the role of science and the need to protect people *and* the diversity of ecosystems and biological species. These seven pillars did not necessarily reflect policy positions of the EPA or any other organization. But the discussion of the seven pillars approach provides us with a steppingstone to our next topic at hand, as the seven pillars help to illustrate the complexity involved with any attempt to protect ecosystems, let alone create an approach wherein the environment may thrive. There are people who want to do their part to save the ecosystem and they heed the warnings of the world's scientists, including climatologists, about the many threats to our planet—the only one (that we are aware of) that can sustain human life. Debates over protecting ecosystems are further complicated by those who refuse to believe that there is any need to intervene with the current course of the environment. Such skeptics will tell us, "The sky is not falling!" and will suggest that environmentalists need to "Calm down!"

Mass Extinctions: The Sky Has "Fallen" in the Past

Ultimately, the discussion on environmental sustainability and thrivability comes down to the following: are we doomed with an inevitable ecocatastrophe(s), or are the problems~~, if any,~~ related to the environment greatly exaggerated and over-hyped (eco-hype)? Let's return to the story of Chicken Little. If one screams, "The sky is falling!" and the sky really is falling, that is not paranoia and such an alarm of pending doom should be heeded. Proclamations akin to the sky is falling by "alarmist" environmentalists could be accurate and perhaps, at the very least, everyone should take the time to consider the consequences of ignoring dire warnings about the compromised environment. In other words, what if the proverbial sky actually falls? What happens to our planet's ecosystems? What happens to humanity? Could humans become as extinct as dinosaurs and countless other species? If you don't think that is possible, consider the fact that the planet has faced mass extinctions in the past. What's to stop it from happening again? And, if you really want to look at the "big picture," realize that during the earth's past 3.5 billion years or so, about 99 percent of the estimated 4 billion species that ever evolved are no longer around (Tarlach 2018).

If an ecosystem is compromised beyond the point of reconciliation any number of species may perish. As alarming as it is to acknowledge the demise of a species, a mass extinction is far more dire. A mass extinction has many characteristics, including the destruction of a significant proportion of the world's biota; substantial biodiversity loss; and the overall destruction of the earth's environment. Barnosky and associates (2011) state that a *mass extinction* (ME) occurs when the planet loses more than three-quarters of its species in a geologically short interval of time, usually during a few hundred thousand to a couple of million years. However, a critical event such as meteorite impact may trigger a mass extinction in a much shorter period of time. According to a group of biologists and paleontologists headed by Anthony Barnosky at the University of California at Berkeley, the planet Earth has already endured five mass extinctions in the past 540 million years. In order of occurrence, these five mass extinctions are known as the "Big Five" and include the end–Ordovician, Late Devonian, end–Permian, end–Triassic, and end–Cretaceous (Wake and Vredenburg 2008; Andryszewski 2009).

The first mass extinction event took place at the end of the Ordovician period (beginning nearly 440 million years ago) when, according to fossil records, nearly two-thirds of all genera of both terrestrial and marine life worldwide were killed off (Cook 2010; Endangered Species International 2011). About 365 million years ago in the Late Devonian period, the environment boasted natural reefs that had existed for at least 13 million years but as the world plunged into the second mass extinction nearly, 85 percent of marine species disappeared. During the end–Permian mass extinction (occurring about 250 million years ago), the planet lost approximately 80 to 95 percent of all marine species. Coral reefs would not reappear for about 10 million years, the greatest hiatus in reef building in all of Earth's history (Cook 2010). The end–Triassic mass extinction (approximately 200 million years ago) is estimated to have claimed about half of all marine invertebrates and approximately 80 percent of all land quadrupeds. The end–Cretaceous mass extinction (65 million years ago) is the most famous because it resulted in the demise of the dinosaurs. Nearly all land animals perished during this mass extinction.

The "Big Five" mass extinctions are named after the geological time period during which they occurred. During these events an average of 80 percent of species were killed off (Hallam and Wignall 1997). Although the math does not calculate precisely, Michael Benton (2003a) states that mass extinctions occur about every 65 million years (with the extinction time period lasting millions of years). (We calculate an ME occurring approximately every 65–70 million years since the date of the first ME.) The past five mass extinctions are the result of forces of nature (worldly or celestial), and the inevitability of mass extinctions results in a life expectancy of any particular species of about five million years.

If you are wondering when the sixth ME will occur, Barnosky and associates (and a slew of other scientists) suggest we are already in the sixth mass extinction period. As evidence, consider that during the time period of 1970–2014 humanity has wiped out 60 percent of mammals, birds and reptiles, leading the world's foremost experts to warn that annihilation of wildlife threatens civilization (*The Guardian* 2018). These foremost experts come from around the world and are a part of the World Wildlife Fund (WWF) and they have found that "the vast and growing consumption of food and resources by the global population is destroying the web of life, billions of years in the making, upon which human society ultimately depends for clean air, water and everything else" (*The Guardian* 2018). The WWF cites deforestation, chemical pollution, the introduction of invasive species and disease as a result of global trade as leading causes of species cleansing. "The worst affected region is South and Central America, which has seen an 89% drop in vertebrate populations" (*The Guardian* 2018). Species extinction did not occur as quickly before humans arrived and mammal extinction was very rare, on average, just two mammal species died out every one million years (*The Independent* 2011). Before human existence, there were any number of species subject to extinction, as this is a part of the natural cycle of events on earth. Human existence, however, has sped the process of the sixth mass extinction and consequently represents a threat to biodiversity, ecosystems, and the environment as a whole. Stuart Pimm, a conservation biologist at Duke University, published a paper in the journal *Science* in 1995 which found that species are currently going extinct at rates 100 to 1,000 times faster than the normal pace of evolution would dictate; fifteen years later, Pimm argues the pace is even higher (Weise 2011). More recently, Pimm and ecologist Mark Urban state that for every species disappearing for natural causes, 1,000 are vanishing because of unnatural man-made causes (*Associated Press* 2015).

To further highlight the point that we are currently in the sixth ME consider the "Doomsday Clock" established by the *Bulletin of Atomic Scientists*, a nonprofit organization. The Doomsday Clock was started in 1947 due to the threat of nuclear warfare. The clock was originally set to seven minutes to midnight to highlight the urgency and dangers the world faced because of advanced weaponry and the threat of nuclear warfare. According to former Irish president Mary Robinson, the Doomsday Clock is now recognized as an indicator of the vulnerability of our existence (Welna 2020). As of late 2020, the official time of the Doomsday Clock was 100 seconds to midnight indicating the amount of time to our impending extinction.

The rapid disappearance of so many species and the contributions of Nobel laureates and leading scientists that determine the time for the Doomsday Clock are among the signs that we are in the sixth mass extinction. Another warning comes with the previously mentioned point that mass extinctions occur about every 65–70 million years coupled with the reality that it has been about that long since the planet's last mass extinction. What distinguishes this mass extinction from the previous five is the realization that this will be the first one to involve humans.

Mass Extinctions and Their Causes

The Earth's environment is constantly under "attack" by human-made forces and nature itself. Long before humans rose to the top of the food chain, nature had begun to, and continues to, bombard the environment in a variety of fashions including glaciation, volcanic eruptions, global warming and cooling, lightning strikes and natural wildfires.

The planet has even been attacked by outer-worldly forces such as asteroids and meteorites. (An asteroid is a minor planet, and a meteorite is any object that survives entering the earth's atmosphere and lands on the surface.)

Scientists have used fossil analysis to determine that there have been five mass extinctions in the past (see Table 1.1). Evidence would indicate that the sixth mass extinction is upon us. The first five mass extinctions were caused by forces of nature, worldly or outer-worldly, while the current one has been accelerated by human behavior. Let's take a quick look at the leading causes of mass extinctions and connect them to the mass extinctions.

Table 1.1: Past Mass Extinctions (ME) and Their Causes

ME #	*Name/When It Began*	*Cause(s)*
1st	End–Ordovian/440 million years ago	Glaciation
2nd	Late–Devonian/365 million years ago	Glaciation and Global Cooling
3rd	End–Permian/251 million years ago	Asteroid and Volcanism
4th	End–Triassic/200 million years ago	Asteroid, Volcanism and Climate Change
5th	End–Cretaceous/65 million years ago	Asteroid and Volcanism

1. Glaciation. Near the end of the end-Ordovician period, approximately 450 million years ago, the second largest and first of the five major extinction events took place. This ME is attributed to a brief glacial interval that produced two pulses of extinction (Sheehan 2001). "The first pulse was at the beginning of the glaciation, when sea-level decline drained epicontinental seaways, produced a harsh climate in low and mid–latitudes, and initiated active, deep-oceanic currents that aerated the deep oceans and brought nutrients and possibly toxic material up from the oceanic depths. Following this initial pulse of extinction, surviving faunas adapted to the new ecological setting" (Sheehan 2011). When this glaciation event suddenly ended, sea levels rose, global cooling occurred, ocean circulation stagnated, and another pulse of extinction followed. This second related glaciation event marked a long stasis of several million years (Sheehan 2001).

The glaciation event of the end–Ordovician era led to the extinction of an estimated 60 percent of marine invertebrates, ranking below only the Permian extinction. All known life was confined to the seas and oceans at this time. The cause of this glaciation event appears to have been the movement of Gondwana, the name of the ancient "supercontinent" that incorporates the present-day Africa, Australia, South America, Arabia, India and Madagascar, into the South Polar Region. Evidence for glaciation was discovered through fossil deposits found in the Sahara Desert.

Although scientists are not as sure about the cause(s) of the Late Devonian mass extinction, the second ME is believed to be the result of the two prime factors that led to the first ME: glaciation and global cooling. We do know that the second ME took place around 364 million years ago. The evidence that does exist indicates that warm water marine species were the most severely affected in this extinction event. This leads paleontologists to believe that an episode of global cooling occurred as a result of another glaciation event on Gondwana, which is evidenced by glacial deposits of this age in northern Brazil (Endangered Species International 2011).

2. Asteroid Impact. As a speck of dust hurtling through space, the planet Earth is statistically bound to get hit by something or run into something. Yes, we have a fairly stable orbit, but orbits are subject to change and to celestial impact violations. Scientists are quite sure that the fifth mass extinction was caused, at least in part, by an asteroid

and reasonably sure that the third mass extinction was also caused by an asteroid. As for the end of the Permian period (the third ME), life on Earth was nearly completely wiped out by an environmental catastrophe of a magnitude never seen before or since (Benton 2003b). In February 2001, a team of scientists led by Luanne Becker of the University of Washington claimed that they had found clear evidence (extraterrestrial helium and argon in rocks from the Permo-Triassic boundary in China and Japan) that the mass extinction at the end of the Permian era was caused by meteorite impact (Becker et al. 2001). The end–Permian extinction, which took place about 251 million years ago, was the Earth's worst mass extinction with 95 percent of all species, 53 percent of marine families, 84 percent of marine genera, and an estimated 70 percent of land species such as plants, insects and vertebrate animals were killed during this catastrophe (Endangered Species International 2011).

Researchers have been quite confident that an asteroid impact caused the fifth ME (Benton 2003b; Gibbons 2011). The huge asteroid that hit Earth 65 million years ago incinerated all the dinosaurs and nearly all other life forms. Research conducted at the University of Colorado at Boulder estimates that the asteroid hit the coast of what is now Mexico. Within a few hours of impact only those organisms already sheltered in burrows or in water were left alive. The six-mile-in-diameter asteroid is thought to have hit Chicxulub, in the Yucatan, striking with the energy of 100 million megatons of TNT. The kinetic energy of the explosion would have turned the normally blue sky red-hot for hours, baking the earth below. According to researchers Douglas S. Robertson and associates, "For several hours following the Chicxulub impact, the entire Earth was bathed with intense infrared radiation from ballistically reentering ejecta. The global heat pulse would have killed unsheltered organisms directly and ignited fires at places where adequate fuel was available. Sheltering underground, within natural cavities, or in water would have been a necessary but not always sufficient condition for survival" (2004: 760). Recent research suggests that the asteroid strike may have been a contributing force to the die-off and that subsequent planetwide volcanism is what actually killed the dinosaurs (Tarlach 2018; Rosen 2019a). Other research suggests that the asteroid strike caused a temporary but devastating "impact winter"—darkening the sky, cooling the Earth and inhibiting photosynthesis (Netburn 2014).

Writing for the journal *Nature*, Roff Smith (2011) proposes that an asteroid strike on earth at the end-Triassic period could have wiped out, or at least had an impact in wiping out, much of that period's flora and fauna, resulting in the opportunity for dinosaurs to spread around the globe during the subsequent Jurassic period. If the earth was to be hit with another asteroid the size of the one that led to the kill-off of the dinosaurs, humanity would be helpless to stop it. The likelihood of that happening again is certainly a possibility.

3. Massive Volcanism. Volcanic eruptions can cause a great deal of harm to the environment and local ecosystems. Lava flow will kill animals and plant life upon impact, while sulphur dioxide emissions harm the atmosphere (Wignall 2004). When the sulphur dioxide gas reacts with water vapor, it forms clouds of sulphuric acid that block out the sun's light, causing changes in the climate while rain from such clouds results in a phenomenon known as acid rainfall (Wignall 2004). Although the evidence seems to support the idea that an asteroid was responsible for the third and fifth ME, there are those who argue that massive volcanic eruptions, sustained over half a million years or more, caused catastrophic environmental deterioration via poisonous gases, which in

turn caused global warming, stripping of soils and plants from the landscape, and mass deoxygenation, and great loss of marine biodiversity (Benton 2003b; Stanley Jr., Shepherd and Robinson 2018). Suffice it to say the end of the Permian was characterized by huge volcanic eruptions.

Massive volcanism may also contribute to global warming and cooling, another contributing cause to mass extinctions. It is believed that the fourth ME, the end–Triassic extinction, which took place roughly 251 million to 199 million years ago, was most likely caused by massive floods as a result of lava erupting from the central Atlantic magnetic province triggering the breakup of Pangaea and the opening of the Atlantic Ocean. Volcanism may have led to deadly global warming (Endangered Species International 2011).

4. Global Warming and Cooling (Climate Change). As we have learned so far, the causes of specific mass extinctions have been tied to natural events such as volcanism, glaciation and asteroid impact; these events, in turn, contributed to either, or both, global warming and global cooling. An asteroid strike is capable of creating huge dust storms that lead to devastating darkness, and cold and warming temperatures can cause mass bleaching of corals (Cook 2010). A consistent pattern of changing atmospheric CO_2 has been linked with every mass extinction. Since oceans do not respond instantly to a CO_2 build-up, and since it can take decades to centuries to develop significant problems, humans may fail to see, or heed, the warning signs of global warming and global cooling.

The main evidence that we have of global warming and cooling comes from fossils (Wignall 2004). For example, scientists argue that the causes of the end–Triassic ME were a combination of climate change and rising sea levels that resulted because of the sudden release of large amounts of carbon dioxide. The release of carbon dioxide from widespread volcanic activity contributes to the global greenhouse effect, which raises average air temperatures around the globe. Other theories suggest that rising carbon dioxide concentrations in the atmosphere could have freed massive amounts of methane trapped in permafrost and under sea ice. Methane is a gas that traps heat in the atmosphere and contributes to global warming. Methane, then, could contribute to global temperature change as much as CO_2 (*Encyclopædia Britannica* 2012).

Citizens of earth have been warned about both global cooling and global warming for the past four decades, at least, and yet many people live in denial of the inevitable conclusion. That is to say, both global warming and global cooling can take place in a same era and ultimately both natural events will contribute to the sixth mass extinction. Unlike past eras, however, humans are contributing to the accelerated pace of global warming and cooling.

5. Human Impact. Tens of millions of years after the fifth ME, the humanoid species began to evolve. For multiple millennia now, humans have, arguably, been the most evolved creatures on the planet. There are those who now worry that humans are slowly, or rapidly, depending upon one's perspective, destroying the Earth. Humans are certainly capable of destroying the earth quickly via nuclear war, but it is a slow death that we are likely inflicting upon our home planet.

Depending on one's definition of what constitutes a "human," archeologists estimate that our species has been on this planet for about 200,000 years. Humans are a member of the bipedal primate species in the Hominidae (the great apes) family, having diverged from apes 6 million to 8 million years ago and a dozen humanlike species removed (PBS 2001; Pickrell 2006). Current day humans are Homo sapiens, a Latin term for "thinking human." Unlike other animal species, humans have developed a high-level brain, a bipedal gait, and opposable thumbs. All evidence to date indicates that the birth of modern humanity can be traced to Africa. Approximately 70,000 years ago, humans migrated out of Africa and began to colonize every corner of the planet. Evidence indicates that humans reached Eurasia and Oceania 40,000 years ago and reached the Americas less than 15,000 years ago (Leakey 1996).

In the grand scheme of things, human existence hardly registers a hash mark on the historic timeline of our planet. As intellectually advanced as humans are, we have done a poor job as stewards of the environment. Our overall blatant disregard for the health of the environment highlights human arrogance. We have treated the planet as a dumping ground for our human-made toxins and trash. We are responsible for the destruction of countless species (animal, plant and marine), and our willingness to pollute the planet with abandon has contributed to the destruction of numerous local ecosystems. Humanity, it seems, has placed its hopes on a God or nature to self-correct the problems it causes. In all but rare occurrences, however, nature does not have an "auto-reset" button so long as humans continue to contaminate the environment.

An atypical example of nature correcting human neglect has taken place on Glass Beach. Located near Fort Bragg in Northern California, Glass Beach was a spot where residents brought all forms of trash, including old cars, kitchen appliances, glass bottles, and so on, to dump at a beach. People, it seems, have always been willing to pollute water (as we shall see in greater detail in later chapters), especially before the days of environmental consciousness spearheaded by social movements. According to the City of Fort Bragg (2012), people first started dumping trash at the beach following the 1906 San Francisco Earthquake when the city's streets were filled with rubble. Using the ocean as a dump made perfect sense to officials in those days because seaside towns around the world had done so for centuries (City of Fort Bragg 2012). Only in the early 1960s did officials at the North Coast Water Quality Board finally begin to regulate what was dumped on the beach, first by forbidding toxins and then everything else (Keneally 2012; City of Fort Bragg 2012). In 1967 the board moved the official dump to a different location.

Today, Glass Beach, which is a part of Mendocino Area State Parks (Glass Beach is in the MacKerricher State Park, which once was a part of the Mendocino Indian Reservation), looks nothing like a toxic dumping ground. After years of thrashing waves pounding away at the debris left behind, a very colorful and almost magical-looking beach has emerged. The beach is now covered with stone-sized pieces of sea glass, coloring the seascape and adding a tourist element to the natural beauty of the spot (Keneally 2012). The colorful sea glass is so valued by local officials and residents that it is a misdemeanor offense to remove any artifacts from the beach (Fort Bragg 2012; Keneally 2012). The beach is about a three-hour drive north of San Francisco and is a regular tourist destination spot.

The Russian equivalent to Glass Beach is Ussuri Bay, once a Soviet era dumping ground for old glass bottles and porcelain. However, like Glass Beach, nature has transformed this bay (within the Peter the Great Gulf of the Sea of Japan) near Vladivostok to one of the most stunning beaches imaginable. Over a period of many years, the crashing waves have washed the broken beer, wine and vodka bottles into millions of smooth and colorful "pebbles" that have turned the area into a tourist attraction that visitors pay to see (Bored Panda 2016; Roxan 2020). After an even longer period of time, these colorful rock pebbles will transform into sand.

Although the Glass Beach and Ussuri Bay stories represent two bright examples of the ability of nature to overcome human errors, there are at least three takeaways. First, the ability of nature to "clean-up" human environmental neglect and harm is very limited and takes a great deal of time. Add to this the constant new forms of devastation caused by humans against the environment and it becomes clear that nature could never keep pace. Second, while the beaches may be pretty to look at, it exists as the result of humans dumping trash and toxins into an ecosystem not equipped to handle such a poisonous invasion. Third, numerous plant and marine species were exposed to toxic substances that may lead to, or have already caused, their extinction.

We not only pollute our oceans and beaches, but we also pollute our rivers and lakes. In fact, humans have been polluting the earth in a wide variety of ways throughout the millennia. Although most commentaries point to industrialization as the beginning era of drastic negative human involvement with the environment, the ancient Romans and Chinese empires were also producers of greenhouse gases. Sapart and associates (2012) found that the methods of burning of plant matter to cook food, clear cropland and process metals utilized by the ancient Romans and Chinese released millions of tons of methane gas into the atmosphere each year for several periods of pre-industrial history. Methane, as we have already learned, is a gas that traps heat in the atmosphere and is a contributor to global warming. The quantity of methane produced then pales in comparison with the emission released today (the amount is roughly 70 times higher today). The findings suggest that man's footprint on the climate is larger than had been realized (Morin 2012). Sapart and his fellow scientists reported their findings in *Nature* and based their conclusions on analysis of ice core samples from Greenland that dates back 2,000 years. They found three distinct periods predating industrialization wherein human activity caused significant environmental harm because of methane production. The first period—roughly 1 CE to 300 CE—encompassed the end of the Roman Empire and the Han Dynasty, when charcoal was a preferred form of fuel. The second period of elevated methane emissions occurred during the Medieval Climate Anomaly (800 to 1200 CE) and the third era was during the Little Ice Age, between 1300 and 1600 (Morin 2012; Sapart et al. 2012).

The research conducted by Sapart and associates indicates that human behavior has already contributed to increased levels of methane release significant enough to affect climate change. The results also indicate that negative past methane-producing behavior pales in comparison to today's production. And yet, only recently, in the last 50 years or so, has there been any concerted interest or effort in doing something about humanity's negative impact on the environment. Concerted interest is spearheaded by collective action and collective action is accelerated via social movements. The first environmental social movement, arguably, was the result of Rachel Carson's landmark warning about the indiscriminate use of pesticides detailed in her controversial 1962 book *Silent Spring*. It could be argued that the environmental social movement began in 1962 as a result of this publication.

In *Silent Spring*, Carson warned the public about the indiscriminate use of pesticides and connected DDT (dichlorodiphenyltrichloroethane, a colorless, tasteless, and mostly odorless chemical compound) spraying to the decimation of bird populations. This book launched the modern environmental movement as well as a ferocious backlash that continues to this day (Souder 2012). Carson's book caused controversy even before it was published as a book, as *The New Yorker* published three long excerpts prior to its book release. An alarmed public deluged the Department of Agriculture and other agencies with demands for action, and the outraged chemical industry with its allies in government fought back, claiming Carson's warnings were overblown.

This pattern of social activists and environmental social movements being countered by big business and in many cases big government has continued right up to the current day. Today, chemical corporations are much larger than the ones Carson attempted to bring down and they are equipped with lobbyists and public relations machines that are well-funded and sophisticated and exert unprecedented influence over regulators, the media, and most critically, elected officials. For example, the chemical industry benefits from the massive amounts of chemicals that are used in hydraulic fracturing, or "hydrofracking."

Carson's critics claimed she was a left-wing nut and tried to lump her with other "radicals" such as organic farmers, food faddists and anti-fluoridationists (Souder 2012). Conversely, with the creation of the Environmental Protection Agency in 1972, the deadly insecticide DDT was banned in the United States. That did not, however, stop the U.S. manufacturing and exportation of DDT to nations across the globe until the mid–1980s. As a deadly neurotoxin, DDT destroyed many species and has been connected to cancer in humans. The residual of DDT remains in our environment. Interestingly, some of the criticism directed at Carson originated from health officials who pointed out that DDT spraying was an effective way to curtail the spread of malaria, especially in developing nations. In 2006, the World Health Organization (WHO) announced a renewed commitment to fighting malaria with DDT, primarily in Africa (Sounder 2012). Rachel Carson died in 1964, long before such data became available. How would she likely have responded to this? What do you think?

Social Movements

The primary focus of this book is to analyze the human impact on the environment and ultimately our contribution to the sixth mass extinction. We will also examine whether or not it is the responsibility of humanity to curtail its harmful activities in an attempt to slow the extinction process. It should be pointed out that there are plenty of people who have formed, or contribute to, social movements that are designed to raise awareness of the negative human impact on the environment. Let's take a quick look at the stages of social movement.

Stages of Social Movements

Carson's *Silent Spring* helped to spur the environmental social movement and the many subsequent variations (e.g., "Going Green"). A *social movement* may be defined as "an organized effort at claims making that tries to shape the way people think about an issue in order to encourage or discourage social change" (Macionis 2010: 6). As described by sociologist John J. Macionis, social movements play a key part in the response to numerous social problems, including the AIDS epidemic, sexual harassment, and family violence. He could have very easily have mentioned environmental issues as well. Macionis states that social movements involve four stages.

1. Emergence—The emergence of a social movement occurs when people (initially just a few) come together and share their concern about the status quo or some specific

social problem. For example, when Carson described the negative effects of unregulated insecticides on the food we eat and on the ecosystem in which we reside, a few people voiced their shared concern and eventually mobilized others into social action. Another example would include the initial concern of a few people that using grocery store paper bags contributed to the depletion of trees, causing them to spearhead the promotion of plastic bags instead of paper. In turn, it was just a matter of time before a few, and then many, people led a social movement to ban single-use plastic bags in grocery stores.

2. Coalescence—The coalescence (to come together, to unite, and form a whole) of a social movement occurs when a new organization begins holding rallies and demonstrations, making public its belief, and engaging in political lobbying. A major step in the U.S. environmental movement occurred in 1970 as people across the nation joined in celebrating the first Earth Day. In 2011, the "Occupy Movement" that started as a protest against Wall Street finance companies, big banks and the 1 Percenters (the very rich) gained coalescence as Occupy protesters set up on college campuses and local cities across the nation and internationally. The social movement to ban plastic bags gained momentum because a large number of diverse interest groups (e.g., environmental groups, save the oceans activists and the million women march campaign) came together to put pressure on governments and businesses.

3. Formalization—Social movements begin with a few highly-motivated individuals who have expressed their displeasure about a certain policy or behavior of others, especially the government, big business, or traditional standards of morality that seem outdated to members of a younger generation. In an attempt to accomplish specific goals (and the establishment of specific goals is critical in the success of any social movement), social movements must become formalized as they attempt to become established players on the political scene. The environmental movement was believed to have become formalized via the establishment of the Environmental Protection Agency (EPA) in 1970. However, as we know today, the EPA is drastically underfunded and viewed by many of today's environmentalists as being inept in its ability to carry out the movement's objectives. The Occupy Movement failed to establish any specific and attainable goals and continues to exist primarily as an example of how not to maintain a social movement. Although, to be fair, the sentiment of the Occupy Movement remains strong among those who want to keep a check on the rich and powerful. For example, "both Elizabeth Warren and Bernie Sanders were convinced to run for president [during the 2020 presidential election] in large part by Occupy activities. And you can hear it. They both rail against big banks. They both talk about the billionaire class and inequality. But Sanders and Warren never really say the word occupy" (Sanders 2020). In early March (2020) Warren ended her presidential campaign and in early April (2020) Sanders did the same. Sanders proclaimed, however, "while this campaign is coming to an end, our movement is not" (Goodman and Moynihan 2020: A4). Indeed, Sanders assured his followers that he would press Democratic presidential candidate Joe Biden to embrace ideals of the progressive movement. Formal organizations such as the Sunrise Movement (a youth-led political movement) vowed to continue Sanders' "visionary solutions" to many contemporary social problems. As another example of formalization, the ban the plastic bag movement was effective (at least in terms of banning certain usages of plastic bags) because it was spearheaded by formalized organizations that were already established (i.e., 1 Million Women Organization; Save the Oceans Foundation; The Ocean Foundation; The Surfrider Foundation; and the Plastic Pollution Coalition).

4. Decline—As argued by Macionis (2010), social movements tend to decline over time, and there are a number of reasons for this. Social movements may decline because they run out of money, because their claims and concerns failed to catch on with the public, or because opposing organizations are more convincing than they are. Social movements may also decline because they actually met their goals or because their goals have been modified. In the case of the latter, the social movement that led to the banning of plastic bags at stores has been expanded to additional goals including banning plastics wherever feasible (e.g., plastic straws and water bottles) and banning polystyrene foam (i.e., Styrofoam). The need to ban plastic straws is evidenced by the fact that in the U.S. alone, more than one-half of plastic straws are thrown away every day; that equals 15,183 miles when you lay them end-to-end. In the U.K, more than 8.5 billion plastic straws are discarded in a year (WBUR 2018).

Throughout this text we will examine a number of environmental social problems that have once been, or currently are, connected to social movements. We will discover that a great number of significant laws designed to help the environment were passed in the 1970s and we will look at current environmental-related concerns among activists and the masses. It will become clear that certain movements, such as recycling have become routine for most people while other issues, such as the social movements to ban hydraulic fracking and combat climate change, remains in the early stages, somewhere between stage 2 (coalescence) and stage 3 (formalization).

Environmental social movements are designed to protect the planet and whether explicitly stated or not to lessen human contributions to the sixth mass extinction. We are already in the sixth mass extinction era, but such eras can last for thousands or millions of years. So how far along are we? Are we merely at the beginning of the sixth ME era, meaning humans and other species have hundreds of thousands of years remaining before they become extinct? Or are we at the point where if we don't take drastic steps to modify our behaviors now, we have already gone past the breaking point? Scientists look for evidence to answer such questions. Open-minded scientists, and there shouldn't be any other kind, will look for evidence, conduct analyses and report the results regardless of their findings. In this, the second edition of *Beyond Sustainability*, the authors add the results of the research they have conducted in the seven years since the first edition.

Popular Culture Section 1

The Planet Earth and a Potential Asteroid Apocalypse

There are no shortages of natural disaster films depicting mass destruction of the planet Earth. Disaster-themed films include such sources of obliteration as planetary geological catastrophes (e.g., avalanches, earthquakes and cave-ins); weather-related storms (e.g., massive lightning strikes, volcanic eruptions, blizzards, wildfires, floods and tsunamis); and outer-worldly sources (e.g., attacks from aliens; and meteorite, asteroid, and comet impacts).

As humans, we prepare the best we can for impending attacks from nature. Buildings and bridges are increasingly designed to absorb the impact of earthquakes; advanced radar tracks the progress of storms so that we can take shelter; warning systems alert us to tornados; fire-fighting crews combat wildfires; and so on. Although we are ultimately at the mercy of nature, we like to think we can handle, or at least minimize, the negative impact of nearly any potential natural catastrophe.

Preparing for meteorite, asteroid and comet strikes, however, represents an entirely different challenge. As a matter of distinction, a meteorite is a meteoroid that has entered into the Earth's atmosphere. Thus, once a meteoroid breaches the atmosphere, they are called meteorites. A major meteorite (say the size of 5 to 10 miles in width) impact could destroy the planet, but even smaller ones can be highly destructive. Asteroids are small, airless rocky worlds (too small to be called planets) that revolve around the sun and are primarily found in orbits lying mostly between Mars and Jupiter (the asteroid belt). It is estimated that there are more than 750,000 asteroids larger than three-fifths of a mile (1 kilometer) in diameter and millions of smaller ones (Space.com 2013). Asteroids are leftovers from the formation of our solar system billions of years ago that are subject to breaking away from their orbit, opening the possibility of their eventual pull into the Earth's atmosphere. According to NASA (2012), comets are cosmic snowballs of frozen gases, rock and dust roughly the size of small towns. Like asteroids and meteoroids, comets are capable of breaking free from their orbits and striking Earth. In fact, since the Earth formed some 4.5 billion years ago, meteorites, asteroids and comets have routinely slammed into our planet. In three instances, the size of these outer-worldly bodies of rock and metals has been large enough to contribute to three of the previous five mass extinctions.

Among the more memorable examples of films depicting the deadly consequences of asteroid and comet strikes on Earth are *Deep Impact* (1998) and *Armageddon* (1998). *Deep Impact*, the more scientifically accurate of the two films, starring Robert Duvall, Tea Leoni, Elijah Wood, Vanessa Redgrave, and Morgan Freeman, describes humanity's preparation for a 7-mile-wide comet on a collision course with Earth to cause a mass extinction. In this film, reporter Jenny Lerner (Leoni) discovers that a government acronym E.L.E. stands for "Extinction-Level Event." Because of her investigation, President Tom Beck (Freeman) is forced to prematurely announce the grim reality that the comet, named "Wolf-Biederman," large enough to cause a mass extinction is on course to impact the planet. American and Russian scientists have built a spacecraft, called *Messiah*, that will carry a team led by Captain Spurgeon Tanner (Duvall) to the comet so that it can be destroyed with nuclear weapons. The crew land on the comet and plant nuclear bombs 100 meters beneath the surface.

Instead of destroying the comet, its splits in two pieces, named "Wolf" and "Biederman" respectively. The wealthy governments of the world have built underground shelters that will house scientists and key officials along with a lottery to allow 800,000 "everyday" citizens a chance to survive. The "Biederman" fragment crashes into the Atlantic Ocean, creating a mega-tsunami that kills millions of people due to the resulting massive flooding. In a last-ditch attempt to stop the "Wolf" fragment from hitting the earth and causing greater damage, the crew of the *Messiah*, still equipped with its remaining nuclear warheads, decides to go on a suicide mission by intercepting and colliding with it in hopes of breaking it up into much smaller pieces that will burn up in the Earth's atmosphere.

While *Deep Impact* did very well at the box office, generating more than $350 million worldwide, *Armageddon* fared better, earning over $550 million worldwide. *Armageddon*, however, was highly criticized by the scientific community for its many scientific flaws. *Armageddon* stars Bruce Willis, Ben Affleck, Billy Bob Thornton, Liv Tyler, Owen

Wilson, Steve Buscemi and slew of other stars. In *Armageddon*, a massive meteor shower destroys the Space Shuttle *Atlantis* and shells New York City and other areas on the Eastern Seaboard of the United States with meteorites. The cause of the meteor shower is a Texas-sized comet that has escaped the gravitational pull of the asteroid belt. In the film, NASA sends a crew led by Harry Stamper (Willis) to land on the asteroid and drill a hole deep enough to plant nuclear devices in hopes that the asteroid can be broken into small enough pieces to burn out in the Earth's atmosphere. After a great deal of drama, the crew is ultimately successful, as the planted nuclear device breaks the asteroid into two pieces that fly past Earth.

In both *Deep Impact* and *Armageddon*, crews are sent to land on a moving asteroid, and charged with the task of setting nuclear devices deep within the surface in hopes of blowing it up into small enough pieces as to not cause harm to our planet. There is ample reason to fear the impact of an asteroid or meteorite here on Earth, as the consequences of Yucatan-size meteorite (like the one that destroyed the dinosaurs 65 million years ago) would cause massive tidal waves, forest fires, acid rain, and high surface temperatures. Smaller meteors that hit the Earth's surface are capable of causing significant climate change, as such impacts create explosions similar to that of atomic bomb explosions, resulting in fire balls and shockwaves that can cause windows to break from buildings (*Deadliest Space Weather* 2013).

There are real life reminders of the potential harm meteors and asteroids can cause. For example, on February 15, 2013, a meteor streaked across the sky of Russia's Ural Mountains and caused a booming shockwave that exploded with the force of 20 atomic bombs, injuring more than 1,100 people as it blasted out windows and spread panic in nearby Chelyabinsk, a city of more than one million people (*The Citizen* 2013). As we shall learn in Chapter 5, this event led to the creation of the annual Asteroid Day, designed in part to warn people of the dangers of asteroid impacts.

Around the same time this meteor hit Russia, the planet was spared the impact of a 150-foot asteroid, dubbed Asteroid 2012 DA14. The asteroid came within 17,150 miles of Earth. To put this in perspective, there are communications and weather satellites orbiting the earth 22,300 miles from the surface. According to NASA officials, Asteroid 2012 DA 14 is very small compared to the one that hit the Yucatan and contributed to the demise of the dinosaurs, but if it had hit the planet the impact would have been equal to the energy equivalent of 2.4 million tons of TNT and would have wiped out 750 square miles (*The Post-Standard* 2013). By comparison, the meteor that hit Russia was about 49 feet wide. Scientists estimate that there are over 10,000 asteroids near enough to Earth to cause us harm. NASA is preparing for what may be an inevitable asteroid or meteorite hit on our planet through its "Near Earth Object" program, which scans the heavens for dangerous objects and the orbiting telescope called "WISE" that specifically hunts for asteroids (*Associated Press* 2013).

Because the devastation that meteors, asteroids and comets can cause is so apocalyptic, scientists are preparing a defense. And the solution is somewhat similar to that of a Hollywood production. One proposed solution involves designing and constructing a type of "Death Star" station that can fire lasers at asteroids as they near our planet. Two California scientists propose developing a station equipped with lasers that could vaporize asteroids from as far away as 93 million miles. They call their device "Directed Energy Solar Targeting of Asteroids and exploRation," or DE-STAR for short (Khan 2013). In popular discourse the term "Death Star" is being used to describe the laser-based defense mechanism.

Since such defenses still do not exist, we are left to hope that a deep-impact or Armageddon scenario as the result of asteroids, meteors or comets does not occur prior to the completion of Hollywood-type solutions. Even then, we have to hope that our technology is sufficient to protect us from outer-worldly impacts. Interestingly, there are any number of major, deep-pocketed corporations also interested in asteroids, not so much because of the possible harm they may cause the planet, but because they see potential opportunities to land on asteroids and mine them for valuable precious minerals. We encourage the governments of the world to create an early warning system and hope that they, and the general public, come to the realization that species annihilation as the result of asteroid impact is a real possibility in the future.

The next predicted close encounters of the asteroid kind are expected to occur in 2029 and 2036. Will we be ready?

Summary

The underlying concern with any discussion on the environment is the degree to which it has been compromised. That is to say, has the environment been so sullied that it is beyond repair or are alarmists simply overreacting when they warn, "The sky is falling?"

In this opening chapter a number of key terms are discussed including, sustainability, thrivability, ecosystems and environment. Sustainability refers to the ability of the environment to hold, endure, or bear the weight of a wide variety of social and natural forces that may compromise its functionality. Many environmentalists promote sustainability, but the authors promote thrivability which refers to a cycle of actions which reinvest energy for future use and stretch resources further; it transcends sustainability by creating an upward spiral of greater possibilities. The environment is the totality of social and physical conditions that affect nature (land, water, air, plants, and animals) and humanity and their influence on the growth, development, and survival of organisms found in a given surrounding. An ecosystem can be thought of a subset of the environment as it is generally applied to limited spaces that contain an ecological network of interconnected and interdependent living organisms (plants, animals and microbes) working in union with the nonliving aspects found in their immediate community, including air, water, minerals and soil.

Thrivability involves incorporating a vision of the world that extends beyond the limitations of a sustainability attitude of the environment by promoting heightened levels of collaboration among individuals, groups, corporations and government. Thus, thrivability requires a major shift both in social consciousness and social action; that is, if we hope to avoid an ecocatastrophe such as those involved in earlier mass extinctions.

The connectedness between ecosystems and ecology is evident as ecology is a branch of science that concerns itself with the interrelationship of organisms and their immediate environment (ecosystems). The ecosystem is constantly changing and may evolve through a process of natural selection and rare genetic mutations. Seven pillars of ecosystem management were discussed within the context of biocentric and anthropocentric approaches.

Environmentalists warn us that the earth's environment risks potentially dire consequences so severe that humanity itself is threatened. There is reason to heed such warnings, as scientists have documented five past mass extinctions. Past mass extinctions occurred because of forces of nature or celestial attacks on our planet. Among the major causes of past extinctions are glaciers, asteroid impacts, massive volcanism, and global warming and cooling (climate change). Scientists have gathered enough data to conclude that mass extinctions are inevitable. We are already a part of the sixth mass extinction era and this one represents the first to involve humans. Although this sixth mass extinction is inevitable, humans are speeding the process of demise in a wide variety of ways.

In an attempt to slow the process of our own demise, many environmental activists have spearheaded social movements in the hope of altering human behavior. The first environmental social movement was spearheaded by Rachel Carson following the publication of her book *Silent Spring* that explored how the indiscriminate use of pesticides had decimated bird populations as a result of DDT spraying of crops. More recent environmental social movements have concentrated on banning or restricting the use of plastics and polystyrene. Environmental social movements are designed to protect the planet and, whether explicitly stated or not, to lessen the human contribution to the sixth mass extinction. How effective we can be in such endeavors will be explored in detail in the following chapters.

Chapter 2

Climate Change and Human Dependency on Fossil Fuels

The focus of this chapter will be on the Earth's limited carrying capacity, an examination of the consequences of human dependency on fossil fuels, and some of the many harmful effects of global warming. Ideally, as "intelligent" creatures, humans will be smart enough to heed the warning signs of impending danger that exist nearly everywhere.

Compromising the Earth's "Carrying Capacity"

Of all of the Earth's organisms, humans place the highest demands on the environment. Humans also represent the greatest threat. The earth has a limited capacity, or "carrying capacity," to support life. *Carrying capacity* refers to the maximum feasible load, just short of the level that would end the environment's ability to support life (Catton 1980). In other words, carrying capacity is tied to the number of organisms that can be supported in a given area (ecosystem) based on the natural resources available without compromising present and future generations. Once the environment is sullied, the carrying capacity shrinks, thus negatively altering its ability to sustain life.

What is the *actual* carrying capacity you may wonder? Scientists seem to agree that when we speak of humans specifically, that number would be between 9 and 10 billion people. The highly esteemed Harvard professor Edward O. Wilson, author of two Pulitzer Prize-winning books, is one such scientist that puts forth the notion that the earth's maximum feasible load is connected to the number of people (as well as other organisms) and in his 2002 book *The Future of Life*, he uses this 9- to 10-million mark (Wilson 2002; Delaney and Malakhova 2020). At the start of 2020, the world population had already exceeded 7.77 billion. The human population is projected to reach the 9 billion mark by 2037 and reach the 10 billion mark in 2057 (Worldometers 2020). The Global Footprint Network (2020a), an international nonprofit organization, states that humans use as much ecological resources as if we lived on 1.75 Earths. The GFN calculates an annual "Earth Overshoot Day," the date that marks when humanity has exhausted nature's budget for the year. This date is typically around August 1 of each year. When we go over budget the debt compounds. "It is an ecological debt, and the interest we are paying on that mounting debt—food shortages, soil erosion, and the build-up CO_2 in our atmosphere—comes with devastating human and monetary costs" (Global Footprint Network 2020b).

Over the generations, but especially recently, the Earth's carrying capacity has been stretched to its limit due to a number of threats to the environment. These threats include, but are not limited to, urban sprawl, the spread of deserts, the destruction of forests by acid rain, deforestation, the stripping of large tracts of land for fuel, radiation fallout, and the many areas where the population is exceeding the carrying capacity of local agriculture. It has been theorized that overpopulation in particular is a leading threat to the thrivability of the earth's environment. Consequently, it has become increasingly important to find a way to sustain the environment.

A number of animal species have been studied with respect to a specific area's carrying capacity. As explained by the Sustainability Scale Project (2011), the carrying capacity formula has two basic scenarios. First, starting from a low population level there are two quite different patterns which describe how various species reach carrying capacity, the sigmoid and peak phenomena. "Populations which exhibit the sigmoid pattern increase rapidly while food and habitat are abundant, and then slow down as regulatory factors such as lower birth rate and reduced food availability come into play. As the rate of population growth slows down to zero, the population reaches a fairly stable level. This pattern is referred to as 'K' (for constant) selected species" (The Sustainability Scale Project 2011). People who hunt deer and other game understand this principle well. There are plenty of deer in any given season when the food supply is abundant, and restrictions are placed on the number of deer each hunter can kill. Conversely, when the population is small to begin with, there is evidence of a lack of sufficient food supply or regulatory restrictions on hunting.

The second scenario of the reaching carry capacity formula is similar in the early stages when populations are still small. "But, in this situation, the same regulatory factors do not come into play and the population increases rapidly to the point where it exhausts the resources upon which it depends. At this point, mortality becomes the primary regulatory factor, and the population collapses to a low level. When resources are replenished the population begins to rise again; this process is repeated in a boom and bust cycle" (The Sustainability Scale Project 2011). This second situation is referred to as the "r-selected" scenario. The r-selected scenario is applicable to species, such as humans, that conserve only when faced with their own mortality. Lax regulatory guidelines allow situations to grow to the r-selected scenario because proper conservation and sustainability standards were not in place or not enforced. In short, the species is dooming itself.

The concept of carrying capacity has been applied to human populations since the 1960s. It has been shown that consumption habits of humans are much more variable than most other animal species, making it more difficult to determine or predict the carrying capacity of earth for human beings. "This realization gave rise to 'The IPAT Equation' which pointed out that capacity for humans was a function not only of population size but also of differing levels of consumption, which in turn are affected by the technologies involved in production and consumption" (The Sustainability Scale Project 2011). The IPAT formula is expressed as I (Environmental Impact) = P (Population) × A (Affluence) × T (Technology). The carrying capacity for any specific area is not necessarily fixed, as food production capacity, for example, can be improved through technology. However, once the environment is sullied, the carrying capacity shrinks, thus negatively altering its ability to sustain life. Biologist Peter C. Schulze (2002) puts forth that "the IPAT equation is particularly useful as a starting point for disentangling the determinants of per capita impact, either of a group of people, such as a nation, or of an individual, such as one's self" (p. 149). To allow for individual calculation of one's environmental impact, Schulze (2002) and his colleagues introduce a new variable of Behavior, expressed as "B" in their modified formula of I = PBAT. Barrett and Odum (2000) demonstrate the link between behavior (B) and affluence (A) by pointing out how wealthy people may consume 100 or more times as much energy and resources as a poor citizen of a developing country. In this reality, affluence reduces the number of people that can be supported by a given resource base. By implication, as affluence across the globe increases the total number of people necessary to stretch the carrying capacity decreases. The affluence variable also helps to explain why some people are climate change deniers as any acknowledgment of human contribution to climate change would lead to policies being enacted to curtail non-essential activities which, in turn, would lead to increase costs on a number of human behaviors (e.g., an increase on gasoline taxes).

Using the traditional IPAT formula, we can see how the more people (P) in a given society, the richer the society, and the more complex the technology (T), the bigger the environmental impact (I) the society has (Macionis 2010). Let's break it down. The idea that more people equals a greater environmental impact is straightforward. In a simple analogy, a household with ten people has far more needs than a household with three people (e.g., food, energy consumption, trash accumulation, and clothing). At the time of industrialization, there were fewer than 1 billion people in the world. Human impact on the environment was not damaging. The technology that accompanied industrialization brought with it higher standards of living, reduced death rates and increased fertility rates. Once again, more people equals greater needs and a higher environmental impact. Furthermore, people living in a communal environment in the woods have less of an environmental impact than those who live in urban areas that are dependent on the social services of others (the delivery of food to local stores, auto and train emissions, trash pick-up, heating and cooling energy consumption, and so on).

Affluence (A) places greater demands on the environment, as people with extra income and wealth will purchase and consume goods and services beyond their survival needs. Affluence fuels the production of consumer goods, such as automobiles, second homes, well-manicured lawns and the creation of golf courses (sometimes in desert areas where water is already a commodity). Most of these goods, arguably, are not necessities and create a throwaway mentality. This is especially true with electronic devices like computers or cell phones and disposable products like lighters and diapers, which, once they are deemed outdated or used up, are thrown away. As we shall see later, electronic waste, or E-waste, causes a great deal of harm to the environment.

The advent of industrialization spearheaded affluence among hundreds of millions of people across the globe. Prior to industrialization, societies were generally characterized by people who relied on hunting and gathering, farming, and guilds and trades to survive. These relatively simple forms of technology not only kept affluence to a minimum, but they also had little impact on the environment. Early humans especially adapted their lives to the "rhythms of nature moving along with the migration of animals and the changing seasons and in response to natural events such as fires, floods, and droughts" (Macionis 2010: 422). The Industrial Revolution changed everything, including human environmental impact as muscle power was replaced by combustion engines that burn fossil fuels such as coal and oil. With this technology, humans learned to manage their existence against that of nature's way. As the years progressed, humans tunneled through mountains, dammed rivers, irrigated deserts, and drilled for oil beneath the ocean floor. Advancements in technology have caused many other environmental problems. As our medical knowledge increases, life expectancies also increase. Many diseases that once thinned our population have been kept at bay and people are living longer, albeit at the cost of a great deal of energy.

High intensity food production has increased in the prosperous areas of the world. And while food provides energy and nutrients, its acquisition requires the expenditure of a great deal of energy. "In post-hunter-gatherer societies, with progressively increasing inputs of extra-somatic energy, the scale of catching, gathering, and producing food has been greatly expanded and methods intensified. Today, relations between energy, food, and health have become complex and multifaceted, raising serious policy concerns at national and international levels" (McMichael, Powles, Butler and Uauy 2007: 1253). The world's agricultural sector, especially livestock production, accounts for about 20 percent of total greenhouse-gas emissions, thus contributing to climate change. There have already been calls for people to eat less meat because of the costs of livestock to the environment.

Technology has drastically improved our ability to mass-produce food. Thanks to refrigeration transport and open markets from international locations, people who live in affluent societies have access to foods once restricted by growing season and geographic location. For example, people in the Northeast and Midwest regions of the United States can still shop for fresh oranges, apples and a wide assortment of other fruits and vegetables at their local grocery store even in the dead of winter. In many areas of the world, modern agricultural practices and the expansion of food production has compromised topsoil and the biodiversity of earth and cover. The reliance on fertilizer for the production of a great amount of food increases the concentration of bioactive nitrogen compounds in the global environment (McMichael et al. 2007). Add to this the previously stated reality that livestock production accounts for a significant amount of the total greenhouse gases emitted into the environment and we begin to see the problem associated with current food production and consumption practices among humans. Assuming there is a forty percent increase in population by the year 2050 and no advances in reducing greenhouse gases caused by livestock, meat consumption per person would need to drop significantly just to sustain our current rate of compromising the environment. If we wish an environment that thrives, we will need to drastically reduce our meat consumption and food production practices.

Poverty causes great environmental strain, but people who are preoccupied with basic survival have little choice but to consume the resources they do have. (Note: According to a 2019 report from the World Bank, there were 736 million people living in extreme poverty worldwide. The World Bank also estimates that an additional 40 million to 60 million people will fall into extreme poverty in 2020 as a result of COVID-19.) For example, while the rest of the world condemns the depleting of the Amazon Rainforest, the impoverished local citizenry has worked with big industry for money enough to feed their populations. Meanwhile, beyond the depletion of thousands of acres of rainforest, numerous species that reside there have become extinct. It should be emphasized, however, that high income nations consume far more energy than those in poor countries. "The typical U.S. resident uses the same amount of energy in a year as 200 people in Ethiopia or the Central African Republic" (Macionis 2010: 423). Thus, the greater the affluence of a nation, the higher its negative impact on the environment.

In affluent nations people generally pay little attention to the negative impact their mass consumption of energy has on the environment. Furthermore, people seem to be oblivious to the source of the energy they consume. Most people seem to ponder the source of energy only when there is a disruption of service in energy flow (e.g., power outages due to storms, rolling blackouts that occur when people place too high of a demand on "the grid," and a "dead zone" where cell phones cannot pick up a signal). And although throughout most of history humans lived without benefit of electricity and most of our other modern-day forms of energy, the vast majority of people today are ill-prepared to survive in a world without readily available energy to fuel our many electronic needs. This reality led the authors of this text to ponder—what happens to humanity when the power grid goes down?

Human evolutionary growth in intellect has allowed us to develop technology that has, in turn, led to our increased consumption of scarce resources. This evolutionary process has caused more environmental damage to the planet in the past two centuries than in all of human history prior to industrialization (Macionis 2010). If humans are truly intelligent creatures, we will need to learn to drastically modify our behaviors if there is any hope of delaying the 6th mass extinction.

In short, over the generations, but especially recently, humans have contributed to the deterioration of the earth's carrying capacity. Our growing population has placed a great strain on our land's topsoil that is used to grow crops and raise cattle, sheep, chickens and other animals for consumption. The Earth's carrying capacity has been stretched to its limit due to a number of threats to the environment as well. We will discuss many of these threats, including our dependency on the conversion of fossil fuels to meet our energy needs (in the form of gasoline, heating oil, and diesel, for example), overpopulation, and the Five Horrorists (see Chapter 3). Our dependency on the burning of fossil fuels has led to a number of serious problems that are being felt around the world, including global warming, the greenhouse effect and a compromised ozone.

As we have mentioned a few times that the Earth's carrying capacity is being stretched to its limit it is important to point out that you can only stretch something so far until it snaps and breaks. Stretch a rubber band beyond its capacity and it will snap back and smack you in the face. Stretch the Earth's carrying capacity too far and many life forms, including humans, may be smacked into extinction.

Human Dependency on Fossil Fuels

For more than a century, humans have become increasingly dependent on fossil fuels, especially for transportation and heating. The major forms of fossil fuels—coal, crude oil and natural gas—were formed many hundreds of millions of years ago, long before the time of the dinosaurs—thus the name. Fossil fuels were formed in the Carboniferous Period (approximately 360 to 286 million years ago and part of the Paleozoic Era), a period named from carbon, the basic element in coal, oil and natural gas. Some deposits of coal can be found dating back to the time of Tyrannosaurus Rex (65 million years ago), but the main deposits of fossil fuels are from the Carboniferous Period (California Energy Commission 2012). Fossil fuels come about as a result of fossilized plants that rotted away over an extended period of time and became solids, liquids and gases.

Measuring human dependency on fossil fuels varies somewhat depending upon the source cited. For example, the World Bank (2019b) estimates that nearly 80 percent of the world's total energy consumption comes from fossil fuels; this is down from the record high 95 percent in 1970 (Delaney and Malakhova 2020). The World Bank estimates that

92 percent of Russia's and 82 percent of the United States' energy sources come from fossil fuels. The U.S. Energy Information Administration (EIA) estimates that 80 percent of energy used comes from fossil fuels (United States Energy Information Administration 2019a). Five nations (Oman, Qatar, Kuwait, Saudi Arabia and Brunei Darussalam) are completely dependent of fossil fuels and five more nations (Trinidad and Tobago, Bahrain, United Arab Emirates, Algeria, and Islamic Republic of Iran) are more than 99 percent dependent of fossil fuels for all their energy needs (World Atlas 2020a).

The relative abundance of fossil fuels and the huge profits made by corporations that control its production, distribution and sale all but guarantees our continued dependence. Along with its relative abundance, fossil fuels are relatively cheap and yield a high net useful energy quotient. "Net useful energy is the total useful energy left from the resource after subtracting the amount of energy used and wasted in finding, processing, concentrating, and transporting it to users" (Harper 2012: 111). Oil, in particular, is easy to transport, but it is also very versatile in that it can be burned to propel vehicles, heat buildings and water, and supply high-temperature heat for industrial and electricity production (Harper 2012). Although coal gets a deservingly bad rap for being "dirty" in terms of mining, transporting and burning it, there still exists huge amounts of it that have yet to be mined. Burning coal also produces a high useful net energy yield and because domestic mining of coal is subsidized, costs to corporations are minimal. Thus, coal is a cheap way to produce heat for industry and generate electricity (Harper 2012). Natural gas deposits are so plentiful that a huge hydro-fracking industry has emerged to tap into this energy source. That's the good news associated with our reliance on fossil fuels for our ever-expanding energy needs.

Now the bad news, and there is plenty of it as we will demonstrate in this chapter and subsequent chapters. (Note: We will take a closer look at the negative environmental impact of natural gas and coal extraction in Chapter 4.) A sampling of problems associated with a dependency on fossil fuels is provided here. While we acknowledged that transporting crude oil (transportation networks transport crude oil to refineries where it is refined in remote locations as usable consumer products and then transported to such places as gas stations for consumption) is relatively easy, mishaps can lead to ecological nightmares. Oil spills, such as the Exxon Valdez (1989) spill in Prince William Sound, an inlet in the Gulf of Alaska, Alaska, which dispersed nearly 11 million gallons (41,640 kiloliters) of crude oil across the sound caused decimation to surrounding ecosystems and may exterminate much of native wildlife (*Encyclopædia Britannica* 2020). In the case of the Exxon Valdez spill, salmon, herring, sea otters, bald eagles, killer whales, and a wide variety of aquatic life were among the examples of ecological damage. Numerous other global oil spills have dumped a total of hundreds of millions of crude oil into oceans, bays, sounds and coastal shorelines resulting in great ecological destruction.

According to the United Nations' Environment Program, oil companies are extracting too much fossil fuel. In its "Emissions Gap 2019" report it was concluded that by 2030, global production of fossil fuels would be more than double what we can safely consume if we hope to limit the most severe impacts from human caused global warming. "In other words, rather than adopting policies and practices to slow the rise in global temperatures, humanity is largely continuing on the same suicidal course it's been on for decades. And, of course, Trump administration policies seeking ever more fossil fuel production in the U.S. are exactly opposite of what we need to do" (*Los Angeles Times* 2019a: A10). The United Nations' "Emission Gap 2019" compares projected global emissions with the stated goals of the 2015 Paris Agreement on climate changed (the agreement was shepherded by the Obama administration, but fossil fuel-friendly Donald Trump pulled the U.S. out of this agreement when he became president, making it the only country to not agree to emissions caps) and is designed to reduce the emission of gases that contribute to global warming. The leading countries in CO_2 producing emissions are China, the United States, India, the Russian Federation, and Japan (United Nations Environment Program 2019). The report indicates that if we rely on the current climate commitments of the Paris Agreement, temperatures can be expected to rise 3.2 degrees Celsius this century.

Another problem associated with the oil industry is the toxic legacy it leaves behind when oil wells are deserted. Deserted wells pose a health risk and across much of California, for example, "fossil fuel companies are leaving thousands of oil and gas wells unplugged and idle, potentially threatening the health of people living nearby and handing taxpayers a multibillion-dollar bill for the environmental cleanup" (Olalde and Menezes 2020: A1). Under federal, state, and local laws, fossil fuel companies are mandated to post funds, called bonds, to ensure that unused wells are capped and remediated. Industry representatives sheepishly admit that their bonds are woefully inadequate to meet the expected costs (Olalde and Menezes 2020). Ultimately, the fossil fuel industry will never fully take care of the problem and nearby residents will suffer health problems for years to come and the ecosystems near the deserted wells will be harmed indefinitely. Such is the reality of the rich and powerful fossil fuel industry versus the health and welfare of the injured people.

We also know that the burning of fossil fuels has compromised the quality of the air that we breathe and has compromised the ozone. The depletion of the ozone, the greenhouse effect, global warming, and climate change, all directly a result of rising CO_2 levels, will be discussed in the following pages.

Climate Change

Before we discuss the evidence that supports climate change as a real phenomenon and demonstrate the link between human behavior and the burning of fossil fuels to climate change, let's first define the word *climate* itself. *Climate* refers to the average or typical weather conditions (e.g., temperature, precipitation, humidity, and windiness) that is prevalent for a particular region in general or over a long period. *Climate change* refers to a periodic modification in the earth's climate, especially due to shifts in average atmospheric temperatures as well as interactions between the atmosphere and various other geological, chemical, biological, and geographic factors within the Earth's system (Jackson 2020). "The atmosphere is a dynamic fluid that is continually in motion. Both its physical properties and its rate and direction of motion are influenced by a variety of factors, including solar radiation, the geographic position of continents, ocean currents, the location and orientation of mountain ranges, atmospheric chemistry, and vegetation growing on the land surface. All these factors change through time" (Jackson 2020). In the 1970s, scientists worried about potential global cooling, while in the past few decades, scientists have become far more concerned with global warming. There is good reason to be concerned with climate change, as history has shown that extreme global warming and global cooling have occurred in the past and contributed to previous mass extinctions (Late-Devonian and End-Triassic). There is overwhelming evidence to indicate that global warming is occurring now, and it is likely to be a contributor to the sixth mass extinction. While humans had nothing to do with climate change millions of years ago, we most assuredly are contributing to it in the present. The only question and corresponding debate about human involvement in present-day climate change in general, and global warming in particular, is just how much harm are we causing? That is, are humans responsible for a fraction, a quarter, half, or nearly all of our current climate change predicaments?

Carbon Dioxide and Global Warming

Humans are increasingly influencing the climate and the earth's temperature by burning fossil fuels, cutting down rainforests, farming livestock, rice cultivation, industrial production and other activities (European Commission 2020; Jackson 2020). Many of these human activities combined with such natural forces as volcanic eruptions have increased the atmospheric levels of carbon dioxide (CO_2). "Humans have increased atmospheric CO_2 concentration by more than a third since the Industrial Revolution began. This is the most important long-lived 'forcing' of climate change" (NASA 2020). We will explore the increase in CO_2 since the Industrial Revolution below.

The prime human contribution to global warming is the burning of fossil fuels which directly increases the output of carbon dioxide (CO_2). Carbon dioxide is measured in terms of parts per million (ppm). Global warming is mainly the result of CO_2 levels rising in the Earth's atmosphere. The measurement of CO_2 is significant because carbon dioxide is the chief greenhouse gas that results from human activities and causes global warming. Both atmospheric CO_2 and climate change are accelerating, and climate scientists say we have years, not decades, to stabilize CO_2 and other greenhouse gases (CO_2 Now 2012).

Scientists calculate that the global average level of CO_2 before the Industrial Revolution was about 280 ppm (Porter 2013). The primary source of CO_2 was from burning fossil fuels—coal, oil and gas—and the level of CO_2 has been climbing ever since with the continued acceleration of industrialization of China and other countries. Scientists warn that the atmospheric CO_2 count needs to be at 350 ppm (maximum) in order to halt global warming and avoid catastrophic weather patterns that could spell the demise of human civilization. This 350 ppm level is referred to as the "safe" level for the continued existence of life on earth. In September 2012, the earth's atmosphere registered 391.07 ppm. Four years later, the earth's CO_2 level officially surpassed 400 ppm with dire predictions that there's little hope of returning them to safe levels (Science Alert 2016). "Even if, by some miracle, we all stopped emitting carbon dioxide tomorrow, it would take decades to get us back below the 400 ppm threshold—and we all know that's never going to happen" (Science Alert 2016). The last time earth reached the 400 ppm barrier was 800,000 years ago (Science Alert 2016). So there you have it, reaching the 400 ppm mark is a sign of impending doom, according to climatologists. As you likely imagined, the CO_2 level has not gone down and in March 2020, the CO_2 level reached 414.11 ppm (CO2.earth 2020). According to CO2.earth (2020), the global land and ocean surface temperature for January 2020 was the highest in the 141 years of record-keeping and an average of 2.05 degrees Fahrenheit (1.14 degrees Celsius) above the 20th-century average.

In addition to examining global warming based on CO_2 ppm levels are the scientific and historical records of both humans and nature around the world. This research is led by the PAGES (Past Global Changes) project, an international effort of over 5,000 scientists from more than 125 countries, who have assembled in order to coordinate and promote past global change research to help us understand our current climate (PAGES 2019a). In their 2017 comprehensive publication, PAGES reports that after examining 648 locations across the globe (including all 7 continents), and after analyzing 2,000 years of detailed records of both nature (e.g., examining tree rings, arctic ice, corals and deep ocean sentiments) and humans, researchers have discovered that the average surface temperature of the Earth has warmed faster in the past few decades than it did in the previous 1,900 years, proving yet again the current warming of the planet is unprecedented in human history. In addition, while volcanic events were the primary cause of global warming in the past, it is greenhouse gases in the atmosphere causing a more dominant role in driving global temperatures during the past 100 years (*Tribune News Service* 2019; PAGES 2019b).

An awareness of CO_2 levels will make the conscientious person think about their own carbon footprint (as we authors, as noted in the introduction, have done). The carbon footprint of a person, group, organization or society is the sum of emissions of greenhouse gases (carbon dioxide, methane and other carbon compounds) emitted due to the consumption of fossil fuels. Generally, a carbon footprint is calculated for a given period of time, such as a calendar year. One's carbon footprint provides the best estimate of the full impact they have on the environment. There are carbon footprint calculators available online (the EPA is one site) that you can use to measure your personal impact on the environment. Daily activities, such as using electricity, driving a care, or disposing of waste-cause greenhouse gas emissions along with other activities such as taking a vacation and the travel involved (e.g., flying on a plane) are among the behaviors that are measured. People can reduce their carbon footprint in a number of ways including by walking instead of driving, carpooling, public transportation, driving a low carbon vehicle, working remotely (e.g., video-conferencing tools), insulating and sealing your home, using energy efficient appliances, turning off lights when you are not using them, using LED lights instead of incandescent, setting the thermostat at a level not too high or not too low, using solar panels on the home, reducing or eliminate the consumption of beef and dairy, reducing food waste, buying products in bulk, hanging clothes to dry following a wash, lowering water usage, reusing and recycling, reducing the amount of plastic you use (e.g., straws), going paperless, and voting for candidates that support green initiatives. Another important way that one can reduce their carbon footprint is by joining social movements (see Chapter 1) that are trying to curtail the power of the fossil fuel industry and that are trying to reduce the amount of CO2 in the earth's atmosphere. Among the latest climate change activists is Greta Thunberg, a Swedish teenager (born in 2003) who has gained international recognition as an advocate of saving the climate from CO_2 emissions. (Thunberg will be discussed again later in this chapter.)

Climate Change and Ozone

Above the surface of our planet are layers of ozone: the Troposphere (0–11 km); stratosphere (11–50 km); mesosphere (50–85 km); thermosphere (85–800 km); and exosphere (800+ km). Exact starting points of these layers and the distance from the earth where "space" begins are approximations. The boundary where space begins (100 kilometers) is known as the "Karman line" (Jessa 2010). NASA's mission control uses 76 miles (122 kilometers) as their re-entry altitude because that's where the shuttle switches from steering with thrusters to maneuvering with air surfaces (Thompson 2009).

What is ozone? *Ozone* is a colorless gas; chemically it is very active and reacts readily with a great many other substances (NASA Goddard Space Flight Center 2020). "Ozone forms when groups of three oxygen atoms bond together into single molecules, which chemists signify as O_3. Most atmospheric oxygen is in the form of two bonded oxygen atoms, or O_2, but a vital layer of O_3 in the upper atmosphere helps protect life on the Earth's surface from the effects of the Sun's ultraviolet radiation" (Bell and Ashwood 2016: 20). Ozone causes chemical reactions near the Earth's surface and can damage people's lung tissues. However, ozone also absorbs harmful components of sunlight, known as "ultraviolet B," or "UV-B" (NASA Goddard Space Flight Center 2020). "Ultraviolet light can cause skin cancer, promote cataracts, damage immune systems, and disrupt ecosystems" (Bell and Ashwood 2016:20). Ozone also varies with latitude and for this reason it affects temperatures on Earth. "Ozone in the tropics, near the Equator, is typically much less concentrated than at the middle latitudes, between the Equator and the poles. Ozone concentration also changes with the seasons" (Somerville 2008: 179).

Without healthy ozone the planet would suffer dramatic environmental change and life would not be the same as we know it. And yet, humans compromise the ozone on a regular basis, especially via our dependency on fossil fuels. Humanity's overuse of coal, oil and natural gas has altered the natural environment. "We are using them [fossil fuels] up at a rate in excess of one million times their natural rate of production, and this too is an unprecedented change in geological environment" (Officer and Page 2009: 152). A compromised ozone increases ultraviolet B Rays (UVB) exposure, which can harm humans (it can also lead to crop injury and damage to ocean plant life) (EPA 2012a). In humans, ozone can cause respiratory symptoms (e.g., coughing; throat irritation; pain, burning, or discomfort in the chest when taking a deep breath; and chest tightness, wheezing, or shortness of breath); decrements in lung function (e.g., decrements such as reductions in forced expiratory volume or FEV1; shallow breathing; and retention of other inhaled substances such as allergens and particle pollution); inflammation of airways (e.g., an increase in small airway obstruction; a decrease in the integrity of the airway epithelium; and a decrease in phagocytic activity of alveolar macrophages); increased hospital admissions and emergency visits for respiratory causes; school absences; increased likelihood of asthma (EPA 2016a).

At its most elementary level, we know that ozone affects climate, and climate affects ozone. A number of variables including "temperature, humidity, winds, and the presence of other chemicals in the atmosphere influence ozone formation, and the presence of ozone formation, and the presence of ozone, in turn, affects those atmospheric constituents" (NASA Goddard Institute for Space Studies 2004). The interactions between ozone and climate first became serious subjects of discussion in the early 1970s when scientists first suggested that human-produced chemicals could destroy our ozone shield in the upper atmosphere (NASA Goddard Institute for Space Studies 2004). The first subject of concern was the human use of chlorofluorocarbons (CFCs). CFCs were widely used as aerosol propellants in consumer products such as hairsprays and deodorants, and as coolants in refrigerators and air conditioners (EPA 2012a). In 1978, the U.S. government banned CFCs as propellants in most aerosol uses (EPA 2012a). As a result of that governmental action, the ozone became healthier. The acknowledgment of the relationship between human activities and climate change intensified in 1985 when atmospheric scientists discovered an ozone "hole" in the upper atmosphere (stratosphere) over Antarctica. Scientists then expected a recovery of the ozone hole because they made the public aware of the seriousness of the situation and because most nations were expected to abide by international agreements to phase out production of ozone-depleting chemicals such as CFCs (NASA Goddard Institute for Space Studies 2004). Among the international agreements that provided scientists with hope was the Montreal Protocol. The Montreal Protocol was agreed on in the 1980s. It was an international commitment to phase out ozone-depleting chemicals that was universally ratified by all countries that participate in the United Nations (Union of Concerned Scientists 2017). The European Commission (2020b) describes a number of other ozone regulations that have taken place since the Montreal Protocol including the Commission Regulation (EU) 537/2011 (placing limits on the allocation of quantities of controlled substances allowed for laboratory and analytical uses); Commission Regulation (EU) No 291/2011 (controls on other substances such as hydrochlorofluoro-carbons); and Commission Decision 2010/372/EU (updates on earlier restrictions of controlled substances/chemicals).

Did all this legislation and scientific efforts help to eliminate the ozone hole over Antarctica? Continued research indicates that the ozone hole became increasingly smaller since the 1980s to a low point in the early 1990s; however, the ozone hole has become slightly larger each year since (with ebbs and flows) and by 2020 it was back to the same level as the late 1980s (but still lower than when it was first "discovered") (NASA Goddard Space Flight Center 2020). We should caution readers that whenever they hear about the size of the ozone hole to bear in mind that it is subject to seasonal changes within a given year. In addition, it needs to be pointed out that there is also a smaller ozone hole above the North Pole (Bell and Ashwood 2016). NASA Goddard Space Flight Center (2020) also points out that there are ozone mini holes that are generally associated with weather systems and the jet stream and occur from late autumn to early spring. As shown here, scientists across the globe agree that climate change, caused in large part by human behaviors, affects ozone.

Climate Change and Greenhouse Effect

As NASA (2020) explains, life on Earth depends on energy coming from the Sun. "About half the light reaching Earth's atmosphere passes through the air and clouds to the surface, where it is absorbed and then radiated upward in the form of infrared heat. About 90 percent of this heat is then absorbed by the greenhouse gases and radiated back toward the surface" (NASA 2020). This process leads to the greenhouse effect.

The greenhouse effect refers to circumstances where the short wavelengths of visible light from the sun pass through the atmosphere, but the longer wavelengths of the infrared re-radiation from the heated objects are unable to escape the earth's atmosphere. The trapped long wavelength radiation (infrared light) leads to more heating and a higher resultant temperature, thus contributing to global warming. The infrared light is felt as heat similar to the heat lamps used by, for example, fast food restaurants that keep foods hot. It also causes the heat we feel from ordinary light bulbs. Since carbon dioxide absorbs the heat, the more carbon dioxide there is in the atmosphere, the warmer the air will be. When the air gets too hot, the balance of life is disrupted to the point where species of plants and animals will die, and the entire food chain may be negatively affected

(National Oceanic and Atmospheric Administration 2004). The human expansion of the greenhouse effect contributes to the global warming observed since the late 20th century. "Certain gases in the atmosphere block heat from escaping. Long-lived gases that remain semi-permanently in the atmosphere and do not respond physically or chemically to changes in temperatures are described as 'forcing' climate change. Gases, such as water vapor, which respond physically or chemically to changes in temperature, are seen as 'feedbacks'" (NASA 2020). The feedback quality refers to the possibility of water vapor in the Earth's atmosphere coming back to earth after the formation of clouds and precipitation.

In addition to water vapor (the most abundant greenhouse gas) are other major greenhouse gases including carbon dioxide (CO_2) (as the result of forces of nature such as respiration and volcano eruptions and human activities such as deforestation, land use changes, and especially the burning of fossil fuels); chlorofluorocarbons (CFCs) (synthetic compounds entirely of industrial origin used in a number of applications); methane (CH_4) (as the result of natural sources and human activities, including the decomposition of wastes in landfills, agriculture, and especially rice cultivation, as well as ruminant digestion and manure management associate with domestic livestock); and nitrous oxide (N_2O) (a powerful greenhouse gas produced by soil cultivation practices, especially the use of commercial and organic fertilizers, and fossil fuel combustion) (Drake 2000; NASA 2020). It is CO_2 that is the most commonly human produced greenhouse gas, and it is responsible for 64 percent of man-made global warming. "Its concentration in the atmosphere is currently 40% higher than it was when industrialization began" (European Commission 2020a).

Human activities on Earth are changing the natural greenhouse primarily because of the burning of fossil fuels like coal and oil which has increased the concentration of atmospheric CO_2. According to NASA (2020) the consequences of changing the natural atmospheric greenhouse are difficult to predict by most assuredly will involve: some regions on Earth will become warmer, others may not; some regions will have an increase in precipitation, others will become dryer; warmer oceans, melting glaciers and rising sea levels; and some crops and plants that will respond favorably, others that will fail, and ecological changes in plant life communities (NASA 2020). "In its Fifth Assessment Report, the Intergovernmental Panel on Climate Change, a group of 1,300 independent scientific experts from countries all over the world under the auspices of the United Nations, concluded there's a more than 95 percent probability that human activities over the past 50 years have warmed our planet" (NASA 2020). Ritchie and Rosen (2020) conclude that the burning of fossil fuels is the most dominant source of local air pollution and emitter of carbon dioxide and other greenhouse gases and recommend that the world must balance its energy needs with lower-carbon energy sources. (Note: We will discuss these lower-carbon energy sources later in the chapter.)

The absolute last thing the planet needs is any legislation designed to empower the fossil fuel industry and that curtails efforts to combat climate change. And yet, that is exactly what U.S. president Donald Trump did on March 31, 2020, when he signed legislation that would weaken one of the nation's most aggressive efforts to combat climate change, the promotion of new fuel-efficiency standards for cars and trucks (Phillips and Mitchell 2020). Trump dismissed strong fuel-efficiency standards as the public was distracted by the COVID-19 pandemic (*Los Angeles Times* 2020a). The new rule relaxes Obama-era fuel-efficiency standards, then a triumph for environmentalists specifically and those who want to assure a future for younger generations of humans. "The final rule is a dialed-down version of the one the [Trump] administration originally planned. Instead of proposing zero improvement in fuel efficiency in coming years, it would require automakers to increase fuel economy across their fleets by 1.5% a year, with a goal of achieving an average of 40 miles per gallon by 2026. That's still a major departure from current rules, which mandate annual increases of 5%, reaching an average of 54 miles per gallon by 2025" (Phillips and Mitchell 2020: A4). Shortly after this major blunder, and still amidst the backdrop of the COVID-19 pandemic, the Trump administration gutted an Obama-era rule that compelled the country's coal plants to cut back emissions of mercury and other toxins produced by coal- and oil-fired power plants. Also, during the pandemic, the EPA, under the "leadership" of Andrew Wheeler, ended an Obama-era drive to regulate a widespread contaminant—perchlorate, a rocket fuel component—in drinking water that is linked to brain damage in infants. The agency rejected warnings that such a move would mean lower IQs for an unknown number of American newborns (Knickmeyer 2020). Under Wheeler, who was appointed the deputy administrator by Trump, the EPA has lost nearly all of its credibility as Wheeler is a former coal lobbyist with continued loyalty to the fossil fuel industry. As impossible as it should be, it is true that a former coal lobbyist (2009–2017) was head of the EPA under the Trump administration. Such moves on the part of a U.S. president are clearly fossil foolish. For Americans who like to drive gas-guzzling, fuel inefficient vehicles and spend more money on gasoline, and for those who do not care about fresh air and safe drinking water, this was a triumph. For those Americans against water pollution, air pollution, pipeline exhaust, and climate change, these moves were a disaster.

The Trump administration has assaulted the environment in so many ways, most of which are documented throughout this text, but to gut environmental protections during a global pandemic seems especially deplorable as the public was concerned about so many other things throughout most of 2020 including the deaths of hundreds of thousands of Americans during the pandemic, a record national debt, a crumbling economy, high unemployment, systemic racism leading to protests and clashes between anti-racism protestors and pro–Trump supporters, and so on. A few other examples of Trump's attacks on the environment during the pandemic include opening up more than 1.6 million acres of pristine wilderness in the Arctic National Wildlife Refuge to drilling; rolling back protections of a marine conservation area off the New England coast; and the rolling back of a foundational Nixon-era environmental law that required oversight of construction of highways, pipelines, chemical plants and so, by federal agencies to consider whether a project would harm the air, land, water, or wildlife, and giving the public the right of review and input (Eilperin 2020; Madhani and Freking 2020). The rollback on the Nixon-era rule which ensured scrutiny and public input "may be the single biggest giveaway to polluters in the past 40 years" said Brett Hartl, government affairs director at the Center for Biological Diversity (Madhani and Freking 2020: A7). All these actions on the part of the Trump administration will contribute to climate change and the greenhouse effect.

The Effects of Climate Change

The negative effects of climate change are numerous. As we have demonstrated already, climate change contributes to a compromise of the Earth's carrying capacity, the extinction of numerous plant and animal species, an increased consumption of scarce resources, a dramatic increase in CO_2 levels, global warming, the creation of ozone holes, and changes in the natural greenhouse. There are a number of other specific negative effects of climate change as well. In the following pages, we shall look at some specific examples of the negative effects of climate change. We begin by examining melting glaciers and ice caps and the thawing permafrost.

Melting Glaciers and Ice Caps and the Thawing Permafrost

The effects of climate change, a compromised ozone and the greenhouse effect have had a profound effect on the Earth's glaciers and ice caps. The World Wildlife Fund (WWF) (2020) explains the distinctions between these forms of ice: Sea ice forms and melts strictly in the ocean whereas glaciers are formed on land. Icebergs are chunks of glacial ice that break off glaciers and fall into the ocean. The importance of ice is that it acts a protective cover over the planet's oceans. The Arctic in particular acts as the Earth's refrigerator, cooling the planet. "These bright white spots reflect excess heat back into space and keep the planet cooler. In theory, the Arctic remains colder than the equator because more of the heat from the sun is reflected off the ice, back into space" (World Wildlife Fund 2020). The World Wildlife Fund estimates that as of early 2020, about 10 percent of the land area on Earth is covered with glacial ice. Nearly 90 percent is in Antarctica, while the remaining 10 percent is in the Greenland ice cap. Greenland's unusually mild summer in 2019 caused the world's largest island to lose 600 billion tons of ice in just two months, rivaling the summer of 2012 for the most since mass lost in a single melt season, according to NASA data (Freedman 2020). The summer of 2019 included the then-hottest month on record, July (Dennis and Freedman 2019). As a very strange sight, wildfires raged across millions of acres in the Arctic during this same month (July 2019) (Dennis and Freedman 2019). Researchers for the National Oceanic and Atmospheric Administration's (NOAA) Artic program report that 95 percent of Greenland's ice sheet surface underwent melting during the 2019 summer. "Melt duration in 2019 exceeded the 1981–2010 mean for most of the ice sheet ablation zone. The entire northern periphery of the ice sheet had at least 20 more days with melt compared to the mean" (Tedesco et al. 2019).

Canadian researchers state that evidence of our changing climate is clear in Canada's north where researchers have spent 20 years studying ice caps in the Canadian Arctic using remote sensing technology and taking measurements in person (Natural Resources Canada 2019). "On the larger ice caps there is a recent increase of massive ice layers just beneath the ice cap surface caused by excessive melting and rain at elevations that normally only see snow and freezing temperatures, even during the summer. On the smaller ice caps, some measurement poles that were drilled into the ice more than 50 years ago have been discontinued as the ice is now too thin to support them" (Natural Resources Canada 2019). In addition to the poles in Canada and Greenland, glaciers are melting in central Europe, the Caucasus region, the lower 48 U.S. states and New Zealand. Temperatures are warming faster at the Earth's higher elevations than in the lowlands, leading researchers to believe that the Andes Mountains ice sheet could be gone within two decades. Scientists say Venezuela will be the first country in South America to lose all its glaciers (Larson and Narancio 2019). In all, the Earth's glaciers are losing, on average, 360 billion tons of snow and ice each year, more than half of that in North America. Since 1961, the world has lost 10.6 trillion tons of ice and snow (Borenstein 2019).

Among the immediate concerns of rapid glacial melt in Antarctica and Greenland is its effects on ocean currents, as massive amounts of very cold glacial-melt water entering warmer ocean waters is slowing ocean currents. Furthermore, as ice on land melts, sea levels will continue to rise. California Assemblywoman Tasha Boerner Hovarth explains, "We know the sea is rising.... This is not something that's out there in the future, it's happening now" (Xia 2019a). Rising sea levels, of course, can have very adverse effects on coastal and low-lying areas. Melting sea ice also has devastating consequences for walruses and polar bears that rely on such areas for hunting and basic survival (WWF 2020). It is important to note that currently about 29 percent of our planet is covered in land (United States Geological Survey 2020a) but that has not always been the case as glaciation is, at least partially, responsible for two past mass extinctions. This points to the obvious—that any level of significant glacial melt can have a tremendous impact on humans both directly (most people live in coastal areas and they will be the first to feel the effects of continued climate change) and indirectly (food supplies and land availability for homes and agriculture would be restricted far beyond the modified carrying capacity of Earth). It is not beyond the realm of possibility that glaciation could contribute, along with climate change, the sixth mass extinction.

Because of the deadly consequences glaciation poses, the study of glaciers is very important. Glaciers around the world may contain ice that is several hundred to several thousands of years old and provide a scientific record of how climate has changed over time. Through the examination of glaciers, scientists gain valuable information about the extent to which the planet is rapidly warming. In this regard, they provide scientists with a record of how climate has changed over time (WWF 2020).

While glaciers provided scientists with valuable information on climate change because of their longevity, their long life on the planet is a cause of concern for other scientists. "As the planet warms and ice thaws, scientists warn we could see the re-emergence of ancient pathogens currently unknown to science. These viruses, which have lay dormant and locked away in glaciers and permafrost—permanently frozen soil—for hundreds if not thousands of years, could 'wake up,' researchers have said" (McCall 2020). This potential reality should really scare us all especially in light of such recent pandemics as COVID-19, SARS, MERS, Ebola, HIV, Zika, H1N1. (Note: We will discuss pandemics later in this chapter.) In 2020, scientists analyzed two core samples from the Guliya ice cap, Tibet, and identified several such dormant viruses. "One of the core samples dated back 520 years, while the other held sediments locked away 15,000 years ago. Four of the virus genera—the taxonomic rank between species and family—were already known, but 28 had never been seen before" (McCall 2020). In August 2016, in a remote corner of the Siberian tundra called the Yamal Peninsula in the Arctic Circle, a 12-year-old boy died and as least twenty people were hospitalized after being infected by anthrax. "The theory is that, over 75 years ago, a reindeer infected with anthrax died and its frozen carcass became trapped under a layer of frozen soil, known as permafrost. There it stayed until a heatwave in the summer of 2016, when the permafrost thawed. This exposed the reindeer corpse and released infectious anthrax into nearby water and soil, and then into the food supply" (Fox-Skelly 2017). The planet was not prepared for the microbes associated with COVID-19, and we certainly are not prepared for glacial microbes and viruses that have been trapped and preserved for tens of thousands of years (McCall 2020).

The ultimate question regarding melting glaciers and sea ice is why—why are glaciers melting? "Since the early 1900s, many glaciers around the world have been rapidly melting. Human activities are at the root of this phenomenon. Specifically, since the industrial revolution, carbon dioxide and other greenhouse gas emissions have raised temperatures, even higher in the poles, and as a result, glaciers are rapidly melting, calving into the sea and retreating on land" (WWF 2020). Fossil fuels are, of course, the primary culprit, and scientists have known this for a long time. Scientists also project that if emissions continue to rise unchecked, the Arctic could be free of ice in the summer months as soon as the year 2040 (WWF 2020). As anyone with even a basic understanding of science can conclude from this, air temperatures will continue to rise but at a much quicker pace than what we are already experiencing. "Even if we significantly curb emission in the coming decades, more than a third of the world's remaining glaciers will melt before the year 2100. When it comes to sea ice, 95% of the oldest and thickest ice in the Artic is already gone" (WWF 2020).

The permafrost is not so permanent any longer. "Right now, the Greenland ice sheet is disappearing four times faster than in 2003 and already contributes 20% of current sea level rise. How much and how quickly these Greenland and Antarctic ice sheets melt in the future will largely determine how much ocean levels rise in the future. If emissions continue to rise, the current rate of melting on the Greenland ice sheet is expected to double by the end of the century. Alarmingly, if all the ice on Greenland melted, it would raise global sea levels by 20 feet" (WWF 2020). Inland Canada is also experiencing permafrost loss. "'Permafrost has been warming and thawing over the past four decades,' says scientist Sharon Smith who is tracking developments. 'When ice-rich permafrost thaws, it results in ground instability and settlement which can impact infrastructure and the surrounding environment'" (NRCAN 2019). Warming temperatures and changes in precipitation patterns are affecting groundwater levels as well. This, in turn, will affect the availability of freshwater for human use (NRCAN 2019).

As the glaciers, ice caps and permafrost continue to melt and oceans warm, ocean currents will continue to disrupt weather patterns worldwide. Industries that thrive on vibrant fisheries will be affected as warmer waters change where and when fish spawn. Coastal communities will continue to face billion-dollar recovery bills as flooding becomes more frequent and storms become more intense (WWF 2020). Furthermore, as the Earth continues to warm, more permafrost will melt, exposing the planet to more bacteria that has laid doormat, perhaps as long as a million years. The temperature in the Arctic is rising quickly, about three times faster than the rest of the world and as the permafrost melts, other infectious agents may be released (Fox-Skelly 2017). As we have learned, the permafrost is a very good preserver of microbes and viruses. This is because the cold temperatures keep out oxygen (Fox-Skelly 2017). The medical world has not developed antibiotics for what lurks beneath the ice surface.

Yes, we are indeed in deep trouble if climate change continues; and it *will* continue unless we greatly lessen our dependency on fossil fuels.

Ocean Acidification

The increased output of CO_2 not only contributes to climate change, but it also contributes to ocean acidification. Ocean acidification should not be an afterthought compared to global warming; instead, it is a twin concern (Ocean Acidification 2012). As the Smithsonian (2018) states, "Ocean acidification is sometimes called 'climate change's equally evil twin,' and for good reason: it's a significant and harmful consequence of excess carbon dioxide in the atmosphere that we don't see or feel because its effects are happening underwater." It is important to realize that at least 25 percent of the CO_2 released by burning fossil fuels doesn't stay in the air but instead dissolves into the ocean (Smithsonian 2018). The Smithsonian (2018) estimates that since the beginning of the industrial era, the ocean has absorbed some 525 billion tons of CO_2 from the atmosphere; at present, the oceans are absorbing around 22 million tons per day. NOAA (2020) concurs with the Smithsonian that since the industrial revolution began, the concentration of CO_2 in the atmosphere has increased due to human burning fossil fuels and changes in the way land is used (e.g., deforestation). "During this time, the pH of surface ocean waters has fallen by 0.1 pH units. The pH scale, like the Richter scale, is logarithmic, so this change represents approximately a 30 percent increase in acidity" (NOAA 2020a). NOAA (2020) states that the ocean absorbs about 30 percent of the CO_2 that is released in the atmosphere, and as levels of atmospheric CO_2 increase, so do the levels in the ocean. "This increase causes the seawater to become more acidic and causes carbonate ions to be relatively less abundant" (NOAA 2020a).

With this background information we can define *ocean acidification* as the increase of acidity of sea water due to the decreasing level of pH over an extended period of time due primarily to the burning of fossil fuels. It was in 2003 when researchers Caldeira and Wickett first used the term "ocean acidification" to describe the impending changes of the ocean acidification levels due to carbon dioxide as a result of the burning of fossil fuels and warned of the adverse consequences for marine biota. Caldeira and Wickett (2003) describe how their research allowed for the quantification of changes in ocean pH as the result of the continued release of CO_2 compared with pH changes estimated from geological and historical records. "We find that oceanic absorption of CO_2 from fossil fuels may result in larger pH changes over the next several centuries than any inferred from the geological record of the past 300 million years, with the possible exception of those resulting from rare, extreme events such as bolide impacts or catastrophic methane hydrate degassing" (Caldeira and Wickett 2003).

Over the past few decades as scientists and enlightened thinkers have come to realize the dangers of burning fossil fuels because of increase level of CO_2 in the air, the dangers of the carbon dioxide in the waters were ignored and, in some cases, deemed preferable. However, with the increased attention paid to ocean acidification in recent years, especially the past decade, we know that CO_2 mixed in the seawater results in more acidity and high pH (Smithsonian 2018). "Such a relatively quick change in ocean chemistry doesn't give marine life, which evolved over millions of years in an ocean with a generally stable pH, much time to adapt. In fact, the shells of some animals are already dissolving in the more acidic seawater, and that's just one way that acidification may affect ocean life. Overall, it's expected to have dramatic and mostly negative impacts on ocean ecosystems—although some species (especially those that live in estuaries) are finding ways to adapt to the changing conditions" (Smithsonian 2018).

The effects of ocean acidification are most obvious with the bleaching and destruction of coral reefs. "Coral reefs provide important ecosystems services, such as sustaining fisheries, coastal protection, and social and cultural services such as recreation and tourism.... Coral biodiversity and cover have been decreasing over the last three decades, and the trend is projected to accelerate as ocean temperatures continue to rise" (Welle,

Small, Doney and Azevedo 2017). What exactly are coral reefs? We must begin by realizing that coral reefs consist, of course, of coral. Reef-building corals craft their homes from calcium carbonate, forming complex and colorful reefs that house the coral animals themselves and provide habitat for many other organisms. "Acidification may limit coral growth by corroding pre-existing coral skeletons while simultaneously slowing the growth of new ones, and the weaker reefs that result will be more vulnerable to erosion. This erosion will come not only from storm waves, but also from animals that drill into or eat coral" (Smithsonian 2018). The harm to coral reefs caused by acidification may manifest itself even before the coral begin construction of their homes, as the acidity may prevent them from reaching adulthood. Some types of coral use bicarbonate instead of carbonate ions to build their skeletons, which gives them more options in an acidifying ocean. Nonetheless, by the end of this century we can expect that all coral reefs may be destroyed because of ocean acidification (Smithsonian 2018).

Coral bleaching is also killing the pristine world of corals. *Coral bleaching* occurs when "corals are stressed by changes in conditions such as temperature, light, or nutrients, they expel the symbiotic algae living in their tissues, causing them to turn completely white" (NOAA 2020b). As the Australian Marine Conservation Society (2020) explains, the stunning colors "in corals come from a marine algae called zooxanthellae, which live inside their tissue. This algae provides the corals with an easy food supply thanks to photosynthesis, which gives the corals energy, allowing them to grow and reproduce. When corals get stressed, from things such as heat or pollution, they react by expelling this algae, leaving a ghostly, transparent skeleton behind. This is known as 'coral bleaching.' Some corals can feed themselves, but without the zooxanthellae most corals starve."

Carbon pollution is causing unprecedented damage to Australia's Great Barrier Reef. We warned of this in the first edition of text as we stated: The world's largest coral reef ecosystem, Australia's Great Barrier Reef, featuring nearly 3,000 individual reefs within 133,205 square miles, has lost more than half of its coral cover since 1985 (De'ath et al. 2012; Eilperin 2012). Sadly, the destruction of the Great Barrier Reef has continued. The Australian Marine Conservation Society (AMCS) states, "In recent years, our Reef has suffered severe mass coral bleaching, faster and more severe than scientists predicted. If we don't act to halt this pollution, we risk the future of precious Reef." In 2016 and 2017, the Great Barrier Reef suffered back-to-back bleaching leaving half of the shallow water corals dead (AMCS 2020). The AMCS (2020) warns that all corals risk, at the least, bleaching and at the most, extinction, unless humans stop using fossil fuels and transition to renewable energy today. (Note: We will discuss renewable energy sources later in this chapter.)

As you might expect, coral reefs around the world including in the United States and the Caribbean are also at risk. The EPA (2018) reports that the very existence of coral reefs may be in jeopardy unless we intensify our efforts to protect them. In North America, coral reefs face many threats from both a planetary level due to rising CO_2 levels and locally from coastal development, dredging, quarrying, destructive fishing practices and gear, boat anchors and groundings, and recreational misuse (touching or removing corals) (EPA 2018). Other sources of threats confronting coral reefs include pollution that finds its way into coastal waters (sedimentation, nutrients such as nitrogen and phosphorous, pathogens, toxic substances, and trash and micro plastics); overfishing; and coral harvesting (EPA 2018).

At the current rate of coral reef destruction, the time will come when the only images of these natural sources of beauty will be from archival photos.

Climate Change and Severe Weather

Climate change has a tremendous impact on the Earth's weather patterns that manifest themselves in many forms (and the multitude of examples are far too numerous to chronicle completely here), including more severe storms (e.g., hurricanes, superstorms, tornadoes and snowstorms), flood deluges, and record heat. In 2018, a congressionally mandated report by 13 federal agencies found that climate change is being felt in communities across the United States. The report projected widespread and growing devastation as increasing temperatures, rising sea levels, worsening wildfires, more intense storms and other cascading effects will harm environmental ecosystems, societal infrastructure, and our very way of life. "The assessment paints a dire picture of the worsening effects of global warming as nearly every corner of the country grows more at risk from extreme heat, more devastating storms, droughts and wildfires, waning snowpack and other threats to critical infrastructure, air quality, water supplies and vulnerable communities" (Barboza 2018: A1). And these are problems in addition to what we have already described in preceding pages. Just as this report was released (and as he has done throughout his presidency) Donald Trump was moving to dismantle Obama-era climate regulations and replace them with fossil-fuel friendly policies that would allow more planet-warming emissions from power plants, cars and trucks (Barboza 2018). Demonstrating his ignorance on the subject (or at the very least, speaking in such a manner as to appease his supporters), Trump mocked climate science and tweeted about cold weather in the Northeast (November 2018) asking, "Whatever happened to Global Warming?" Anyone with an elementary school-level education in science can answer that question.

Among the most intense examples of severe weather are hurricanes. The most widespread damaging storms on earth, hurricanes are getting worse, and climate change is a big reason why (Denchak 2018). *Hurricanes* are named after the ancient indigenous god Hurican (who in turn was derived from the Mayan god Hurakan) and can be defined as an intense low-pressure weather system with winds of 74 miles (119 kilometers) per hour or greater that gain energy from warm tropical waters and are usually accompanied by rain, thunder, and lighting and that sometimes moves into temperate latitudes. Tara Rava Zolnikov (2018) provides a first-hand account of the effects of Hurricane Irma after she decided not to heed the mandatory evacuation orders and wait out Hurricane Irma in Florida: "the worst part of the hurricane is not the assailing rain pellets shooting against the window, it is the bass-like drone of the raging wind tearing at the roof of the house; every angry gust is a relentless effort to disassemble the structure offering the most protection and last line of defense against the storm … the house starts to feel unsafe once the whistling wind and accumulating pools of water seeping under doors and windows reaffirm that this man-made shelter is imperfect" (p. 27).

Depending on where this weather event takes place, the hurricane (in the Atlantic and in the central and eastern Pacific) is also known as a typhoon (northwestern Pacific) or cyclone (Indian Ocean and South Pacific) (Denchak 2018). Monsoons take on the characteristics of hurricanes but refer to seasonal change rather that a single weather event. Hurricanes come in categories based on wind speed (not the storm's potential for damage) ranging from Category 1 (74–95 mph), very dangerous to Category 5 (157+

mph), catastrophic hurricane. The power of the wind and rain of a hurricane can cause harmful consequences directly to those in its path. Hurricanes may also lead to floods which can cause great harm to local ecosystems, humans and property.

Denchak (2018) explains how climate change affects hurricanes: "Over the past 50-plus years, the earth's oceans have absorbed more than 90 percent of the extra heat generated by man-made global warming, becoming warmer as a result. Since warm sea surface temperature fuel hurricanes, a greater temperature increase means more energy, and that allows these storms to pack a bigger punch." Goodman and Moynihan (2018) add, "Fossil fuel extraction from the Alberta tar sands drives global warming, which in turn increases the destructive power and frequency of storms like Hurricane Michael" (p. A4). As we documented earlier in this chapter, the burning of fossil fuels and other human activities have caused a global increase in temperature. This results in a hotter atmosphere and because a hotter atmosphere can hold—and then dump—more water vapor, a continued rise in air temperature would result in wetter storms and an even greater capacity to generate flooding.

Combine wetter storms with continued sea rise as a result of glacier and ice sheets melt, and the threat of storm surge also increases. Consider, for example, Hurricane Katrina's 28-foot storm surge that overwhelmed the levees around New Orleans in 2005 and unleashed a devastating flood across much of the city (Denchak 2018). (Note: Hurricane Michael hit North Carolina, South Carolina, Virginia, Georgia, and Florida in October 2018.) In addition, research suggests that global warming is weakening atmospheric currents that keep weather systems like hurricanes moving, resulting in storms that linger longer. "Sluggish storms can prove disastrous—even without catastrophic winds—since they can heap tremendous amounts of rain on a region over a longer period of time. The stalling of Hurricane Harvey over Texas in 2017, as well as the slow pace of Hurricane Florence, helped make them the storms with the greatest amount of rainfall in 70 years" (Denchak 2018). Due to climate change, the forward speed of hurricanes has decreased by 16 percent over land in North America in past years and rain rates are expected to increase by median of 14 percent in the coming years (*CBS This Morning* 2020).

In addition to hurricanes producing heavy rains, research indicates that heavy rainstorms are becoming more common in the U.S. Northeast. Howarth, Thorncroft and Bosart (2019) state that rainfall events that can cause significant flooding are increasing in frequency and magnitude and that the frequency of extreme events has increased over the course of span of years (1979–2014) of their study. The number of severe rainstorms (defined as those that drop more than 1 inch in 24 hours) has increased 74 percent in the Northeast, more than any other region of the country. The increasing trend in extreme precipitation were most robust during the fall months of September, October, and November, and particularly at locations further inland (Howarth et al. 2019). In Central New York, severe rainstorms occur twice as often as they did 25 years ago (Coin 2019).

In some instances, storms become so powerful they attain the label of "superstorm." Such was the case with Superstorm Sandy in 2012. (A "superstorm" is a subjective term for any storm which is extremely and unusually destructive but is not classified as a hurricane or blizzard or some other distinct meteorological classification.) Sandy helps to further illustrate what can happen when a huge storm occurs in low-lying areas and such places experience a corresponding rise in sea levels. On October 29, 2012, Hurricane Sandy, which was downgraded to a tropical storm when it hit the mainland, hit the

Northeast with New York City getting the brunt. The storm, which included elements of a hurricane with a nor'easter, hit Atlantic City and New York City around 8:00 p.m. just as the high tide was coming in. To make matters worse, it was a full moon, meaning a lunar high tide (higher than normal tides). Televised news reported that the Battery (Lower Manhattan) was hit with a high-water mark of 13.88 feet. Other area high-water marks included Sandy Hook, New Jersey, 13.31 feet; Bridgeport, Connecticut, 12.74; Cape May, New Jersey, 8.17; Atlantic City, 8.04; and Newport, Rhode Island, 6.35.

The effects of Superstorm Sandy are hard to quantify and qualify but consider that within the first 48 hours of the storm 46 people had died, 18 in New York City alone, and 8 million people in the 15-state area affected by the storm were without power. The death count rose to 92 later in the week as more bodies were found amid the devastating ruins left behind by Sandy. After the storm subsided, power outages compounded problems: refrigerated food was ruined; people were stuck in elevators; emergency responses increased in volume, but rescue efforts were hampered; sewer systems failed; crime and looting occurred; work productivity was compromised; and so on. Rising sea waters in Atlantic City ripped apart parts of the famed Boardwalk. A beachside roller coaster was completely surrounded by the Atlantic Ocean.

In New York City, rising sea waters flooded the subways and the salt water corrupted the rails and short-circuited underground power transformers. The storm caused the tunnels to be flooded; the bridges connecting Manhattan and the outer boroughs were closed; mass transportation (trains and buses) halted; airports closed; fires started; beaches were ruined; marine life was destroyed; and well over 125,000 trees were destroyed or damaged (113,000 in New Jersey alone and 10,000 in New York City). Health issues arose due to problems such as raw sewage dumping back into fresh water supplies because power outages stopped pumps from working. The storm brought loss of life, billions of dollars in damage and billions of dollars in lost productivity. Entire neighborhoods were ravaged, including Breezy Point, Queens, a beachside community of mostly Irish Catholic policemen and firefighters. The Breezy Point neighborhood was destroyed by an out-of-control fire assisted by 80 mph winds. Firefighters were unable to find hydrants (NBC News 2012; Fitzgerald 2012). For a time, much of New York City looked like a larger version of a submerged New Orleans following Hurricane Katrina.

Just as climate change affects the intensity of rainstorms, so to does it affect snowstorms. The EPA (2016b) explains that the Earth's surface contains many forms of snow and ice, including sea, lake, and river ice; snow cover, glaciers, ice caps, and ice sheets; and frozen ground. "Climate change can dramatically alter the Earth's snow- and ice-covered areas because snow and ice can easily change between solid and liquid states in response to relatively minor changes in temperature" (EPA 2016b). The National Centers for Environmental Information (NCEI) (formerly the National Climatic Data Center), an affiliation of NOAA, reports that "Years with heavy seasonal snow and extreme snowstorms continue to occur with great frequency as the climate has changed. The frequency of extreme snowstorms in the eastern two-thirds of the contiguous United States has increased over the past century. Approximately twice as many extreme U.S. snowstorms occurred in the latter half of the 20th century than the first" (NCEI 2020). Warmer-than-average ocean surface temperatures in the Atlantic are contributing to unusually high ocean surface temperatures which in turn leads to more moisture that can fuel bigger snowstorms. The authors of this text can attest to the effect of living nearby two of the Great Lakes (Ontario and Erie) and the fact that the warmer lake water temperature routinely leads to the region getting hammered by huge "lake-effect" snow events. This helps to explain why the cities of Syracuse, Rochester (NY), and Buffalo are routinely three of snowiest cities (population of 100,000+) in the country. We also know that climate change sometimes leads to less snow in any given season than when compared to average snowfall years. This natural variability is due to global surface temperature increases as the temperature at any given time is higher than it would be without climate change (NCEI 2020). Studies have also shown that natural variability associated with the presence of El Niño conditions has a strong influence on the incidence of severe snowstorms in the eastern United States (NCEI 2020).

Climate change has a direct affect on the record heat experienced in the United States and across the planet. It seems as though every time some record for the hottest day is set that it is broken a short time later. During a one-week span in the summer of 2018, *The Washington Post* reported that all-time heat records were set all over the world, and this includes such milder climate countries as Ireland, Scotland and Canada to the hotter climates of the Middle East and Southern California (Samenow 2018). The heat was blamed for at least 54 deaths in southern Quebec. In Northern Siberia, along the Arctic Ocean, temperatures soared 40 degrees (Fahrenheit) above normal to over 90 degrees on July 5 (2018). During this same week, Africa also baked with its hottest temperature ever reliably measured occurring in Ouargla, Algeria (124.3 F; 51.3 C) (Samenow 2018). Data provided by NASA and NOAA indicates that the 2010s was the warmest decade in modern human history and that each of the last five years were among the warmest years on record (Khan 2020). NOAA chief of climate monitoring, Derek Arndt, reported that the warming trend probably won't be slowing down anytime soon either. Arndt stated, "Notwithstanding some sort of major, major geophysical event, it would be almost certain that the [coming] decade [2020s] will be warmer than the previous" (Khan 2020). With this report in mind, it is all but certain that readers of this text in the years to come will have witnessed even warmer record high heat.

One of the consequences of high heat is the increased use of air conditioning. Air conditioning was invented in 1902 by Willis Haviland Carrier. In that same year, the first air conditioner unit was installed in a Brooklyn printing plant, not to cool the air but to control humidity that was playing havoc with the paper and ink (*The Post-Standard* 2012). Almost ironically, the spread of air-conditioning around the world is contributing to the growing problem of global warming. Air conditioning harms the environment by releasing poisonous gases such as chlorofluorocarbons and hydrochlorofluorocarbons into the environment. Air conditioning systems also consume electricity which puts a strain on the electrical power grid. This is why people are asked to turn their air conditioning off during peak summer hours. Running the air conditioner in your car decreases the gas mileage, which is another negative impact on the environment.

The NCEI reports that in 2019 there were 14 weather and climate disaster events in the United States with losses that each topped the $1 billion mark. That made it the fifth consecutive year with 10 or more billion-dollar weather or climate disasters (Khan 2020). "Between 1980 and 2019, the average number of such events was 6.5 per year; in the last five years, that figure has shot up to 13.8" (Khan 2020: A2). These natural disasters include flooding, severe storms, tropical cyclones (a storm or system of winds that advances at a speed of about 20 to miles an hour) and wildfires. Wildfires will be discussed in greater detail in Chapter 5 (as a natural occurrence) so our coverage here will be limited.

While the eastern part of the U.S. often has to contend with severe, water-drenched storms, the western part of the country often deals with warmer, drier conditions, increased drought, and a longer fire season that increases in wildfire risk. "Climate change has been a key factor in increasing the risk and extent of wildfires in the Western United States. Wildfire risk depends on a number of factors, including temperature, soil moisture, and the presence of trees, shrubs, and other potential fuel. All these factors have strong direct or indirect ties to climate variability and climate change. Climate change causes forest fuels (the organic matter that burns and spreads wildfire) to be more dry, and has doubled the number of large fires between 1984 and 2015 in the western United States" (Center for Climate and Energy Solutions 2020). The National Park Service (NPS) (2016) warns that climate change places the overwhelming majority of parks at extreme fire risk and predicts longer fire seasons. The NPS concurs with other research (Stephens et al 2013) that indicates climate-mediated changes in plant regeneration may lead to novel vegetation patterns in parks. For example, fire-intolerant and moisture-demanding trees may be replaced by fire- and drought-tolerant species, and low elevation dry forests may convert to shrublands or grasslands, depending on the magnitude of change in climate and fire. In California, a state synonymous with wildfires, Governor Gavin Newsom in 2019 declared a 12-month state of emergency throughout the state of California in an effort to protect 200 of California's most wildfire-vulnerable communities (Office of Governor Gavin Newsom 2019). Governor Newsom has good reason to be concerned as the wildfires in 2020 were even worse than the record-setting fires of 2019. By early October, more than 4 million acres were burned by wildfires in California—prompting the creation of the term "gigafire" to describe massive wildfires. A *gigafire* refers to fires that burn through one million or more acres of land in a single go. The calamity of out-of-control fires, increased heat, and air pollution predicted for the future (circa 2040 or 2050) has already besieged the western states, and climate change is a primary contributing factor.

The relationship between climate change and wildfires is certainly not limited to the United States, as many areas across the planet face fire seasons. The fire 2019–20 fire season in Australia was particularly brutal as bushfires spread unabated throughout a large portion of the island nation. More than 13 million hectares (the size of England) were burned in the early days of the season and firefighters from around the world appeared in Australia to help out (just as their firefighters often help the U.S. and other nations fight their wildfires) (Chang 2020). The Amazon fires of 2018 claimed 800,000 hectares (Chang 2020). In 2019, wildfires in the Amazon raged at a higher rate than the previous year. In Brazil alone, there were a recorded 76,720 wildfires, and 85 percent rise over the previous fire season (De Souza 2019). Climate scientists in Brazil state that the "tipping point" for this dramatic increase in wildfires is deforestation (to be discussed in greater detail in Chapter 4). This man-made influence on climate change has scientists believing that if unusually dry seasons continue, leading to droughts in parts of Brazil, Uruguay, Paraguay, and Argentina, coupled with deforestation, weather patterns could dramatically change for the worse across South America and far beyond (De Souza 2019). The election of far-right president Jair Bolsonaro, known uncomplimentarily as "Brazil's Trump" is another blow to Brazil and their attempt to curtail natural disasters that are directly or indirectly caused by climate change as he is a pro-fossil fuel proponent. Valuing old interpretations economy over environment and the future of humanity, Bolsonaro campaigned on promises to loosen protections for indigenous lands and nature reserves, arguing that they were helping to choke Brazil's now-struggling economy by stifling its major agricultural and mining sectors (De Souza 2019).

Fossil Fools and Greenhouse Asses

As demonstrated in the previous pages, the negative effects of our dependency on fossil fuels have contributed heavily to climate change, led to the destruction of many ecosystems, and threatens the very viability of the Earth's environment. The increased levels of atmospheric CO_2 concentration have compromised the ozone and contributed to the greenhouse effect in general; and has also led to melting glaciers, ice caps and permafrost, poisoned oceans and waterways, increased presence of severe storms and wildfires, specifically. And yet, there exists financially greedy fossil fuel power elites and unenlightened masses that have chosen to ignore, discard or hide from scientific realities. Such people have come to be known as "fossil fools" and "greenhouse asses." *Fossil fools* are people who support energy from, or fail to recognize the negative impact of, fossil fuels. *Greenhouse asses* are people who insist that the environment is in fine shape and believe that attempts to conserve and preserve the environment are unnecessary. Perhaps only time will tell who the real "fools" and "asses" are, but if the environmentalists and world's scientists are correct, we need to take drastic steps now to save our planet.

The concept of "fossil fools" has become so popular that throughout North America people celebrate "Fossil Fools Day" on April 1 as a way to connect eco-unfriendly behavior with foolish behavior. Fossil Fools Day (FFD) hopes to raise awareness of the hazards caused by harvesting, processing and utilizing fossil fuels (Anyday Guide 2020). The first FFD was held in 2004. It has become increasingly popular on college campuses and communities throughout the United States (especially in the Northwest) as well as in Canada. Proponents of "Fossil Fools Day" encourage people to kick the oil habit and avoid acting like a fossil fool via such eco-friendly means as walking, cycling, taking public transportation, or car-pooling whenever possible, buying fuel-efficient vehicles (e.g., hybrids), living as close to work as possible, shopping at local stores, becoming vegan, promoting the use of renewable energies, and buying regionally and seasonally produced food whenever possible.

The *Ecologist*, which describes itself as "the journal for the post-industrial age" and has as its primary aim "to foster a greater connection to nature in order to enhance personal wellbeing, support resilient communities and inform social change towards regenerative societies that enrich rather than deplete our natural environment," warns that many climate-conscious consumers could be unknowingly investing 16 billion (euros) in fossil fuels through their ISAs (individual savings account). "Europe's leading ethical and sustainable bank, Triodos Bank, is today [March 14, 2020] launching a new UK campaign for ISA season, called 'Don't be a fossil fool,' encouraging and enabling people to shift their money away from fossil fuels" (*Ecologist* 2020). Triodos Bank (2020) reports that "nearly two thirds (63%) of people across the UK agree that money needs to be moved away from fossil fuels and invested in clean energy instead, in response to the climate emergency." Nearly the same percentage (61 percent) "acknowledges that fossil fuels are undoubtedly damaging the planet, and four in 10 believe having any money invested in the fossil fuel industry is morally wrong" (Triodos Bank 2020). Research conducted by Triodos Bank (2020) indicates that half of the people with ISAs think that fossil fuel investments are foolish given the climate emergency. The UK declared a climate emergency in 2019 and much of the financial focus since then has been on divesting pension funds are large organizations' deposits. Among the problems associated with the good intentions of those who wish not to be foolish by supporting the fossil fuel industry is the chronic lack of transparency in the financial sector (Triodos Bank 2020). "Two thirds of British people admit to having no idea if their money or savings currently go towards supporting climate-changing fuels, such as oil gas or coal. And 70% say that banks and savings providers need to be more transparent about where their money is invested" (Triodos Bank 2020). While enlightened people are fully aware of the damage caused by human dependency on fossil fuels, "a third of people polled do not realize that moving their money out of fossil-fuel-supporting banks is one of the most effective ways to take action on climate change" (Triodos Bank 2020).

The data compiled by Triodos Bank (2020) suggests that while some people (pro-fossil fuel industry proponents) are fossil fools on purpose others are unwilling participants in fossil foolish behavior. But even fossil fools and greenhouse asses that do not care about the future of humanity and the Earth's environment should care enough to recognize the facts that are a consequence of climate change. For example, fossil fools who are not concerned with the health of the planet should care about their own health and the health of those they care about and love. We have long known that climate change not only harms the natural environment, but it also harms the health of humans. In 2016, a report from the Obama administration found that "as the world warms, exploding populations and greater urbanization could increase the number of people exposed to extreme heat, which already kills thousands of Americans each year" (Dennis 2016). A warmer future will result in thousands to tens of thousands of additional premature deaths per year in the U.S. and around the world. Then-U.S. Surgeon General Vivek Murthy stated (in 2016) that "climate change is going to impact health, and it's not a pretty picture" (Dennis 2016: A 10). The 2016 report also warned that there would be more deaths from extreme heat, longer allergy seasons, increasingly polluted air and water, diseases transmitted by mosquitoes and ticks spreading farther and faster, and overall weaker immune systems as the nutritional value of some crops become increasingly compromised as a result of the accelerated use of pesticides; these are among other the health risks, as a result of global warming (Dennis 2016). In 2018, the United Nations Intergovernmental Panel on Climate Change concluded that we need to cut global greenhouse gas emissions in half by 2030 and entirely by 2040 to avoid the most catastrophic effects of climate change (Solomon and LaRocque 2019). Solomon and LaRocque, both physicians and public health experts, stress the importance of curtailing gas emissions as essential in the healing mission of physicians and warn that "dramatic increases in both sickness and death are projected for the foreseeable future as the world's continued reliance on fossil fuels results in more air pollution, infectious diseases, malnutrition, wildfires, extreme heat and increasingly powerful weather events" (Hardy 2019: A2). Solomon and LaRocque (2019), echoing the sentiment expressed by Triodos Bank, suggest financial divestment as a strategy to lessen the power of the fossil fuel industry, stating that it "has been an effective tool in other health movements, including efforts to thwart the tobacco industry. Much like the tobacco industry, fossil fuel companies have used their vast resources to sow disinformation and influence policymakers against the public interest."

In the same issue of the *New England Journal of Medicine* in which the Solomon and LaRocque article appeared, Haines and Ebi (2019) also warn against the continued reliance on the global burning of fossil fuels and its devastating effects it is having on our planet. In addition to the threats on physical health (some of which have already been mentioned but also including selected vectorborne diseases in some locations) that comes with global warming other threats include psychological stress, political instability, forced migration, and conflict. As an example of forced migration and conflict, consider the situation in Nigeria where Fulani herders, a semi-nomadic, mostly Muslim ethnic group that herds cattle primarily in northern Nigeria, among Christian land-owner farmers. The farmers and herders have clashed over land for years, but it has become increasingly accelerated due to climate change. "As grasslands have degraded in northern Nigeria, semi-nomadic herders have starting moving their herds into densely populated farming areas to the south. At the same time, as Nigeria's population has boomed, farmers have expanded their fields, often into the herders' traditional grazing routes" (Mahr 2019: A3). Disputes between the two groups used to be handled via traditional mediation but violence has escalated as that system has broken down. Both sides now routinely attack and kill each other in fierce battles over the scarce resource of land, although sometimes the conflict has taken on religious and ethnic dimensions. The conflict in Nigeria is of special importance due to the realities of climate change coupled with the realization that Nigeria is the seventh most populous nation in the world (in 2020), but with its projected annual increase in population it is predicted by some to become the world's third-most populous country by 2050 (Mahr 2019).

Returning to the research of Haines and Ebi, the doctors also point out that even when/if our dependency on fossil fuels is reduced the lasting physical health problems will continue for some time as carbon dioxide remains in the atmosphere for centuries, with about 20 percent persisting for more than 1,000 years (Haines and Ebi 2019). Haines and Ebi (2019) add, "The magnitude and nature of health risks depend not just on the hazards created by a changing climate but also on the sensitivity of the people, communities, and natural systems that are exposed to those hazards and on the capacity of people, communities, and health systems to prepare for and manage the increasing risks." Americans are in trouble as the United States must contend with Donald Trump, who, at every possible turn, has ended Obama-era regulations aimed at addressing climate change and protecting the environment in favor of the fossil fuel industry. His ineptness in handling the COVID-19 pandemic is a sure sign that the U.S. is not prepared to handle the magnitude of health problems confronting its residents.

One thing Trump and fossil fuel-proponents do care about is wealth, primarily their own. This concern alone should prompt action on their part as the loss of property due to climate change is measured in hundreds of billions of dollars a year and this number will continue to rise year after year. The U.S. Geological Survey estimates that by 2100, rising sea levels alone will cost California two-thirds of its beaches and coastal cliffs will continue to erode 130 feet further inland (Xia 2019a). Coastal property is highly treasured and valued correspondingly. In 2013, a committee formed by then-Assemblyman Richard Gordon warned that rising sea levels would not only damage coastal property but also, agriculture, tourism, fishing and the infrastructure (e.g., roads, sewers, power grid). The longer officials and policymakers wait to take preventive measures to save property as a result of climate change the more costly it will be. In the case of Superstorm Sandy, "If $15 billion of infrastructure improvements had been done prior to the storm, it would have mitigated most of the $60-billion costs that accrued to taxpayers after the storm" (Xia 2019a). While pro-fossil fuel politicians turn their backs to the reality of climate change's negative impact on real estate value, the real estate investing industry has to heed the realities of global warming. "To that end, big real estate firms are pouring significant resources into calculating climate risk and its likely effect on property portfolios—everything from increasingly extreme weather to a rise in sea levels" (Olick 2019). Real estate firms have no choice but to heed the warnings of the loss of real dollars associated with natural disasters. "Damage to U.S. real estate from extreme storms hit a record high in 2017. Natural disasters, including floods, mudslides and wildfires, cost more than $300 billion in damage, the bulk of it to residential and commercial real estate" (Olick 2019).

For most of us, property damage due to climate change will come in the form of damage to our homes (over 40 million Americans are at risk of flooding from rivers, and over 8.6 million people live in areas that already experience coastal flooding from storm surges during hurricanes); more expensive home insurance (insurance companies do not like to lose money); outdoor work will become unbearable as a result of heat waves; higher electric bills and more blackouts (as a result of the grid going down); rising taxes (someone will have to pay to repair the infrastructure); more allergies and other higher risks (will cost people more money out of pocket to treat acute and chronic health problems); food will become more expensive as agriculture takes a beating from climate change-weather related problems (e.g., extreme heat and flooding that destroys food supplies); water quality will suffer (e.g., storm runoff, pesticides, animal waste and chemical fertilizer entering water supplies); outdoor exercise and recreational sports will become more difficult; and disruptions in travel (as temperatures rise, it may get too hot for some planes to fly, and low-lying airport runways may be prone to flooding—Superstorm Sandy flooded LaGuardia Airport for three days) (Cho 2019). Adding to the disruptions in air travel is the realization that climate change will increase turbulence which in turn will increase flight times and fares (Soo 2016). University of Reading professor Paul Williams states that while the impact of aviation on global warming has long been recognized, "it is becoming increasingly clear that the interaction is two-way and that climate change has important consequences for aviation" (Soo 2016).

In the first country-to-country study on how much each country around the world will suffer in future economic damage from each new ton of CO_2 pumped into the atmosphere it was calculated that the United States that would suffer the second most, behind only India (Morford 2018). In the 2018 study (published in the journal *Nature Climate Change*) lead author Kate Ricke states, "Our analysis demonstrates that the argument that the primary beneficiaries of reductions in carbon dioxide emissions would be other countries is a total myth" (Morford 2018). It makes sense that the larger a country's economy is, the more they have to lose, said Ricke. The costs, caused by emissions, appear in the form of energy costs and damage to property and agriculture. Utilizing cost-benefits analysis of the median social cost of carbon per metric ton of new CO_2 released into the atmosphere, the Obama administration estimated it would be $42 per metric ton for 2020. The Trump administration, however, altered the cost-benefits formula and not surprisingly came up with a much lower figure, close to $3 per ton. This low-ball figure helps fossil fools to justify their continued dependency on fossil fuels.

Trump is not the first Republican president to question science in favor of a political agenda. We can recall the actions of former president Ronald Reagan who started this trend in earnest. In the months just prior to Reagan taking office, the Carter administration had begun negotiations with Canada to control air pollutants that cause acid rain (see Chapter 3) and was moving toward a treaty that would have severely limited air pollution from power plants. "Reagan's response upon winning the White House was to instruct his science advisor to alter the findings of a major report on acid rain to make the science seem more uncertain, and therefore justify delay. Some of Reagan's advisors and Cabinet members disputed the science behind stratospheric ozone depletion too, though Reagan eventually agreed to sign an international treaty—the Montreal Protocol—that controlled ozone-destroying chemicals. But his administration established a precedent: to be a conservative hostile to federal power and regulation meant skepticism toward environmental science" (Oreskes and Conway 2020: A11). Reagan's successor, George H.W. Bush, did not completely buy into Reagan's anti-science and anti-environment sentiments. He championed the 1990 Clean Air Act amendments and also established the U.S. Global Climate Research program to improve our understanding of climate change and agreed to a complete ban on production of ozone-depleting chemicals (Oreskes and Conway 2020). As we know, George H.W. Bush was a one-term president, in part because his moderate and fact-based positions were not in step with an emerging Republican ideology that did not support the science behind climate change analysis. Ever since, most Republican leaders "have downplayed, misrepresented or rejected the scientific evidence of man-made climate change, and some have gone as far as to ridicule and harass climate scientists for telling the truth" (Oreskes and Conway 2020: A11). Their ever-lasting support of the fossil fuel industry, coupled with an ideological commitment to "limited government," has caused conservatives to drag their feet on climate change, healthcare, the opioid crisis and slew of other social problems that are too big for individuals, or even states, to fix. "This ideology has caused conservative leaders to discard scientific findings, even when delivered by their own experts and even when lives are at stake. The tragic consequence, as we are seeing now [COVID-19 pandemic and a worsening of the negative effects of climate change], is that lives are lost that could have been saved" (Oreskes and Conway 2020: A11). This ideology, to put as mildly and politely as possible, is foolish.

The lesson here is quite clear: Don't be a fossil fool or a greenhouse ass.

Plant and Animal Reaction to Global Warming

Humans are not the only species affected by climate change, as evidence of global warming exists in the natural world of plants and animals. Let's first take a look at the world of plants as climate change affects a number of variables that determine how much and where plants can grow; minimum, maximum, and average temperatures affect plant growth and distribution; and seasonal growing cycles. In some instances, plants will benefit from a declining average number of freezing days and later first fall frosts. According to Professor William Cox, soybeans are not labor intensive, and they command a high market price (*The Citizen* 2012a). Historically, New York has been a marginal region for soybean cultivation because of its cold climate, but with global warming a new opportunity presents itself to farmers. Northern areas of the planet, such as Russia, China and Canada have gained growing days (Worland 2015).

Conversely, extreme temperatures, a decrease in water availability and changes to soil conditions will make it more difficult for plants to thrive and will likely stunt plant growth (Worland 2015). "Declining plant growth would destroy forests and dramatically change the habitats that are necessary for many species to survive. And, if conditions get bad enough, forests could actually produce carbon instead of removing it from the atmosphere, exacerbating the root cause of climate change" (Worland 2015). While traditionally colder areas may benefit from a warmer climate, the regions that already experience great warmth and heat will potentially lose 100–200 growing days per year. "In total, 3.4 billion people would live in countries that lose nearly a third of their growing days. More than 2 billion of those people live in low-income countries" (Worland 2015). Such a worst-case scenario could be altered if humans become more actively engaged in reducing their CO_2 output. As we demonstrated in this chapter, however, the CO_2 output has continued to increase and has not decreased.

There exists a relationship between CO_2 and plant growth that will have a negative impact on water supplies. "Plants are the primary regulators of the water cycle, responsible for 60 percent of the flow of water from the land to the atmosphere" (Leahy 2019). As CO_2 levels increase, plants need less water to do photosynthesis, which would lead us to think that would equate to more fresh water available for soil and streams. However, a warming world means longer and warmer growing seasons, which gives plants more time to grow and consume water. "Plants in this hotter, carbon dioxide-rich environment grow bigger, with more leaves. That means when it rains there will be far more wet leaves creating more surface area for more evaporation to occur. Computer modeling shows that such enhanced leaf evaporation has a large effect on runoff and soil moisture" (Leahy 2019). The combined effects of increased CO_2 and warmer temperatures will increase water consumption by vegetation and will decrease water in rivers and streams (Leahy 2019). By 2100, it is likely that the Southwest and central Great Plains of the U.S. will experience a 35-year or longer "megadrought" if immediate action to curtail CO_2 output is not put into place. Water scarcity is already a major concern, as four billion people suffer from severe water scarcity at least one month a year (Leahy 2019).

There are many other issues connected to an increasing level of CO_2 and plant life as well. As anyone who suffers from certain types of allergies can tell you, longer plant seasons equates to more pollen and longer allergy seasons. Changing climate patterns can also alter soil type, affecting which plants thrive and which ones cannot in any given region. Invasive species (see Chapter 5) adapt more quickly to changing environmental conditions where native species might struggle (Arcadia 2017). While environmentalists promote the idea of "going green" (see Chapters 7 and 9), the greening effect that has been seen across the northern landscapes has a negative effect on the environment. Vegetation absorbs sunlight rather than reflect it like snow and ice does, thereby causing more warming (Arcadia 2017). The thawing of the tundra, as discussed earlier in this chapter, may also lead to the release of methane, a greenhouse gas. Warmer temperatures can also potentially kill off tropical forests, releasing more gases that can contribute to atmospheric warming (Arcadia 2017). Over a decade ago, scientists at Duke University reported that in their 18-year study of trees they found that trees' ability to grow and produce seeds is more sensitive to climate than was thought. Trees in the Southeast are showing signs of risk factors from earlier springs to summer droughts, making some species especially vulnerable. Climate change poses a serious potential threat to the vast Eastern forests in the form of stress to the genetic integrity of the trees and their ability to cope with exotic insects (invasive species) (Henderson 2011). The EPA (2017a) provides us with a further analysis of the harm that climate change may cause forests as rising levels of CO_2 can alter the frequency and intensity of forest disturbances such as insect outbreaks, invasive species, wildfires, and storms. Insect outbreaks often defoliate, weaken, and kill trees. Pine beetles, for example, damaged more than 650,000 acres of forest in Colorado and spruce beetles damaged more than 3.7 million acres in southern Alaska and western Canada. Invasive species that are sensitive to the cold weather are flourishing in the warming regions. Warmer temperatures and drought conditions have contributed to an increase in wildfires, and this includes the extent, intensity, and frequency of wildfires. We discussed the devastation that hurricanes and storms can cause, primarily in the context of coastal areas, but they can also cause significant damage to forests as well (EPA 2017a).

Climate change not only affects humans and plant life, but it also affects birdlife and animals. It only takes a small change in temperature to threaten the survival of many animal species. Research conducted by Warren and associates (2018) indicates that a global warming of approximately 2 degrees (Celsius) above pre-industrial levels (and we are nearly at that level) could make 24–50 percent of birds, 22–44 percent of amphibians, and 15–32 percent of corals vulnerable to extinction. The risk of extinction grows if the temperature continues to rise even higher. Up to half of the animal and plant species in the world's most naturally rich areas, such as the Amazon and Galapagos, could face extinction by 2100 due to climate change if the 2015 Paris Agreement is not enforced (Warren et al 2018). The National Weather Service (NWS) (2019) states, "Unless greenhouse gas emissions are severely reduced, climate change could cause a quarter of land animals, birdlife and plants to become extinct." The NWS (2019) adds, "Climate variability and change affects birdlife and animals in a number of ways; birds lay eggs earlier in the year than usual, plants bloom earlier and mammals are coming out of hibernation sooner. Distribution of animals is also affected; with many species moving closer to the poles as a response to the rise in global temperatures." The continued rise in sea levels could cause sea turtles to lose their nesting beaches. A mere 50-centimeter rise could cost over 30 percent of sea turtles in the Caribbean to lose their nesting season. The already endangered Monk Seals need beaches to raise their pups; whales and dolphins need shallow, gentle waters to rear their small calves; and migratory fish and mammals are at risk because of rising sea levels (NWS 2019).

The impact of rising sea levels is one example of habitat disruption. Habitat disruption as a result of climate change upsets the local ecosystems of animals, often a place where animals have spent millions of years adapting. This is very problematic as animals rely on instincts. They are more or less programmed to respond instinctively to nature. As the world warms, many species have been forced to alter their habitats and instinctive way of survival. "Affected wildlife populations can sometimes move into new spaces and continue to thrive. But concurrent human population growth means that many land areas that might be suitable for such 'refugee wildlife' are fragmented and already cluttered with residential and industrial development. Cities and roads can act as obstacles, preventing plants and animals from moving into alternative habitats" (Earth Talk 2020).

Scientists suspect that nearly half of the world's species have changed their habitants as the temperatures have become too hot (Animal Network 2018). The U.S. Department of the Interior (2015) provides us with a number of examples of species moving farther north because of rising temperatures and among them are moose, salmon, snowshoe hare, American Pikas and polar bears and walruses. Rising temperatures have led to booming parasite populations in northern U.S. and Canada and up to tens of thousands of these parasites can gather on a single moose to feed on its blood—weakening the animal's immune system and often ending in death, especially among the calves. As a result, moose are expected to move further northward or they risk extinction. Salmon require cold, fast-flowing streams and rivers to spawn and with the changing stream flows and warming waters in the Pacific Northwest already impacting some salmon species and populations they will have to move farther north. Like the moose, salmon are confronted by a growing parasite invasion. Snowshoe hares have evolved to turn white in winter to blend in with the snow in order to hide from predators. However, with climate change, snow is melting earlier than the hares have grown accustomed to over tens of thousands of years, leaving them exposed in snow-less landscapes. Moving farther north will be important for their survival. Snowshoe hares play an important role in forest ecosystems, so their movement northward has a domino effect on other ecosystems. American Pikas, about the size and shape of a hamster, typically live in high elevations where cool, moist conditions prevail; but their populations are disappearing from numerous areas that span from the Sierra Nevadas to the Rocky Mountains. Reduced snowpacks and warmer summers as a result of global warming puts this species at risk as they have few options as temperatures continue to rise. Polar bears are already listed as "vulnerable" to extinction under the Endangered Species Act because of their forecasted population declines from the effects of climate change. The primary cause of their predicted demise as a species by 2100 is the loss of sea ice habitat attributed to Arctic warming. Walruses and other Arctic species are facing similar challenges as summer ice continues to retreat (U.S. Department of Interior 2015).

A number of animal species are having reproductive problems caused by climate change and this has a lot to do with their instinctive response to environmental cues, such as the arrival of spring or increased rainfall that triggers their reproductive behaviors. "For example, many reptiles exhibit a phenomenon known as temperature-dependent sex determination (TDSD). This essentially means that the temperatures surrounding their eggs will determine the sex of their offspring. If the global temperatures continue to rise, this will ultimately leave many of these animals with populations that are comprised exclusively of individuals of one sex" (Animal Network 2018). TDSD occurs in a variety of reptile species, including most aquatic turtles and several geckos (Animal Network 2018).

The migration patterns of many species have been altered due to global warming. While some animals have been able to cope with rising temperatures and changing precipitation patterns, others have forced to modify their migration patterns, or they have been forced to move. "Birds are migrating and arriving at their nesting grounds earlier, and the nesting grounds that they are moving to are not as far away as they used to be and in some countries the birds don't even leave anymore, as the climate is suitable all year round" (NWS 2019). The natural order of food chains is becoming disrupted by climate change as well, which in turn will lead to further modifications in migration. The red knot shorebirds travel to the Arctic each year to feed on hatching insects but because these insects are hatching weeks before they used to, there are rarely enough insects to feed all of the birds when they arrive (Animal Network 2018). The changing migration patterns have other consequences as well. For example, the northern exodus from the United States to Canada by some types of warblers has led to a spread of mountain pine beetles that destroy valuable balsam fir trees. In the Netherlands, the northward migration of caterpillars has eroded some forests there (Earth Talk 2020).

Birds, in general, are especially in trouble, as new data indicates that in the period of time between 1970 and 2019, North America has lost nearly 3 billion birds, or 29 percent of the total bird population (Pennisi 2019). This total number is derived from a wide variety of bird species, perhaps from as many as 529 in total, according to data sets provided by such sources as the North American Breeding Bird Survey (Khan 2019). Among the causes of this dramatic loss in the bird population are human intrusions (e.g., the use of pesticides and building construction in the natural habitants of birds) and cats. The loss of birds can cause profound consequences for the ecosystems they inhabit.

The earlier hatching of insects is just one example of shifting life cycles caused by global warming. The study of shifting seasonal life cycles events is called phenology. "Many birds have altered the timing of long-held migratory and reproductive routines to better sync up with the warming climate. And some hibernating animals are ending their slumbers earlier each year, perhaps due to warmer spring temperatures" (Earth Talk 2020).

It is more important than ever before to protect animals, especially those that are endangered. Alarmingly, the Trump Administration continued to push to eliminate or soften the Endangered Species Act. For example, in September 2019, California and 16 other states filed a lawsuit against the Trump Administration's weakening of this landmark law in an attempt to ensure the survival of the California condor, the grizzly bear and other animals close to extinction (Phillips 2019). "California Attorney General Xavier Becerra has sued the Trump administration more than 60 times over its agenda of dismantling Obama-era environmental and public health regulations" (Phillips 2019: B2). To make matters worse, the Interior Secretary (at this time), David Bernhardt, was a former oil lobbyist who actively worked to gut the Endangered Species Act both in his former life as fossil fool and as a member of Trump's Cabinet. Bernhardt's clear conflict of interest should have disqualified him as the Interior Secretary but that is not the case under the Trump administration. Is there any wonder why endangered species are in more peril than ever before? Becerra states, "As we face the unprecedented threat of a climate emergency, now is the time to strengthen our planet's biodiversity, not to destroy it. The only thing we want to see extinct are the beastly policies of the Trump administration putting our ecosystems in critical danger" (Phillips 2019: B2).

Another concern with endangered species is the realization that in many areas of the world hunters are allowed to hunt endangered species primarily so that they mount the dead animal's head (or some other part of the dead animal) on a wall. "For instance, most conservation groups consider African elephants and lions to be threatened or endangered. The amount of protection afforded the animals, however, depends on the country they happen to be roaming in, and some continue to sell the rights to hunt and kill them" (*Los Angeles Times* 2020b). This helps to explain why a California man could legally kill an elephant in South Africa in 2019. And, even though lions in southern and eastern Africa have been listed as threatened under the Endangered Species Act since 2015 (because their numbers had plummeted 43 percent over the previous two decades), the Trump administration allowed a hunter's lion trophy to be brought into the U.S. from Tanzania in 2020 for the first time since 2016 (*Los Angeles Times* 2020b). To address this issue, California State Senator Henry Stern (D–Canoga Park) introduced a bill (Senate Bill 1175) in 2020 that would forbid the possession of such "trophies" from any of 13 iconic African species, including African lions, African elephants, leopards, black rhinoceros, white rhinoceros, giraffes, two species of zebras and baboons (*Los Angeles Times* 2020b). The intent of this bill is to preserve endangered species, not their trophies.

This review of plant and animal reaction to global warming represents a mere sampling of the negative effects of climate change on nature. Combine these problems with those confronting humans and it is very obvious that we need to drastically reduce our dependency of fossil fuels in favor of renewable forms of energy.

The Need for Renewable Energy

Throughout this chapter ample evidence (and it is a mere sampling of available facts) has been provided of the damage that the reliance on fossil fuels as our primary source of energy has caused the Earth's environment. Humans need to immediately change this dirty, carbon dioxide-producing reliance as climate change is greatly harming us, plants and animals, and countless ecosystems. The extremely high level of CO_2 in the atmosphere suggests that it may already be too late to save the planet and all the species that inhabit it. However, if it's not already too late, we should make huge strides to erase our collective negative carbon impact. To break ourselves away from fossil fuel dependency we need to educate those who somehow remain in denial of the realities of world (this includes our policymakers, politicians and the general public) on ways in which can reduce our fossil fuel consumption. The primary manner in which to lessen our dependency on fossil fuels is to switch to renewable sources of energy.

Renewable energy refers to "energy from sources that are naturally replenishing but flow-limited; renewable resources are virtually inexhaustible in duration but limited in the amount of energy that is available per unit of time" (United States Energy Information Administration 2019b). The concept of "flow-limited" means that while renewable energy sources are inexhaustible and replenish naturally, the amount of energy available per unit of time is limited (Zaman 2019). In the past, renewable energy sources were described as "clean energy," but today is en vogue to combine these meanings as one. As the World Atlas (2019) explains, "renewable energy is considered clean energy as it does not cause adverse environmental pollution." Because these energy sources are cleaner, they are healthier for all living creatures—plants, ocean life and life on land. Renewable energy sources include wind (via windmills which help to generate electricity); solar (the sun's radiation is a significant source of heat); water flow (e.g., hydropower energy); geothermal energy (energy from the heat within the earth); natural and human-made biomass (e.g., wood and wood waste; municipal solid waste; landfill gas and biogas; ethanol; and biodiesel); and tidal (energy obtained from tides) (Delaney and Malakhova 2020). "The potential for renewable energy resources is enormous because they can, in principle, exponentially exceed the world's demand; therefore, these types of resources will have a significant share in the future global energy portfolio, much of which is not concentrating on advancing their pool of renewable energy resources" (Ellabban, Abu-Rub, and Blaabjerg 2014: 748).

In the United States, 11 percent of its total energy comes from renewable energy sources including wind, solar, biofuels, geothermal, biomass waste, wood and hydroelectric (EIA 2019b). Delaney and Malakhova (2020) provide a review of the steps that some of the U.S. states, such as New York, Maine and California are taking toward embracing renewable energy sources. In New York, legislators want 70 percent of the state's energy to come from renewable sources by 2050; Maine has a much loftier goal—its governor has pledged to move to 100 percent renewable energy by 2050; and, in California, the governor has pledged that all of its electricity will come from renewable energy sources by 2045. Many states are working toward diversifying their energy resources and have enacted policies that have driven the nation's $64 billion market for wind, solar and other renewable energy sources (National Conference of State Legislatures 2020). A growing number of states have established Renewable Portfolio Standards (RPS) in an attempt to promote a diversified energy resource mix and encourage deployment of certain energy technologies such as offshore wind or rooftop solar (NCSL 2020).

In Russia, the percentage of current total energy from renewable energy sources is much lower than in the U.S., with about 3 percent of its energy coming from wind, water, solar biofuel and waste. However, the Russian Federation has set out to increase and diversify its use of renewables, particularly for power generation (Delaney and Malakhova 2020). "Under current plans and policies, renewables would reach nearly 5% of total final energy consumption by 2030. Accelerated deployment, however, could boost Russia's renewable energy share to more than 11% in the same timeframe" (International Renewable Energy Agency 2019). In particular, the potential is particularly large for wind, biomass, small hydro, geothermal and solar, depending on the geographic regions. Energy experts stress that Russia has abundant resources for all types of renewable energy and thus, as IRENA (2019) states, Russia is a potential "green giant." Nonetheless, the development of renewable energy sources other than hydropower and bioenergy has progressed slowly in Russia as its national energy balance is still dominated by the traditional sources of coal, oil, and natural gas (Delaney and Malakhova 2020).

Additional sources of energy, other than fossil fuels, are nuclear energy and hydrogen fuel cells. As for nuclear energy, some people consider it as renewable energy as it has low carbon emission but, it is more accurately described as *sustainable source energy*. Nuclear energy is not renewable as its primary fuel source is uranium and it is not unlimited; although using breeder and fusion reactors makes it possible to produce other fissionable elements. In this regard, nuclear energy becomes a sustainable source of energy (Delaney and Malakhova 2020). In the U.S., 8 percent of its energy sources come from nuclear power. Globally, nuclear energy accounts for 13.3 percent of the total energy needs, up from less than 3 percent of the total in 1960 (The World Bank 2019b). Russia is moving steadily forward with plans for an expansion in nuclear energy, including the development of new reactor technology (World Nuclear Association 2020). Nuclear energy, however, comes with its own set of potentially deadly and catastrophic consequences both to humanity and the environment (as we shall see in Chapter 3). Hydrogen is an energy source that causes almost no pollution when it burns and can be used to power batteries and fuel cells. "A fuel cell is a device that converts chemical potential energy (energy stored in molecular bonds) into electrical energy. A PEM (Proton Exchange Membrane) cell uses hydrogen gas (H_2) and oxygen gas (O_2) as fuel. The products of the reaction in the cell are water, electricity, and heat. This is a big improvement over internal combustion engines, coal burning power plants, and nuclear power plants, all of which produce harmful by-products" (Hydrogenics 2020). Experiments are being conducted to use solar power to produce Hydrogen and if successful, it could become a very important energy source (Zaman 2019).

There are a number of nations around the world that are far less dependent upon fossil fuels for their energy needs than the U.S. and Russia. According to the World Atlas (2019) the nations least dependent of fossil fuels for energy are (the percentage provided refers to the percent of energy that comes from alternative and nuclear energy sources): Iceland (89.0 percent), Tajikistan (64.1 percent), Sweden (48.5 percent), France (47.0 percent), Switzerland (39.5), Costa Rica (38.7 percent), Norway (34 percent), El Salvador (33.8 percent), New Zealand (31.5 percent), and Kyrgyzstan (29.5 percent).

In their research, Delaney and Malakhova (2020) provide a number of detailed reasons as to why the U.S. and Russia lag behind in lessening their dependency on fossil fuels. A synopsis is presented. The first explanation rests with the realization that while supplies of fossil fuels are finite, advanced technology has led to the discovery of an abundance of fossil fuel reserves. In 2015, 76 percent of the oil consumed in the U.S. was domestic, the remaining 24 percent was imported from foreign countries with the top six being: Canada, Saudi Arabia, Venezuela, Mexico, Columbia and Russia (Institute for Energy Research 2019). American coal production is currently the second highest in the world, behind only China. It is used primarily as fuel for power plants that generate electricity for residential and commercial customers. "In fact, it accounts for over 30 percent of U.S. electricity generation" (IER 2019). Adding to the abundance of fossil fuels in the U.S. is natural gas. The United States is the largest producer of natural gas (Russia is second), producing 27.1 trillion cubic feet in 2015 (IER 2019). The controversial hydraulic fracturing process (see Chapter 4) is an environmental nightmare, but the abundance of natural gas is used as justification by the fossil fuel industry and consumers who want cheap energy supplies. In sum, the U.S. is endowed with huge quantities of fossil fuels and this abundance is likely to keep the nation committed to it for its chief energy needs for decades to come. Nonetheless, at our current rate of consumption of fossil fuels, the CO_2 levels will climb so high that most forms of life will be wiped out before we run out of oil, coal, and natural gas (Delaney and Malakhova 2020).

The second reason the U.S. lags in developing renewable energy is a general disbelief in science. As scientists across the globe have concluded that a reliance of fossil fuels is a threat to the very survival of the human species, ignorant people duped by conservative politicians have allowed themselves to embarrassingly question the legitimacy of the harmful effects of climate change in general, and to naively accept the assertion that further dependency on fossil fuels is the best course of action in the pursuit of energy and economic success. Conservative politicians since the Reagan era have equated anti-science with pro-religion, an ideological commitment to "limited government," and a defense of material consumption and prosperity. To acknowledge that climate change is caused mostly by human dependency on fossil fuels runs counter to the pro-fossil fuel mantra. Collomb (2014) believes that climate change denial in the United States is related to two variables: "First, climate denial stems from the strong ideological commitment of small-government conservatives and libertarians to *laissez-faire* and their strong opposition to regulation. Second, in order to disarm their opponents, US climate deniers often rest their case on the defense of the American way of life, defined by high consumption and every-expanding material prosperity" (p. 1). Securing energy needs fuels feelings of well-being. Concurring with this premise, Wood and Roelich (2019) state that energy plays a role in the attainment of well-being and this feeling of well-being attainment is currently powered by the comfort of easily attainable energy via fossil fuels.

Social movements lacking in impact is the third explanation for the U.S. lagging behind in renewable energy development. To be fair, a number of social movements, especially in the early 1970s, were successful in bringing about environmental changes as evidence by the passage of the Clean Air Act of 1970, Clean Water Act of 1972, and the first Earth Day (1970). However, the activists of the 1970s—the present-day baby boomers—learned a hard lesson about the true strength of the *power elites* (which includes the fossil fuel industry power brokers). The current generation is energized by such climate activists as Greta Thunberg, a Swedish teenager who initially became known for her activism in August 2018 (then aged 15) when she first stood outside the Swedish parliament to protest climate change and calling for stronger action on global warming by holding up a sign that read: "School strike for the climate." Throughout 2019 she led a youth-spirited environmental campaign, in an attempt to halt the negative forces contributing to climate change, which became a global social movement. The ability of Greta Thunberg to stimulate social action led some to refer to her young followers as the "Greta Generation" (McKibben 2020). Thunberg's message has been so inspiring that *Time* magazine named the teenaged climate activist the 2019 "Person of the Year" (she is the youngest person to receive this prestigious honor). However, Thunberg and her peers are also learning first-hand of the resistance among conservatives and the power elite who wish to maintain the status quo. President Trump, for example, tweeted (upon learning that Thunberg received the *Time* honor), "So ridiculous. Greta must work on her Anger Management problem, then go to a good old fashion movie with a friend! Chill Greta, Chill!" (Smith and Stein 2019). Brazilian right-wing president Jair Bolsonaro and Russian leader Vladimir Putin were among the list of leaders who patronized Thunberg following her *Time* honor (Smith and Stein 2019). These grown men have a history of being intimidated by strong women but now they have demonstrated a fear of a teenage girl and her message of trying to protect the environment. This is the type of ignorance that members of the current younger generation must overcome if they hope to be more successful than the previous generations.

Greed among the power elites is the fourth explanation why the U.S. lags in the development and utilization of renewables as its primary source of energy. It is a fundamental proposition that the power elites have an inexorable thirst for power and the rest of us should never underestimate the financial greed of such people. Who are the power elites? Sociologist C. Wright Mills (1956) describes the *power elite* as those persons who hold positions to make decisions that have major consequences over others. Mills argued that the power elites are a select few people who are at the top of major corporations, sometimes referred to as "captains of industry," influential government persons and top influencers of the military. Combined, big business, government and the military form what Mills described as the "tripartite elite" and warned that this triangle of power was an increasing threat to American democracy (Delaney 2014b; Delaney and Malakhova 2020). The power elite make the major decisions regarding energy sources to be consumed, and because they already monopolize the fossil fuel industry, they are not about to give up the unbelievably huge sums of money they receive at the expense of the environment and dying plant and animal species.

Russia is making only small strides toward embracing renewable energy sources. Three of the primary reasons why they lag behind include the presence of large reserves of fossil fuels, comparatively low energy prices, and the lack of sufficient environmental pressure on the State.

Similar to the United States, Russia enjoys large reserves of fossil fuels. Russia is the world's largest producer of crude oil and gas exporter and is the second-largest producer of dry natural gas. Russia also produces a significant amount of coal. As a major producer of hydrocarbons, it is not a surprise to learn that more than one-third of Russia's federal budget revenues come from the production of fossil fuels (EIA 2017). The presence of large reserves of fossil fuels makes Russia a major producer and exporter of oil and natural gas. "Russia's economic growth is driven by energy exports [which accounted] for 36% of Russia's federal budget revenues in 2016" (EIA 2017). *Forbes* reports that Russia derives 40 percent of its revenue from oil and gas sales, making it a "de-facto petro-state" (Cohen 2019). The high revenues associated with fossil fuel use will likely mean that Russia will continue to promote their use for as long as possible.

Primarily as a result of its large reserves of fossil fuels and its desire to assist the continued growth of business, the Russian government has kept energy prices relatively low for consumers and businesses. Electricity prices are much lower in Russia than in the European Union (as much as 60–80 percent lower) (Enerdata Intelligence 2019). In June 2019, the price of electricity in Russia was 0.068 U.S. Dollar per kWh for households and 0.092 U.S. Dollar for businesses which includes all components of the electricity bill such as the cost of power, distribution and taxes (Global Petrol Prices 2019). "For comparison, the average price of electricity in the world for that period is 0.14 U.S. Dollar per kWh for households and 0.12 U.S. Dollar for businesses" (Global Petrol Prices 2019). Tariffs on the electricity market vary from one region of Russia to another but overall, Russians appear to be thankful for the comparatively low energy prices as total energy consumption has been increasing rapidly since 2016, by almost 3.5 percent per year (Enerdata Intelligence 2019). In addition to the low cost of electricity in Russia, Russians enjoy a comparatively low price for gasoline. Data compiled by Global Petrol Prices (2020) indicates that gasoline prices per liter, octane-95, in Russia from 11 November 2019 to 17 February 2020 was 45.79 Russia Ruble, while the average price of gasoline in the world for this period was 92.68 Russian Ruble. The comparatively low energy prices result in no immediate significant motivation for investing into alternative energy sources (Delaney and Malakhova 2020).

The low energy prices in Russia also results in a corresponding lack of sufficient environmental pressure placed on the State by either businesses or consumers to promote renewable energy (with the exception of very remote regions where renewable energy can be considered as a solution to energy security). As a result, Russia has not invested an extensive program for the development of renewable energy. The fact that the State generates a huge amount of its overall revenues from fossil fuels highlights why there is a lack of systematic federal funding in the development of renewable forms of energy (Delaney and Malakhova 2020). Some Russian business owners have shown an interest in the development of energy, but their activity should be overestimated as the main players on the renewables market (i.e., Rusnano and RusHydro) are focus on lobbying and developing their own local projects with vested interest. State-run corporations are not interested in renewables either and in fact, they are currently reducing their investments in this sphere (Delaney and Malakhova 2020).

We would like to conclude this chapter on a high note, but as the evidence presented in this chapter implies, that will be a tall order. One glimmer of hope is the bipartisan "Energy Innovation and Carbon Dividend Act of 2019" (H.R. 763) introduced in the House of Representatives by Rep. Theodore E. Deutch (D-FL) and currently making its way through the long legislative process. "This bill would impose a fee on the carbon content of fuels, including crude oil, natural gas, and coal, or any other product derived from those fuels that would be used so as to emit greenhouse gases into the atmosphere" (Congress.gov 2020). The fee would be imposed on the producers or importers of the fuels and is equal to the greenhouse gas content of the fuel multiplied by the carbon fee rate. The fee rate begins at $15 in 2019 and increases by $10 each year. The idea behind the bill is that the cost of using fossil fuels would off-set the profits and thus lead to a decrease in the usage of fossil fuels in favor of renewable forms of energy. The fees collected would be distributed, as dividend payments, to all U.S. citizens or lawful residents (Congress.gov 2020).

Most climate change deniers are merely uniformed of the seriousness of this environmental disaster on the brink of leading us through the sixth mass extinction.

Researchers at the Yale Program on Climate Change Communication recommend as a viable tactic to reach these uniformed people that we simply talking about it. The Yale team believes the more we talk about global warming, the more we might move the needle on public opinion. The Yale findings "suggest that climate conversations with friends and family enter people into a proclimate social feedback loop" (Rosen 2019: A2). Instead, the topic of climate change has increasingly been treated as a taboo subject something akin to politics and religion to never be discussed in polite company. This attitude must stop.

It is possible, perhaps, that the reason some people are reluctant to discuss the direness of climate change and its immense negative effects on the environment is because of the feelings of despair they experience when contemplating that their environment is under assault. Australian professor of sustainability Glenn Albrecht and his wife Jill first thought of the concept "nostalgia" to describe the profound sense of loss, isolation, and feelings of powerlessness that accompanies the realization that climate change has dramatically compromised the environment. Seeking a more descriptive term, Albrecht coined the term "solastalgia"—a neologism that combines the words nostalgia, solace and desolation (Wick 2020).

Solastalgia aside, it is our intention, with this second edition of *Beyond Sustainability* to more strongly emphasize the seriousness of climate change. If nothing else, after reading this chapter, hopefully you will, at the very least, discuss climate change with your friends and family. Move the needle of social awareness.

Popular Culture Section 2

When the Grid Goes Down

The alarm goes off to start the day. You turn on lights, check your phone for messages, and turn on the computer and television. Next, it's time to start the coffee maker before taking a hot shower and brushing your teeth with an electric toothbrush. You plug in your devices to charge. Maybe you put some clothes in the washer or dryer. Chances are, either the heat is on or the air conditioner. Within ten minutes of waking, you have tapped into the grid and assume there will be enough power to run everything. Electric power, after all, is like air, you assume that when you take a breath there will be air, and the same assumption applies to turning on light switches and every other electrical device. In reality, you have been tapped into the grid while sleeping, as all sorts of household appliances are connected to the grid 24 hours a day, 7 days a week. Only during a power outage do we realize how much we take electrical power for granted and just how dependent we are on the grid. The grid provides us with heat, cooling, refrigeration, light, sound, entertainment, and a slew of electrical devices that would be inoperable without it.

Just what is this *grid* that we are so dependent upon? According to the Trustworthy Cyber Infrastructure for the Power Grid's (TCIPG) Office for Mathematics, Science and Technology Education (MSTE) (2013), the power grid is the system of producers and consumers of electricity. It includes the sources of power or power generators (e.g., wind farms; coal, natural gas plants; and nuclear power plants), the system of substations, power lines, and transformers that deliver electricity, and the consumers that consume electricity (TCIPG 2013). Commonly, local communities will have a generator to provide power to customers and this generator may be able to vary its production as the usage by customers fluctuates. When the demand is too high for the local generator, the community buys electricity from another source; at other times, the generator may be making more electricity than the community is using, so it is sold to other providers (TCIPG 2013). A number of homes, schools and businesses supplement the grid and lessen their dependence on power being transported to them by adding features such as solar panels. All of these activities are connected and accomplished on "the grid."

If you live near a substation or an electrical transformer, you may experience more disturbances in the flow of power through the grid than other customers. (The concept of transformer box locations shutting off for the greater good is similar to that of electrical surge protector strips that are designed to protect electrical devices from abnormal flows of energy.) While most electrical devices power back up and automatically adjust to the correct time and previously programmed settings following power outages, some older devices (e.g., certain alarm clocks) still need to be manually updated. But such inconveniences are minor compared to what people would experience if the grid went down and never powered up again. Imagine if you will, what life would be like without access to the grid, or if the grid did not operate, for any extended length of time. People who experience natural disasters that knock out power for days at a time certainly have a limited idea of just how difficult and challenging life can be without access to the grid.

Now imagine if the grid never came back on. What would happen then? Would humanity be able to survive? Would there be chaos? Would Herbert Spencer's doctrine of the *survival of the fittest* come into effect? It's only intriguing to contemplate if you feel confident that such a situation would never actually occur. If you cannot visualize the idea of what society would be like without the grid, we can turn to popular culture for possible scenarios as many television shows (i.e., "The Walking Dead") and popular films (i.e., "The Day the Earth Stood Still") depict life without the grid.

In the time since the first edition of *Beyond Sustainability* (2014) was released two books of particular note have been published that describe scenarios related to the grid going down. The first book we will draw attention to is *The Grid: The Fraying Wires Between Americans and Our Energy Future* (2016) by Gretchen Bakke. Bakke describes, much as we did, our dependency on an energy source—the grid—that nearly all of us take for granted and this, despite the fact, that the average American is without electric power six hours a year. Bakke states that America has the highest number of outage minutes of any developed nation and this does not include outages caused by extreme weather or other "acts of God." Comparably, Korea averages 16 outage minutes a year, Italy, 51 minutes, Germany 15, and Japan at 11 minutes. Bakke (2016) warns that for every minute of these blackouts, money is lost, and national security is at risk; and, of course, there is a great deal of inconvenience to all those affected. Bakke, in particular, points to America's aging grid system of transmission lines and transformers along with the average age of an American power plant as the primary culprits of blackouts and power outages. In many instances, better maintenance could help alleviate the stress the grid is under, for example, tree limbs often knock down power lines and pests such as squirrels have been known to chew on wires—including one time when both the New York Stock Exchange and the NASDAQ exchange were shut down by squirrels chewing on wires. Among other lessons that Bakke attempts to teach us is the need for a smarter grid that connects cleaner energy. Delaney and Madigan recommend a public works program to help keep the grid running more efficiently, such a program would not only create countless jobs it would help to maintain the security and economic productivity of the United States.

The second book of relevance to our discussion on the power grid is *Lights Out: A Cyberattack, A Nation Unprepared, Surviving the Aftermath* (2016) by Ted Koppel. In this book on the grid, Koppel warns us that the United States government is not prepared for the power grid being knocked out for any significant length of time, let alone, forever. Koppel's ominous warnings begin immediately with a one word opening sentence—"Darkness." Indeed, darkness would be among the key characteristics of a prolonged power grid failure. No electricity, batteries lose power, cellphones, portable radios, and flashlights all quickly go dark. The emergency generators that some people and businesses possess will provide pockets of lights and power for an indefinite amount of time (and draw attention to others without such conveniences that electric power represents), but there is little running water anywhere and the taps will soon go dry, and toilets no longer flush. All emergency supplies become scarce quickly and suddenly we are living in a *Walking Dead*–type of world where scavengers seek out any supplies remaining. Water, food, and fuel will all disappear quickly and for those in large cities, the demands for basic survival items will soon lead to deadly unrest. As scary as this scenario is, and the fact that people like Koppel and Delaney and Madigan are well aware of this all-too-real possibility, the government, according Koppel, is completely unprepared to step in with a plan of action to save the day. In fact, according to Koppel, the government is not even prepared to protect the grid from likely sources of attack—cyber threats and drone attacks on power stations. With the grid down, Americans will not only have to worry about their fellow citizens turning on them, but also the vulnerability to foreign military intrusions.

Doomsday is upon us if the grid goes down.

Summary

In this chapter, we learned about the concept of "carrying capacity" and the realization that the Earth has a limited capacity to support life. Over the generations, but especially recently, the Earth's carrying capacity has been stretched to its limit due to a number of threats to the environment.

Human dependency on fossil fuels (oil, natural gas, and coal) has directly led to climate change and a slew of very specific problems including global warming, as a result in the dramatic increase in carbon dioxide (CO_2) the atmosphere since the time of industrialization; a damaged ozone (which affects the very air we breathe); and an unnatural rise in the greenhouse effect (which contributes to the global warming observed since the late 20th century).

The negative effects of climate change are numerous and contribute to a compromise of the Earth's carrying capacity; a dramatic increase in CO_2 levels; melting glaciers, ice caps and the permafrost; ocean acidification; severe weather (e.g., hurricanes, superstorms, tornadoes, flooding; and snowstorms); and an increasing imbalance in the natural world of plants and animals (in some cases leading to their extinction). The topic of fossil fools and greenhouse asses was presented to highlight the ignorance of climate change deniers and the harmful consequences of climate change denial.

In order to lessen our dependency of fossil fuels it is necessary to develop viable alternatives, known as renewable energy or clean energy. Renewable forms of energy include wind (via windmills which help to generate electricity); solar (the sun's radiation is a significant source of heat); water flow (e.g., hydropower energy); geothermal energy (energy from the heat within the earth); natural and human-made biomass (e.g., wood and wood waste; municipal solid waste; landfill gas and biogas; ethanol; and biodiesel); and tidal (energy obtained from tides). The key obstacles impeding our switch to renewable forms of energy were also presented.

Chapter 3

Overpopulation and the Five Horrorists

In this chapter, we will examine the role of human population on the environment in an attempt to answer the question, "Are there too many people in the world?" That is to say, can the planet handle the ever-increasing number of people that make demands for food, water and energy? Concern about overpopulation is not a new one, as many scholars and scientists have pondered such consequences for centuries. We believe that overpopulation has led to the "Five Horrorists"—an updated, and more deadly, version of the "Four Horsemen" concept.

Overpopulation: Are There Too Many People in the World?

Overpopulation is generally defined as an excessive number of people in an area to the point of overcrowding, depletion of natural resources, an impaired quality of life, or environmental deterioration. Since the advent of early industrialization, a number of social thinkers have pondered whether there were too many people on the planet. Scholars such as Thomas Malthus did not use the concept of "carrying capacity" in their calculations, but the rudiments existed. For example, Malthus, in his *An Essay on the Principles of Population* (1798), claimed that the world's population was growing too quickly in proportion to the amount of food available. Compared to such formulas as IPAT, his theory was rather simple and harshly criticized by subsequent scholars, such as Karl Marx and Friedrich Engels, who called his conception "false and childish." But is there any merit to Malthus's concern about the total number of people and the ability of societies around the world to provide an adequate food supply for its people?

Before we answer that question let's first take a brief look at Malthus's theory from Marx and Engel's point of view. In their opinion, Malthus "stupidly" reduces two complicated issues—human reproduction and natural reproduction of edible plants (or means of subsistence)—as two natural theorems, one (population) as geometric and the other (food supply) as arithmetic (Marx and Engels 1978: 276). Marx and Engels are correct in that human population does not grow geometrically, nor has our ability to grow food increased arithmetically. In fact, our ability to produce food has grown dramatically since the time of Malthus through such means as advanced machinery to replace human and animal labor, technology that allows for irrigation and a slew of other improvements, and large corporate farms which produce far more food than traditional family farms.

Marx and Engels (1978) were, however, also concerned about overpopulation and state, "There are *too many* people. Even the existence of men is a pure luxury; and if the worker is 'ethical,' he will be *sparing* in procreation" (p. 97). Marx and Engels' concern for population control was directed at those who could not afford (based on such criteria as the worker's ability to provide adequate food and shelter for his children) to bring children into the world, suggesting that it is unethical for people to have children if they cannot provide for them. This was an interesting perspective for Marx to take, as he was reliant on handouts from Engels; not only that, but four of Marx's seven children died in childhood largely due to their impoverished circumstances. Like many moral theorists, Marx did not necessarily practice what he preached. A working-class agitator, Marx rarely earned enough money to take care of his own children. Conversely, Engels could have easily afforded to raise children, financially speaking.

Whether or not overpopulation is a legitimate concern comes down to the issue of the Earth's carrying capacity (discussed in Chapter 2) in general, and specifically with the answer to two interrelated questions: "Is the number of people in the world too large for the earth to sustain itself?" and the second, "Is there any merit to Malthus's concern about an adequate food supply?" With regard to the first question, as we stated in Chapter 2, the world's population in early 2020 was in excess of 7.77 billion. The U.S. population was over 330 million. The Worldometers (2020) website provides running totals (counters, or meters) of the number of people in the world and a slew of other interesting demographic data including "Births today," "Births this year" (a running tabulation since January 1), "Deaths today," "Deaths this year," "Population growth today," and "Population Growth this year." It is somewhat fascinating to see how quickly the numbers of the current world population meter and births today increase faster than two per second. It is perhaps equally disturbing to see that more than one person dies per second as well. The net world population growth on a given day exceeds 220,000.

At the dawn of agriculture (circa 8000 BCE), the world's population was approximately 5 million. Over the 8,000-year period up to 1 CE the population grew to approximately 200 million, although some estimates are higher. One thousand years later and the total world population was estimated at 275 million and by 1750, a date just prior to industrialization, there were approximately 750 million people (Worldometers 2020). Thomas Malthus, who lived from 1766 to 1834, was alive when the world's population reached its first one billion people (1804). Since the time of industrialization, the world's population has not grown geometrically, but it certainly has sped up its pace. While it took all of human history to reach one billion in 1804, the 2nd billion was reached in just 123 years (1927); the third billion in less than 30 years; the fourth billion in 15 years; and the fifth billion in only 13 years. It is estimated that we will reach 9 billion in 2037 and 10 billion in 2057 (our current annual rate of population growth has slowed to 1.05 percent compared to its peak rate of 2.07 percent in 1970) (Worldometers 2020). Are these figures indicative of overpopulation? The sheer number of people in the world unquestionably puts a strain on the environment in totality and especially on any number of localized ecosystems.

If the environment can flourish, or at least sustain itself, with 9–10 billion people, the issue of overpopulation becomes somewhat moot, for now. However, the second aspect of Malthus's concern (and a concern shared by all social thinkers who worry about overpopulation) resides with our ability to feed all the people of the world so that no one goes hungry. While hunger is applicable in the context of each of us craving breakfast, lunch, dinner or a snack, in the environmental context we are concerned about world hunger.

World hunger addresses one of Malthus's chief concerns associated with an exploding world population, our ability to feed all the people of the world. In this context, hunger refers to the life-threatening lack of food. *World hunger* refers to hunger aggregated to the global level (World Hunger Education Service 2018).

While there are people in affluent nations that literally throw away perfectly good food and the number of overweight and obese people continues to expand, there are more than 820 million people facing hunger worldwide (WHO 2019). "Hunger is increasing in many countries where economic growth is lagging, particularly in middle-income countries and those that rely heavily on international primary commodity trade ... income inequality is rising in many of the countries where hunger is on the rise making it even more difficult for the poor, vulnerable or marginalized to cope with economic slowdowns and downturns" (WHO 2019). The United Nations reported in July 2020 that world hunger would increase by as many as 130 million people due to the COVID-19 pandemic.

The dual issues of income inequality and the lack to access to adequate supplies was evident in the United States during the COVID-19 pandemic as millions of Americans relied on food banks and handouts from others in order to feed themselves and their families. Most of these Americans temporarily faced food insecurity. (There are millions of Americans who regularly face food insecurity, but the numbers dramatically increased during the global pandemic.) *Food insecurity* refers to the limited or unreliable access to foods that are safe and nutritionally adequate (World Hunger Education Service 2018). "People experiencing moderate food insecurity face uncertainties abut their ability to obtain food and have had to reduce the quality and/or quantity of food they eat to get by" (WHO 2019). Furthermore, it is estimated "that over 2 billion people, mostly in low- and middle-income countries, do not have regular access to safe, nutritious and sufficient food. But irregular access is also a challenge for high-income countries, including 8 percent of the population in Northern America and Europe" (Who 2019). This fact highlights the necessity for a more efficient form of food distribution and for a profound transformation of food systems to provide sustainably produced healthy diets for a growing world population (WHO 2019). (Note: We will discuss harmful agricultural practices in Chapter 4.)

Another issue related to hunger is malnutrition. *Malnutrition* is a condition resulting from insufficient intake of biologically necessary nutrients (World Hunger Education Service 2018). Although malnutrition includes both overnutrition and undernutrition, when it comes to a discussion on global hunger, the focus is undernutrition. The number of undernourished people in the world has been on the rise since 2015 (Food and Agriculture Organization 2019). There are two basic types of malnutrition/undernutrition: protein-energy malnutrition (PEM) and micronutrient (vitamin and mineral) deficiency. PEM is of far greater concern to those who study hunger. With PEM, people are not eating enough calories and protein. Protein is necessary for key body functions, including the development and maintenance of muscles (World Hunger Education Service 2018). CARE (2020) states that malnutrition can severely affect a child's intellectual development and physical growth. They point out what we said above that "there's enough food in the world for everyone, but one person in eight goes hungry every night, and 2.3 million children die each year from malnutrition. Hunger and malnutrition kill more people each year than AIDS, malaria and tuberculosis combined" (CARE 2020).

When 820 million people worldwide face hunger and malnutrition and another 2 billion people do not have regular access to safe, nutritious and sufficient food, is it safe to say that there is not enough food to feed all the people in the world, as Malthus predicted? Or, is the underlying problem the lack of concern to distribute the food evenly across the globe? The answer could be a combination of both. Regardless, the primary question—is there enough food to feed the people of the world?—is very relevant from the perspective of overpopulation. Furthermore, when over 2 billion people face nutritious and sufficient food shortages, we have sown the seed of desperation that could trigger global social unrest. Nineteenth century social thinker Herbert Spencer believed that the scarcity of resources would lead to conflict between individuals, groups and entire societies.

Spencer coined the term "survival of the fittest" to describe the adaptation process that people go through in order to survive their environment. The very evolution of a species or of societies is ultimately a matter of the survival of the fittest. According to this notion, the survival of the fittest doctrine serves as an evolutionary process wherein the "weak" (those who fail to adapt) die off and the "fit" (those who successfully adapt to the environment) survive (Spencer 1862). Modification of behavior and adaptation, then are the key to survival, according to Spencer (Delaney 2004; Peel 1972; Spencer 1908). While some scholars have labeled Spencer strictly as an evolutionary theorist, he did in fact explain the opposite outcome in his theories on dissolution. *Dissolution* means to dissolve, or fall apart, and is applicable to entire societies that once flourished but have since disappeared. Spencer believed that dissolution occurs when a society ceases to evolve and remains stagnant, thus rendering it incapable of competing in the marketplace and of providing the critical scarce resources needed by its citizens. In addition, when a society ceases to evolve and or becomes stagnant, it risks being conquered or overwhelmed by more dominant and evolved societies. Spencer believed that the history of mankind has generally been one of evolution and dissolution. In this matter, societies that fail to provide subsistence for its people risk dissolution. The understanding of the terms "evolution" and "dissolution" is applicable to hunger because any society characterized by widespread hunger, malnutrition, and the lack of access to safe, nutritious and sufficient food risks dissolution. It is also important to point out that when people face the lack of basic necessities, they potentially become a danger to society as they are most likely to do whatever it takes to survive.

Cities and Urbanization

The advent of industrialization changed the course of human history in a vast array of ways, including our role in shaping the environment. Generally speaking, human impact on the environment has been quite negative and has increasingly led to simplified-growth ecosystems because of our producing virtual monocultures (areas where primarily one type of crop grows). Whether we are talking about massive cutting of trees to build homes and businesses, the plowing of prairies for crops and the allocation of huge tracts of land for ranching and farming, cultivating natural grass growth to conform to well-manicured lawns, and especially, the mass migration to cities, humans have greatly reduced the biological diversity of living things that exist in the "wild" ecosystems. We have created a number of monocultures (e.g., a field of corn, wheat, or soybeans; and the weeding, spraying of herbicides and pesticides to maintain lawns and gardens) at the expense of biodiversity. This loss of biodiversity in monoculture environments has its price, not only by the introduction and addition of pesticides and other chemicals that are very difficult for nature to recycle, but by the realization that monocultures are less robust and stable than more diverse ecosystems. It is our manipulation of the natural environment in conjunction with our ever-expanding population that makes environmentalists wonder whether there is a "saturation point" for the earth's carrying capacity.

We have already suggested that the total number of people in the world right now may be in excess of the earth's carrying capacity. At the very least, human activity has placed the environment in immediate danger and the diversity of many ecosystems is beyond repair. That humans are congregated in urban areas is certainly more problematic than if we were spread across the vast open lands living within the local carrying capacity confines of diverse ecosystems. Our quick review of city life merely sheds a light on the environmental problems associated with urban living. Among the many problems associated with cities are congestion and its associated problems including transportation problems like automobile emissions; increased air pollution as a result of more automobile driving (AAA reports that Americans spend an average of 17,600 minutes of driving each year), which contributes to poor health and smog problems; the transportation of goods and services into cities (especially food) and the transportation of by-products out of the cities (especially trash); high energy costs (e.g., heating and cooling); urban sprawl that infringes on valuable biodiversity and displaces many species—both plant and animal—native to that area; all sorts of pollution (from water and air to noise); and a slew of social problems (e.g., high crime, poverty, homelessness, slums, public housing, and racial segregation).

The United States Census Bureau first defined urban places in reports following the 1880 and 1890 censuses (U.S. Census 2012). An urban area in 1880 and 1890 was defined as any incorporated area that had a minimum of either 4,000 or 8,000, depending on the report. In 1950, the minimum population for an urban area was changed to 50,000 or more. The term "urban cluster" was introduced in 2000 for any incorporated area of at least 2,500 and less than 50,000 (cities over 50,000 are known as urbanized areas). A rural area continues to be defined as any population, housing, or territory outside urban areas (U.S. Census 2012). In the 2010 Census there were 3,601 identified urban areas, of which 3,104 fell under the urban cluster classification. There were nearly 250 million Americans (80.7 percent) residing in urban areas in 2010 (U.S. Census 2012). Four out of five Americans are urbanites. The 2020 U.S. Census data was not available at the time of this writing but data from other sources indicate that in 2019, approximately 84 percent of the U.S. population lived in urban areas, up from 64 percent in 1950. By 2050, is predicted that 89 percent of the U.S. population and 68 percent of the world population will live in urban areas (Center for Sustainable Systems 2019). There are more than 300 urban areas in the United States with populations above 100,000; New York City, with 8.4 million inhabitants, is the largest (Center for Sustainable Systems 2019).

As urbanization increases, many local ecosystems are lost. Consider that the changing of land from forest or agricultural uses to suburban and urban uses is increasing; between 2000 and 2010, urban land area in the U.S. increased by 15 percent. Urban land area is 106,386 square miles, or 3 percent of the total land area in the U.S. and is projected to triple from 2000 to 2050 (Center for Sustainable Systems 2019). Urban living is characterized by high density. In the U.S., the average population density is 87 people per square mile. In contrast, the population density of metropolitan areas (MSA) is 283 people per square mile; in New York City, the population density is 27,012 people per square mile

(10,429 per square kilometer). The most densely populated city in the world is Manila in the Philippines, where 107,524 people reside per square mile (41,515 per square km). Rounding out the top five most densely populated cities in the world are Mumbai, the capital of the state of Maharashtra, India (28,508 individuals per sq. km./73,835 per sq. mi.); Dhaka, the largest and capital city of Bangladesh (28,410 people per sq. km./73,582 per sq. mi.); Caloocan, the Philippines (27,916 people per sq. km./72,302 per sq. mi.); and Chennai, capital of Tamil Nadu, India (25,854 individuals per sq. km./66,961 per sq. mi.) (World Atlas 2020b).

Urbanites live in an environment with a limited ecosystem. Despite the existence of parks and trees and scattered lawns and gardens, cities are mostly concrete. There is so much concrete in the world that the only substance more widely used is water. Concrete is the material for the very foundation of modern development; it is how we try to tame nature, one slab at a time. While it provides us with structural well-being, it entombs vast tracts of fertile soil, constipates rivers, chokes habitats and is responsible for 4–8 percent of the world's CO_2 (Watts 2019). "If the cement industry were a country, it would be the third largest carbon dioxide emitter in the world with up to 2.8bn tonnes, surpassed only by China and the US.... All the plastic produced over the past 60 years amounts to 8bn tonnes. The cement industry pumps out more than that every two years" (Watts 2019). Concrete causes other environmental problems as well, including sucking up almost 10 percent of the world's industrial water use, which often strains supplies for drinking and irrigation. Concrete also adds to the heat-island effect by absorbing the warmth of the sun and trapping gases from car exhausts and air-conditioning units. Concrete and its airborne particles worsens the problem of silicosis and other respiratory diseases, while also becoming a choking hazard for all life forms (Watts 2019).

The animal species that live in cities have learned to adapt; telephone and electrical poles and wires have replaced tree branches as resting places. Pigeons and rats eat trash. Family pets are kept indoors except for the daily walking of the dog and perhaps a trip to the park. The larger the city, the more pollution there is in the air, streets and sidewalks. Noise, especially in cities like New York, is a constant irritation; that is, until residents simply learn to tune it out. One of the authors of this text teaches at a State University of New York (SUNY) university in a small, college town setting and constantly hears from his New York City students that it is "too quiet," and they complain of sleepless nights because of the lack of noise. Perhaps urbanites have simply adapted to their noisy environments. While urbanites seemed to have adapted to noise and high-density living, they often chose to ignore how dependent they are on the burning of fossil fuels for energy, rural people to grow their food, other people to deliver it to them into the cities, a lack of biodiversity, the constant search to find an area willing to take their trash, and in many cases, water being piped to them from hundreds of miles away.

City life brings with it a diversity of people and cultural opportunities and experiences, but this costly mono-ecosystem also lacks in environmental diversity.

Malthus and the Four Horsemen

Let's revisit Malthus' theorem of overpopulation. His fundamental principles have been proven to be incorrect—population, while it continues to grow rapidly, does not grow geometrically and our ability to produce food has improved more than arithmetically—nonetheless, we have already shown that he was correct in his basic assumption that there is not an adequate food supply for all the people of the world. Malthus provides us with one other important contribution to the study of the population and its effects on the environment, as he introduced the "Four Horsemen" concept as one of academic relevance.

Malthus's use of the Four Horsemen was, in part, an adaptation of the Bible's religious and spiritual description of the Four Horsemen of the Apocalypse. In the Book of Revelation (the last book in the Bible), chapter six, the story of the Four Horsemen is unveiled: conquest, war, famine, and death, each of whom rode a different colored horse. Malthus described his version of the Four Horsemen—war, famine, pestilence and disease—as forces of nature designed to keep the human population at a manageable number. He argued that when there were too many people the Earth would be confronted by these deadly forces. In short, if war, famine, pestilence and disease exist in the world it's because there are too many people fighting for scarce resources. It doesn't take a news junkie to realize that all four elements are taking place in the world today. From a Malthusian perspective, the answer to the question, "Are there too many people in the world?" would be "yes!" as evidence by the Four Horsemen.

Once again, Malthus provides us with an interesting concept that needs to be revised substantially. Years ago, Tim Delaney proposed that the four horsemen are not really forces of nature, but instead, they are mostly human made. By this reasoning, we are the creators of our own potential future demise. Delaney expands upon Malthus's theory of the "Four Horsemen" by introducing his concept of the "Five Horrorists" (Delaney 2005).

The Five Horrorists

The "Five Horrorists" represent an advanced evolutionary interpretation and development of the "Four Horsemen" concept. Breaking from the religious roots and de-emphasizing (but not excluding it) the primacy of the forces of nature component of the four horsemen, the Five Horrorists emphasize the social forces (the role of humans) that will lead to the destruction of humanity and the environment and hasten the arrival of the sixth mass extinction, if left unchecked by global socio-political powers. Along with an updated version of Malthus's four horsemen—war, famine, pestilence and disease—Delaney introduced a new, and equally deadly, partner—the "enviromare." *Enviromares* are environmental-related threats, or nightmares, to humanity and the environment, and if left unchecked, will spell the doom of the human species along with countless others. In other words, the Five Horrorists will lead the charge toward the sixth mass extinction. That is why we refer to the Five Horrorists as the updated version of the "destroyers of life."

War

One of the most devastating and consistent destroyers of life is war. *War* can be defined as a period of collective fighting between large groups or countries via the use of armed combat, usually in an open and declared manner. The United States has a long history of engaging in war beginning with the American Revolution and continuing through the present day. Today, the U.S. is known around the world as a warring nation and there is good reason for this perception as the nation has been at war for over 93 percent of its history (realizing that many of these years include wars with indigenous populations). Americans born after September 11, 2001, (the vast majority of college students) have not lived in peaceful years. In 2019, the Institute for Economics and Peace (IEP) ranked the U.S. as only the 128th most peaceful nation out of the 163 nations it tracks (Ventura 2019). The bottom five peaceful nations are Afghanistan (163); Syria (162); South Sudan (161); Yemen (160); and Iraq (159) (Ventura 2019). At any given time, nearly all the nations of the world are not at peace, for example, the IEP identified all but 11 countries in 2014 and all but 10 countries in 2016 as not peaceful. Put another way, the World Atlas (2020c) named the world's most war-torn countries (in 2020) as Syria; Iraq; Afghanistan; Mexico; Somalia; Nigeria; Sudan; South Sudan; Libya; and Egypt. To illustrate the preponderance of war, especially in these most war-torn countries, battles between combatants at war in Syria, Iraq, and Yemen persisted, and intensified in Libya, even as the COVID-19 pandemic spread throughout the Middle East (and the rest of the world) in 2020 (Bulos 2020).

The United States not only wages war against other combatants, but it has have also engaged in a "War against Poverty," "War on Drugs," "War on Terrorism," and "War against an Invisible Enemy" (COVID-19). The U.S. has lost its war against poverty and drugs and the war on terrorism has been going on for twenty years, so far. Ideally, it will at least win the war against COVID-19.

The United States certainly did not create war; in fact, war has been waged throughout history between clans of people and entire nation-states. As sociologist C. Wright Mills (1958) states, "To reflect upon war is to reflect upon the human condition" (p. 1). And what does it say about the human condition that war is so commonplace that there exist "laws" of war? In essence humanity is resigned to the fact that there will always be war and that rules must be abided by, rather than having the nations of the world adhere to an agreement not to go to war. In other words, why not have a law that forbids war? Instead, the exact opposite is the case. "The law of war is usually divided into *jus ad bellum* (the right to resort to war) and *jus in bello* (the law during war)" (Wallace 2002: 1699). The first law of war states that nations have the *right* to go to war. And if they have that right, why not exercise it whenever deemed necessary? But it better be a "just" war or the rest of the world will condemn the aggressor. The human condition, influenced by political motivation, dictates that other nations must take sides during war. In extreme cases this results in "world wars." Can we survive another world war? What if nuclear weapons are used by both sides? Do the acts of "terrorists" justify going to war against a nation from where the insurgents originated, even if the masses have nothing to do with terrorism? The second law of war dictates that warriors must abide by certain protocols; otherwise, they risk being accused of "war crimes." *War crime violations* include the ill-treatment of prisoners, mass murder of an indigenous population, subjecting a population into slavery, killing hostages, and wanton destruction of property not justified by military necessity (Wallace 2002). The wanton destruction of cultural landmarks and heritage by the Islamic State of Iraq and Syria (ISIS) of various places of historical, architectural, and religious importance in Iraq, Syria, and Libya in 2015 would qualify as a violation of the law of war (Delaney 2017).

The very meaning of war involves willful destruction of people and the physical environment. With the advent of nuclear and chemical warfare, the environment risks permanent forms of destruction. War is clearly a human-made condition and is not caused by such forces of nature as wind, heat, or snow. Humans create war and, as a result, it is not a force of nature as once argued by Malthus.

War affects more than just the killing of humans and humanly constructed forms of property; it also contributes to numerous environmental problems and these problems are so severe we could say that war's impact on the environment is equal to ecocide. *Ecocide* refers to the extensive damage to, or destruction of, large areas of the natural environment, leading to the loss of ecosystem(s) as a consequence of human activity. Wars are a major cause of ecocide. Among the earliest recorded attempts of war-related ecocide involved the Romans in 146 BCE who salted the fields around Carthage to impede food production (Sierra Club of Canada 2012). As the Sierra Club of Canada (2012) points out, "Although ecological disturbances brought on by war have been occurring for thousands of years, modern day warfare has made its impact increasingly severe. Recognizing the long-term and widespread impacts caused by such degradation, experts have coined the term ecocide, literally meaning the killing of the environment." Modern warfare contributes to land contamination, water contamination, deforestation, natural resource depletion, social resource depletion (wars require equipment, machinery, fuel, vehicles and so on), and landscape and ecosystem obliteration (Priddy 2020). The carbon footprint of war contributes to a rise in the CO_2 level, which results in damage to the ozone and acceleration of the greenhouse effect. The effects of war on the environment also includes habitat destruction (e.g., during the Vietnam War when U.S. forces sprayed herbicides like Agent Orange on the forests and mangrove swamps); the mass movement of refugees (they are forced to convert a variety of ecosystems into sources of survival); the introduction of invasive species (e.g., military ships, cargo airplanes, and trucks often carry more than soldiers and munitions—they may contain non-native plants and animals); infrastructure collapse (e.g., water treatment plants destroyed, comprising the quality of water; and the bombing or chemical manufacturing facilities that release toxins into the air, water and land); scorched earth practice (a tragic wartime custom); the use of biological, chemical and nuclear weapons (causing devastation to ecosystems) (Lallanilla 2019a).

The wreckage caused by war in general, and ecocide in particular, has led many activists to push for acts of ecocide to be recognized as a crime against humanity. "With the stakes of global heating intolerable, and the fanglessness of international climate agreements undeniable, it is little wonder that activists are calling for the major perpetrators of the environmental decimation to be seen as guilty parties in mass atrocity, on par with war crimes and genocide. The demand that ecocide—the decimation of ecosystems, humanity and non-human life—be prosecutable by The International Criminal Court has found renewed force in a climate movement increasingly unafraid to name its enemies" (Lennard 2019).

There have already been attempts to investigate acts of ecocide as war crimes, including the formation of Green Cross International. The environmental group Green Cross International was created to investigate war-related ecological tragedies. Its roots can be traced back to President Mikhail Gorbachev's time in office as head of state (1985–1991) of the Union of Soviet Socialist Republics (USSR). Gorbachev was concerned about human threats to the planet via the nuclear arms race, use of chemical weapons, unsustainable development, and the human-induced decimation of the planet's ecology (Green Cross International 2012a). In October 1987, five years before the first Earth Summit in Rio de Janeiro, Gorbachev addressed a gathering in the Arctic city of Murmansk and spoke on the link between possible global environmental problems and the use of nuclear weapons in war. On January 19, 1990, Gorbachev suggested creating an international Green Cross that would offer assistance to nations in ecological trouble. This organization would respond to crisis similarly to that of the Red Cross, only on ecological issues. The organization officially emerged as Green Cross International in 1998 (Green Cross International 2012a). The mission of Green Cross International is to respond to the combined challenges of security and environmental degradation to ensure a sustainable and secure future. It attempts to provide resolution to conflicts and war that contribute to environmental degradation and provide assistance to people affected by the consequences of wars, conflicts and man-made calamities (Green Cross International 2012b).

In the years since its creation, the Green Cross International organization has investigated the environments of war-torn areas, especially those in the Gulf and Africa. Unfortunately, and sadly not unpredictably, the nations of the world continue to engage in ecocide in particular and war in general. It seems C. Wright Mills was correct in his assumption that war is a key aspect of the human condition. While war itself is one of the five horrorists, it also contributes to the other horrorists, including famine.

Famine

Famine (from the Latin *fames*, meaning "hunger") is the second horrorist and is the result of both social and natural forces. The word "famine" is used to describe conditions in which a large number of people are drastically affected by a severe scarcity of food, resulting in mass malnutrition, starvation, and death. Although forces of nature, such as floods, droughts, hurricanes, and earthquakes contribute to famine, this horrorist is primarily the result of humans. Wars and deliberate crop destruction are among the leading social forces that cause famine.

Tom Keneally, author of *Three Famines* (2010), argues that another social force, certain types of governmental systems (e.g., imperialist nations that negatively influence the subjected native peoples, and dictators) impact the creation of widespread famine. In an interview with the Australian *Peninsula Living* (2011), Keneally states, "Food shortages in natural disasters are not always the real reason for famines—but one thing I can assure you is that there has never been a famine in a liberal democracy" (p. 9). In *Three Famines*, Keneally examines three of the most devastating famines in the past 200 years that had a huge impact on not only their own societies, but on international relations. The first was the Gorta More, the great hunger of Ireland in 1845–1851. This, together with other Irish famines in the mid-1800s, killed approximately three million people. The second famine detailed by Keneally is well-known but equally deadly, the famine that struck Bengal in 1943, and the third is the Ethiopian famine that decimated the country in the 1970s and again in the 1980s.

Keneally seems to have ignored countless other examples of major famines in the past. Here are a few examples in the past 150 years: Great Persian Famine of 1870–71 (1.5–2 million lives lost), caused by drought; Java, under Japanese occupation in 1942 (an estimated 2.4 million people lost their lives due to famine); Russian famine of 1921 (approximately 5 million lives in Bolshevik from 1921 to 1922); Soviet famine of 1932–33 (estimates of lives lost range from 3 to eight million); and Persian famine of 1917–1918 that was blamed for the death of up to one-quarter of the total population inhabiting northern Iran (8–10 million) World Atlas (2020d). At present, most famines are taking place in the Sub-Saharan Africa region, but with a myriad of problems such as limited food supplies, lack of quality drinking water, wars, and economic failure, famine is a horrorist that exists in many places in the world leading to widespread malnutrition, hunger and death. In 2019, the United Nations warned that 10 million Yemenis are just one step from famine as they face great food insecurity (*UN News* 2019). In 2020, the COVID-19 pandemic caused great fears of widespread famine. "Vulnerable parts of the developing world, particularly in Africa, are at risk of sliding into famine as a result of the Coronavirus pandemic, while humanitarian relief efforts are being hindered by lockdowns and travel restrictions, according to the UN" (*The Guardian* 2020).

While war may cause famine conditions, famine itself often leads to civil unrest—looting in urban areas and the raping of land to gather whatever morsel may be edible in rural areas—which can cause havoc on the environment (Gargulinski 2011). When people face famine, they must eat something and generally, this "something" is counter to their normal diet. Wildlife, livestock, and other animals such as dogs, cats, mice, rats, and birds may be eaten by starving people. Humans may eat tree buds, bark, corncobs, sawdust, or flour pounded from dried leaves (Gargulinski 2011). It was reported that some people resorted to cannibalism during the Russian famine of 1921 (World Atlas 2020d). Disease runs rampant during times of famine due to people's weakened immune systems. Mass migrations are another result of famine, as people flee areas where land now rests barren and devoid of life (Gargulinski 2011).

In Chapter 2, we promoted renewable forms of energy to lessen our dependency on fossil fuels. One small-scale example included ethanol. The benefits of ethanol are that it is proven to reduce combustion emissions, creates byproducts that are also useable, and it doesn't need to be made from just corn. However, while people around the world face hunger, starvation, malnutrition and famine, there are nations, such as the United States, that grow crops like corn for fuel, not for food for people to eat. The U.S. production of corn for fuel purposes is staggering. In 2020, the Department of Agricultural and Consumer Economics of Illinois reported that corn used for ethanol production was 5.425 billion bushels for the past year. The U.S. production of corn ethanol is a contributing force to rising food prices around the world (Goldenberg 2011). Corn is not the only biofuel crop that we use for fuel production, others include rapeseed/canola (fuel made from this crop fare well in cold climates, unlike most other biodiesels); sugarcane (a biofuel cash crop popular in Brazil); palm oil (an energy-efficient biodiesel fuel; popular in Malaysia and Indonesia but the growth of these palm trees have come at the expense of burning thousands of acres of rainforest); Jatropha (popular in India, it is a poisonous weed that grows quickly and does not demand large amounts of water); soybeans (soybean diesel yields more energy than corn ethanol); and switchgrass (far more energy efficient than corn, could represent the future of biodiesel fuels) (McFadden 2017).

What can we conclude about the use of crops used for biofuels? For one thing, we know that the use of biofuels is better than burning fossil fuels. Secondly, the promotion of certain biofuels comes with their own costs to the environment. Thirdly, and importantly, there are variations of biofuels that do not come with the high cost of loss of food supplies for humans. Fourthly, and most relevant to our discussion on horrorists, is that the diversion of grains such as corn to the production of biofuels instead of global distribution for those facing famine and hunger is a human-made force, not one from nature.

Pestilence

The third horrorist is pestilence. *Pestilence* refers to death caused by plagues. The source of these plagues can be swarms of locust, grasshoppers, and other insects that cause widespread ecological damage or a contagious epidemic disease that spreads out of control.

Swarms of locusts causing pestilence have been well-documented and stories of feeding frenzy locusts exist in biblical writings. The potential for pestilence as the result of swarms of insects exists at nearly all times. They have been known to appear "out of nowhere" to devastate crops and contribute to disease. Weather patterns can signal their sudden appearances. Desert locusts—which are usually solitary—lay their eggs in dry sand, where they can stay for years until seasonal rain causes them to hatch, breed rapidly and gather in such large numbers that swarming behavior is triggered (United Nations Environment Programme 2010).

The plague of locust consuming the Horn of Africa in 2020 serves as a case study example of the role of weather in their development. Scientists blame a pair of 2018 cyclones in the Indian Ocean; a region they say is likely to experience more such storms as global temperatures rise. These cyclones produced unusually large amounts of rain on the locusts' primary breeding ground, a swath of the Arabian Peninsula known as the Empty Quarter. That led to optional breeding conditions as locust lay their eggs in the sand, and when that sand is moist, the eggs hatch in as little as two weeks and the larvae can mature in as little as two weeks and then start reproducing again. In just nine months and three rounds of breeding, the population exploded by a factor of 8,000 (Wadekar 2020). "From the Empty Quarter—which encompasses parts of Saudi Arabia, Oman, the United Arab Emirates and Yemen—the swarms crossed the Red Sea to the coast of Somalia in the summer of 2019 and laid more eggs there before carving a path of destruction west and south, deeper into Africa" (Wadekar 2020: A3). This process of laying eggs with a quick turn-around in the breeding cycle as a result of the storms of 2018 occurred in Somalia, Ethiopia, Eritrea, Djibouti and Uganda. "Schistocerca gregaria, better known as desert locust, has smitten East Africa like a plague from the Bible" (Wadekar 2020: A3). The United Nations Environment Programme (UNEP) (2020a) warns that the rising numbers of desert locusts present an extremely alarming threat to food security and livelihoods in the Horn of Africa. Studies have linked a hotter climate to more damaging locust swarms than normal as the abnormal rains were caused by the Indian Ocean dipole, a phenomenon accentuated by climate change (UNEP 2020). As this review indicates, nature with an assistance of humans, contributes to the swarms of locusts devastating crops in Africa.

For the farmers who grow crops in this region and the consumers that depend on the food, this plague of locusts is devastating. "On a continent that has suffered repeated outbreaks of Ebola and is now fighting the coronavirus, the locusts may pose no less of a challenge. Hundreds of billions of them have conquered territory in at least eight countries, overwhelming governments and aid organizations while consuming crops and grazing land so rapidly that experts fear widespread famine" (Wadekar 2020: A3). The swarm of locusts is particularly problematic in Somalia where 2.6 million people are already displaced because of armed conflict and drought. The low level of socioeconomic development in this region of the world hampers the most effective way to combat the infestation—spraying pesticides from airplanes. The farmers are therefore left to defend for themselves resorting to such methods using sticks and clothing to try and fend off the invaders, but such efforts are ineffective. And how would it be possible for farmers to shoo away billions of locusts especially when all they do is consume vegetation? A locust can consume its body weight in vegetation in one day, which may not sound like much for one 2.5-gram locust but 40 million of them—considered a small swarm—can devour as much food as 35,000 people (Earth Observatory 2020). Realize that the swarm in Africa in 2020 is in the billions and it is easy to ascertain the damage caused by this plague of pestilence.

During spring 2020, Iran was struggling to cope with the outbreak of coronavirus and swarms of locust, some of which consists of two billion hungry insects (Iran Front Page 2020). The country utilized biological control mechanisms (e.g., aerial application of pesticides) in their war against the locust. More that 32,000 hectares were sprayed. It has become an annual event for several swarms of locusts, mainly coming from the Arabian Peninsula and Africa, to inundate Iran farmland (IFP 2020).

Locust plagues destroy crops, vegetation and ecosystems, leading to famine, starvation, and death to the populations depended upon such food supplies. And we provided just two of the contemporary examples of locust pestilence.

Pestilence is also viewed as a sudden onset, widespread plague characterized as a highly contagious and fatal epidemic disease. "Humans usually get plague after being bitten by a rodent flea that is carrying the plague bacterium or by handling an animal infected with plague" (CDC 2019a). This horrorist appeared as one of the Four Horsemen in Biblical tales. The Biblical connotation of pestilence indicates that such plagues were divine visitations. "The word is most frequently used in the prophetic books, and it occurs 25 times in Jeremiah and Ezekiel, always associated with the sword and famine" (Macalister 2020). There are many other Bible passages that describe an angry vengeful God who despises sin so much that he sends a deadly plague unto Earth. For example, "I have sent among you the pestilence after the manner of Egypt: your young men have I slain with the sword, and have taken away your horses; and I have made the stink of your camps to come unto your nostrils: yet have ye not returned unto me, saith the LORD" (from Amos 4:10).

Today, we know that the sources of pestilence—in the form of deadly diseases—are a combination of man-made and natural sources. Roos (2020) explains, "As human civilizations flourished, so did infectious disease. Large numbers of people living in close proximity to each other and to animals, often with poor sanitation and nutrition, provided fertile breeding grounds for disease. And new overseas trading routes spread the novel infections far and wide, creating the first global pandemics." These social elements become a breeding ground for pandemics.

There have been numerous pandemics throughout history, and we shall describe a few, beginning with the Plague of Athens, 430 to 426 BCE. The Plague of Athens decimated the city in the second year of the Peloponnesian War with the plague presumed to have entered the city through the port of Piraeus, having begun in Sub-Saharan Africa and moved along trade routes (Powell-Smith 2020). The plague would recur several times over the course of a few years and in total would kill between one-third and two-thirds of the population of Athens. The high-density living of Athenians contributed to the high death toll. The Antonin Plague, 165–180 CE, sometimes called the Plague of Galen, was carried by marching Roman troops and reached Rome from the Near East (Chinese records show the existence of plague around the same time, so they may have been related) (Powell-Smith 2020). The plague spread along Roman roads, reaching Gaul (modern-day France), as well as other parts of Europe. The far-reaching Roman Empire helped to easily spread the illness. Over the course of the 15-year period, approximately 30 percent of the population of the Roman Empire had been killed by the Antonin Plague (Powell-Smith 2020).

Among the deadliest pandemics in recorded history, three were caused by a single bacterium, *Yersinia pestis* (formerly known as *Pasteurella pestis*) (Roos 2020). The first of these plagues is the Plague of Justinian, arriving in Constantinople, the capital of the Byzantine Empire, in 541 CE. "It was carried over the Mediterranean Sea from Egypt, a recently conquered land paying tribute to Emperor Justinian in grain. Plague-ridden fleas hitched a ride on the black rats that snacked on the grain" (Roos 2020). The plague would wipeout Constantinople and rapidly spread across Europe, Asia, North Africa and Arabia killing an estimated 30 to 50 million people, perhaps half of the world's population (Roos 2020). The plague would never completely disappear and returned in full vengeance 800 years later in 1347 and was known as the "Black Death." In a period of just four years, 200 million lives would be lost (Roos 2020). Not sure how to treat the plague, officials in the Venetian-controlled port city of Ragusa decided to keep newly arrived sailors in isolation (generally for 14–21 days) until they could prove they weren't sick, and thus, the roots of quarantine (*quarantino*) and physical distancing. The Black Death would continue to resurface roughly every 20 years from 1348 to 1665 (40 outbreaks in 300 years) in London and with each new plague epidemic, 20 percent of the men, women and children living in the capital city were killed. By the early 1500s, England imposed the first laws to separate and isolate the sick; also, cats and dogs were believed to carry the disease, so there was a massacre of hundreds of thousands of animals (Roos 2020). The Great Plague of 1665 was the last of these series of pandemics and it was responsible for killing 100,000 Londoners in just seven months (Roos 2020). Smallpox and cholera were the cause of the next two major pandemics. Europeans brought the diseases to the New World with them. Without any prior opportunity to build immunity an estimated 90–95 percent of the indigenous population in the Americas was killed off (Roos 2020). Centuries later, smallpox became the first virus epidemic ended by a vaccine; and thankfully, we have vaccines today to kill off most major sources of pandemics from the past. Cholera has largely been eradicated in developed countries thanks to vaccines and access to clean drinking water and proper sanitation efforts; although, it is still persistent in many third-world countries.

In the period from 1889 to 1970, a number of major modern flu pandemics occurred beginning with the Russian Flu of 1889–1890. Following the path of major roads, rivers and railway lines, it took the Russian flu just 4 months to circumnavigate the planet, peaking in the United States 70 days after the original peak in St. Petersburg (Valleron, Cori, Valtat, Meurisse, Carrat and Boelle 2010). The cause of the Russian flu was some variation of the influenza virus A, subtype H1N1 or H2N2. More than one million died as a result of this pandemic. The Russia flu didn't just disappear, instead, it returned several times in subsequent years but by that time, most Europeans and Americans had built up immunity.

The Russian Flu is almost an afterthought today as this global pandemic was followed by the far more devastating Spanish Flu, 1918–1919. This epidemic was caused by an H1N1 virus with genes of avian origin. While there is a lack of consensus as to the origin of this pandemic, it spread quickly at the end of World War I and the subsequent return of soldiers to their home countries, including the U.S. It is estimated that about 500 million people globally became infected with this virus and 50 million worldwide (675,000 in the U.S.) died (CDC 2019b). The high mortality rate is attributed to the lack of a vaccine to protect against influenza infection and no antibiotics to treat secondary bacterial infections and the lack of such intervention strategies as physical isolation, quarantine, good personal hygiene, use of disinfectants, and limitations of public gatherings (CDC

2019b). The Asian flu, 1957–58, and Hong Kong Flu, 1968–70, would round out the global pandemics to kill at least one million people. The Asian flu was caused by H2N2 and the Hong Kong Flu was caused by H3N2.

In more recent times, the world has faced a number of plagues including Human Immunodeficiency virus (HIV) and the most advanced form of the disease Acquired immunodeficiency syndrome (AIDS), 1981–present. The virus attacks the body's immune system, specifically the CD4 cells, often called T cells but the body cannot fight off the infections and disease (CDC 2019c). While strides have been made to combat this deadly plague, it is estimated that nearly 40 million people have died from this global epidemic. Other recent global diseases include the Swine flu, 2009–10, an influenza (H1N1) in pigs that was spread to humans. The Mayo Clinic (2020a) adds that this virus is actually a combination of viruses from pigs, birds and humans. Severe acute respiratory syndrome (SARS-CoV) was a short-lived (2003) viral respiratory illness caused by a coronavirus that spread to more than two dozen countries around parts of the world. The origin of SARS is believed to be bats that spread it to civet cats, which then infected humans. Following the introduction of SARS, is the Ebola Virus Disease, 2014–2016, and MERS, 2015–present. Ebola, a rare and deadly disease that affects people and nonhuman primates (such as monkeys, gorillas, and chimpanzees) and is spread by bats and nonhuman pirates, is found mostly in sub-Saharan Africa. MERS is short for Middle East Respiratory Syndrome and is a coronavirus (MERS-CoV), also known as camel flu, thought to have originated from Saudi Arabia.

This brings us to 2019 when the world would face a new global pandemic known as the novel coronavirus (COVID-19), caused by the virus SARS-CoV-2. As with most of the emerging infectious diseases already discussed here that affect humans, COVID-19 is believed to have come from animals. "Pathogens have leaped from animals to humans for eons, but the pace of this spillover has increased rapidly over the last century. As 7.8 billion people on this planet radically alter ecosystems and raise, capture and trade animals at an unprecedented scale, 'the road from animal to microbe to human pathogen' has turned into a 'highway,' as journalist Sonia Shah was written" (Morris 2020). Scientists around the world seem to agree, the manner in which humans interact with animals in their habitats leads to an expected consequence—pandemic diseases. The socio-environmental aspect of global pandemics is highlighted by human actions. "By changing the nature and frequency of human-animal interactions, our actions—through the wildlife trade, deforestation, land conversion, industrial animal farming, the burning of fossil fuels and more—propel the emergence and transmission of novel and known human infectious diseases.... Humans have altered three-quarters of terrestrial environments and two-thirds of marine environments. Our ecological domination, aside from risking mass extinctions, makes humans more vulnerable to disease" (Morris 2020: A11). Consider the case of deforestation (a topic of further discussion in Chapter 4); not only does this human practice contribute to the loss of biodiversity, increase the release of greenhouse gases, and lead to higher global temperatures, the mosquitoes that transmit malaria become more abundant and infect people at a higher rate (Morris 2020). The bio-catastrophes that are modern factory farms, wherein most of the world's livestock animals live in crammed living conditions that are hotbeds for viral and bacterial pathogens, create an environment perfect for antibiotic-resistant pathogens to develop (Morris 2020).

COVID-19 began on the last day of 2019 and is believed to have originated in

Wuhan, China. (Note: In April 2020, Chinese officials blamed the U.S. Army stationed throughout Asia for the introduction and spread of COVID-19.) Wuhan is the epicenter for COVID-19, and it is a place known for live animal markets, known as "wet markets." The authors have visited wet markets in China. They are foul open marketplaces with stalls of vegetables and meats, live animals and dead animals, all intertwined in one area for tourists like us to visit but also for locals to purchase their daily food items. Shoppers can purchase all sorts of animals and insects on sticks (think of a corn dog) including bats, beetles, scorpions and so on. At these markets it is possible for one animal, like a bird, to poop onto another animal's food and then that animal poops on a civet, and the disease spreads onward. These markets are literally breeding grounds for new viruses to emerge.

As those of you that have just read this brief review have concluded, the authors believe that pandemics will continue to exist along side human populations, at least as long as human overpopulation remains a fixture, unhealthy contact between animals and humans persists, and acts against the environment continue. Consequently, it is critical that every nation in the world have a readily available pandemic-response team in place to immediately jump into action upon the first sign of an emerging pandemic. Unfortunately, the leaders of many countries, including the United States, failed to heed the lessons learned from history that pestilence is inevitable. President Trump was the beneficiary of eight-years of science-driven, environmental-friendly policies from President Barrack Obama. In addition to a robust and upward trending economy, Obama left Trump with a fully staffed White House pandemic team consisting of many top health and science specialists. But, in 2018 Trump dismantled the team, gutting it by 80 percent, leaving the country ill-prepared to handle any new threat. Trump's ineptness and discounting of the seriousness of the spread of the coronavirus led to a full-blown epidemic, stay-at-home directives, closing of schools and colleges, a collapse of the stock market, and record-setting increases in the numbers of unemployed. Trump would blame all sorts of entities (e.g., the Democrats, the media, China, the World Health Organization, and so on) rather than take a leadership role akin to the "the buck stop here" approach popularized by U.S. president Harry S. Truman to refer to the notion that the president, as the leader of a nation, has to make decisions and accept the ultimate responsibility for those decisions.

The politics of how the U.S. handled the early and continuing stages of the COVID-19 is not of primary concern here; rather, it is the effect of the disease on the environment. We have already highlighted a few anti-environment decisions made by the Trump administration as the country was distracted by COVID-19; another harmful act was the "sweeping decision by the Environmental Protection Agency to suspend enforcement on a range of health and environmental protections" (Rust, Sahagun and Xia 2020: B1). This relaxed enforcement of federal environmental laws designed to protect public health and safety is the exact opposite of what we should be doing. The coronavirus pandemic underscores why it is so important to safeguard our safety; in fact, it is more important than ever before. Gina McCarthy, president of the Natural Resources Defense Council and a former Obama-era EPA chief, states, "No one has every seen anything like this. This is a complete pass for every industry. It basically says that if somehow it's related to COVID-19, then you don't have to worry—and this is retroactive to earlier in the month—about monitoring keeping records" (Rust et al 2020: B4).

Many lessons have been learned from past and present pandemics; physical isolation and quarantines make for an effective course of action to fight the spread of the plague. The Spanish flu also taught us that the second wave of plague can result in a higher death count than the original outbreak. Today, we have the technology to develop testing techniques to let us know when it is safe for people to return to work, school, and most daily activities. Modern technology allows us to employ "contact tracing," a technique that helps to prevent the spread of a virus by proactively finding people at higher risk than others due to potential exposure, notifying them if possible, and quarantining them (Coldewey 2020). Contact tracing is accomplished via health officials working with Big Tech firms and the hiring of many more contact tracers. Contact tracing could help stop the spread of a plague but it's also of concern for privacy advocates.

Another thing we have learned about pandemics is that we are far more likely to be killed by a microbe than a missile. It is time to reallocate a large percentage of the military budget to environmental protection and human health and wellness. "Instead, Washington foreign policy has focused on a standard list of threats, most of which are exaggerated and disconnected from Americans' day-to-day lives. At the heart of our foreign policy has been a commitment to sustaining American primacy through frequent use of military power worldwide while deprioritizing domestic needs" (Menon and Ruger 2020). If our political leaders do not heed the warnings of pestilence, humans are at risk of mass extinction.

This is the time to introduce a term developed during the early stages of COVID-19 to describe those people who refused to abide by stay-at-home orders, failed to heed the requirement of wearing masks in public, chose not to avoid large gatherings (e.g., at beaches and church services), and did not maintain proper physical distance from others—*covidiot*. The term "covidiot" is in the tradition of fossil fool and greenhouse ass. To ignore the seriousness of the coronavirus pandemic (and the inevitability of future pandemics) and to hoard essential goods as to deny your neighbors equal access is to be a covidiot. Don't be a covidiot.

Disease

Disease, the fourth horrorist, is also the result of natural and social forces. Despite modern technology, disease is as prevalent today as any time in the past—although specific diseases that kill people have changed. A *disease* is any harmful deviation, or disorder of structure, from the normal functional state of an organism that manifests itself by distinguishing signs and symptoms and results in physical injury or death. According to the Centers for Disease Control and Prevention (CDC) (1994), societal and environmental changes such as a worldwide, explosive population growth, expanding poverty, urban migration, and a dramatic increase in international travel and commerce, are all factors that increase the risk of exposure to infectious agents.

A couple of generations ago, the medical profession believed that mass death due to disease would be a thing of the past. Unfortunately, infectious disease remains a leading cause of death and disability worldwide, and this horrorist is likely to remain a destroyer of humanity and the physical environment in the future. The inevitability of this horrorist has led health officials to utilize the phrase "emerging infectious diseases," or EIDs. EIDs refer to infectious diseases that have emerged since the mid–1980s or are likely to emerge in the near future.

Despite the threat of EIDs, long-existing diseases continue as among the leading causes of death in the United States. According to the CDC, heart disease (647,457) and cancer (599,108) are the two leading causes of death, accounting for nearly half (approximately 46 percent) of all deaths in 2017 (CDC 2017a; Heron 2019). Accidents (unintentional injuries) (169,936) is the third leading cause of death followed by chronic lower respiratory diseases (160,201), stroke (cerebrovascular diseases) (146,383), Alzheimer's disease (121,404), diabetes (83,564), influenza and pneumonia (55,672), nephritis, nephrotic syndrome, and nephrosis (50,633), and rounding out the top ten causes of death is intentional self-harm (suicide) (47,173).

With both heart disease and cancer there exist a wide number of subcategories. Heart disease includes such things as acute rheumatic fever and chronic rheumatic heart disease, hypertension heart disease, and hypertensive heart and renal disease, while the variations of cancer include breast, lung, prostate and colorectal. It would be impossible to discuss in any details the many examples of heart disease and cancer. "The term heart disease is often used interchangeably with the term cardiovascular disease, although cardiovascular disease generally refers to conditions that involve narrowed or blocked blood vessels that can lead to a heart attack, chest pain (angina) or stroke" (Mayo Clinic 2020b). The causes of heart disease can include heredity, infections (e.g., bacteria, viruses and parasites), and behavioral lifestyle choices (e.g., smoking, physical inactivity, poor diet, poor hygiene, excessive use of alcohol or caffeine, drug abuse, stress, some over-the-counter medications, prescription medications, dietary supplements and herbal remedies) (Mayo Clinic 2020b). A number of studies have reported associations between air pollution and heart disease (CDC 2017b). Because of the variety of types of cancer, there are many possible causes but in general, they are similar to heart disease as genetics, infections and behavioral lifestyle choices (e.g., smoking and tobacco, improper diet, lack of physical activity, and exposure to radiation and too much sun) (American Cancer Society 2020). Environmental factors such as solvents (e.g., from paint thinners, paint and grease removers, and dry cleaning); fiber, fine particles, and dust (e.g., asbestos and ceramics); dioxins (by-products of chemical processes); polycyclic aromatic hydrocarbons (PAHs) (burning carbon-containing compounds); metals (e.g., from drinking water); diesel exhaust particles (carcinogens from diesel exhaust); pesticides; toxins from fungi (those that grow on certain foods such as grains and peanuts); vinyl chloride (used almost exclusively in the U.S. by the plastics industry in manufacturing); and benzidine (from chemical compounds used as dyes for textiles, paper, and leather products), can also contribute to cancer (U.S. Department of Health and Human Services 2003).

Chronic lower respiratory disease is an ongoing condition affecting the airways and other structures of the lower respiratory tract (including the lungs) and comprises a number of subcategories including COPD, or chronic obstructive pulmonary disease (a progressive disease which increasingly hampers breathing and is a major cause of disability); chronic bronchitis (a progressive, recurring inflammation of the lower airways of the lungs); and emphysema (characterized by difficulty with breathing) (Right Diagnosis 2015). Chronic lower respiratory disease is usually the result of smoking or long-term inhalation of irritants into the lungs, such as air pollution, chemical fumes, or dust. Globally, more than 1.5 million deaths occur annually from respiratory infections attributed to the environment, such as indoor air pollution and tobacco smoke, including at least 42 percent of lower respiratory infections and 24 percent of upper respiratory infections in developing nations (WHO 2006).

According to the Mayo Clinic (2020c), "A stroke occurs when the blood supply to part of your brain is interrupted or reduced, preventing brain tissue from getting oxygen and nutrients. Brain cells begin to die in minutes." There are a number of lifestyle risk factors that may cause a stroke including being overweight or obese, physical inactivity, heavy or binge drinking, and use of illegal drugs such as cocaine and methamphetamine. Medical risks include high blood pressure, smoking or exposure to secondhand smoke, high cholesterol, diabetes, obstructive sleep apnea, cardiovascular disease, and personal or family history of stroke, heart attack or transient ischemic attack (Mayo Clinic 2020c). Environmental air pollution has emerged as a leading risk factor for stroke worldwide (Institute for Health Metrics and Evaluation 2016).

Alzheimer's disease is a type of dementia that causes problems with behavior, memory, thinking and other cognitive abilities serious enough to interfere with daily life (Alzheimer's Association 2020). It is a progressive brain disease and currently, Alzheimer's has no cure. Older age does not cause Alzheimer's, but it is the most important risk factor for the disease. Among the known causes are genetics (genetic risk factors can influence the chance of getting the disease); health lifestyle (vascular conditions increase the chances of getting the disease whereas a nutritious diet, physical activity, social engagement, sleep and mental stimulating pursuits decrease the chances); and early-life factors (higher levels of education decrease the risk) (U.S. Department of Health & Human Services 2019). A number of environmental factors such as infections (e.g., pathogens and bacteria), toxic chemicals (e.g., aluminum and mercury), and food antigens (certain modified food extracts) may also trigger Alzheimer's (Vojdani 2019).

As explained by the Mayo Clinic (2020d), "Diabetes mellitus refers to a group of diseases that affect how your body uses blood sugar (glucose). Glucose is vital to your health because it's an important source of energy for the cells that make up your muscles and tissues. It's also your brain's mail source of fuel." There are two general types of diabetes, Type 1 and Type 2. Type 1 diabetes (an autoimmune disease) is often diagnosed in childhood, is not associated with excess body weight, often associated with higher than normal ketone levels at diagnosis and is treated with insulin injections or insulin pump. Type 2 diabetes is usually diagnosed in people 30 or older, often associated with excess body weight, high blood pressure and/or cholesterol levels at diagnosis, usually treated initially without medication but may require insulin injections later in life (Diabetes.co.uk 2019). Similar to most diseases discussed here, diabetes is caused by any of the three categories: hereditary, lifestyle (especially unhealthy food consumption and large quantities of refined carbohydrates, sweets, breads and pasta; physical inactivity; stress and social isolation) and environmental (e.g., chemicals in food, water, plastic packaging, and cleaning and beauty care products have all been found to increase insulin resistance and diabetes) (Spero 2015).

Globally, it is estimated that nearly 1 in 4 (approximately 12.6 million) people die as a result of living or working in an unhealthy environment. Environmental risk factors such as air, water, and soil pollution; chemical exposures; climate change; and ultraviolet radiations, contribute to more than 100 diseases and injures (WHO 2016).

The most straight-forward conclusion we can draw from our review of the first four horrorists (the original Four Horsemen), is that a healthier environment equals healthier people. Humans need to practice proper environmental management to assure a healthier future for the next generations.

Enviromares: Environmental Nightmares

As described earlier, an "enviromare" is an environmentally produced nightmare which causes great harm to humanity and the physical environment. The enviromare is the fifth horrorist—a new destroyer of life separate from Malthus's conception of the Four Horsemen. In Malthus's era, it would appear that there was little concern about protecting the environment, as it was in fairly good shape and had been barely exposed to the harmful effects of industrialization. However, in the nearly 250 years of industrialization, humans have come perilously close to altering the earth's fragile ecosystem in a very harmful manner (Delaney 2005). Enviromares are all connected to various forms of pollution and although they are connected to nature (the ecosystem, the biosphere, etc.), they are influenced by a number of social forces.

Each of the enviromares to be discussed in the following pages are worthy of their own chapters and in many instances entire books have been written on these various forms of pollution. We will provide a brief review of each to demonstrate their collective inclusion as enviromares.

1. Air Pollution. The most important need for all living species is oxygen. Oxygen is a colorless, odorless, tasteless gaseous chemical element which appears in great abundance on Earth because it is trapped by the atmosphere. Oxygen is critical for the human respiration process and without oxygen humans and most other organisms would die within minutes. As we breathe in air it enters our blood system and is transported throughout our body to help convert the food we eat into energy. Common sense alone dictates that it is in our best interest and the best interest of our environment to keep our air clean. If we breathe polluted air, we can expect a number of health-related problems, and if our environment's atmosphere becomes polluted entire ecosystems may become compromised. As Brown (2014) explains, air pollution is linked to heart disease, stroke, and can cause or worsen asthma. Research also indicates that Americans in communities with higher smog levels are at greater risk of dying from COVID-19; this is because of the fine-particle pollution associated with smog.

That breathing air in our compromised environment can cause health problems is very bad news as the average person takes between 17,800 and 23,040 breaths a day and each one is an opportunity to put pollutants into your lungs and body and to increase health risks if you are exposed to air pollution (Brown 2014). The air we breathe may be polluted from natural sources such as smoke caused by wildfires, toxins spilled into the air because of volcanic eruptions, vapors emitted into the air from sulfur springs and the natural smells of decay from dead vegetation and animal species. Most air pollution is caused by humans via such activities as driving automobiles (because of the emissions), burning fossil fuels, and manufacturing chemicals. Air pollution is a problem for all humans, whether directly (wherein one comes in direct contact with polluted air) or indirectly (because of the overall threat to the environment).

A reliance on fossil fuels is the most obvious culprit that comes to mind when discussing air pollution (see Chapter 2) but even when people try to do the "right thing" by using mass transit there is still exposure to deadly toxics. An examination of New York City's subway system reveals that diesel fumes stink up the subway air with a toxic brew of pollution (Guse 2019). The trains run on electricity but diesel fumes from the MTA work trains spew so much soot that subway platforms contain about twice as much carbon diesel pollution as the air above ground (Guse 2019). And things are not about to change any time soon as to revamp the subway system would entail literally digging up the system at astronomical costs. Asthma has become increasingly common with subway workers as they are forced to breathe in the diesel fumes and steel dust for hours at a time.

As discussed in Chapter 2, air pollution is correlated with climate change and global warming and the worsening of the greenhouse effect and it can cause haze, smog and degrade visibility. Air pollution may cause acid rain. The term "acid rain" was coined in 1872 by an English chemist named Robert Angus Smith who observed that acidic precipitation could damage plants and other materials (EPA 2012c). *Acid rain* is a broad term describing acid rain, snow, fog and particles caused by sulfur dioxide and nitrogen oxides released by power plants, vehicles, and other sources. Acid rain harms plants, animals, and fish and erodes human-made building surfaces and national monuments (EPA 2012). "Acid rain can dissolve certain more soluble elements from the soil, like aluminum. The dissolved aluminum begins to accumulate and can reach toxic levels as it enters local streams and wetlands. Acid rain also removes important nutrients from the soil, such as calcium, potassium, and magnesium. The lack of nutrients can negatively affect the health of plants and animals. Lastly, the combination of reduced calcium and excessive aluminum can make forests more susceptible to pests, disease, and injury from freezing and drought, as a proper balance of these nutrients is vital to forest health" (USGS 2020b). Acid rain was not considered a serious environmental problem until the 1970s when scientists observed the increase in acidity of certain lakes and streams that were not in the same local area as the cause of the pollutant. This realization led to the study of long-range transport of atmospheric pollutants such as sulfur dioxide as a possible link to distant sources of pollution. Many power plants use coal with a relatively high concentration of fuel and scientists realized that this toxin could be transported throughout entire regions, thus making acid rain a federal issue (EPA 2012c). As a result, the U.S. Congress passed an Acid Deposition Act in 1980 which established, among other things, a 10-year research program that involved monitoring sites to determine how much acidic precipitation actually exists and to determine long term trends. In 1990, the U.S. Congress passed a series of amendments to the Clean Air Act to control the amount of emissions of sulfur dioxide and nitrogen oxides of factories (EPA 2012c).

It is not just the quality of outdoor air that concerns us as indoor air is often compromised as well. Indoor Air Quality (IAQ) takes into account the air quality within and around buildings and structures, especially as it relates to the health and comfort of building occupants (EPA 2020a). Poor indoor air quality may cause of a number of respiratory infections, and it is not uncommon for indoor air quality levels to be anywhere from 2 to 100 times higher than outdoor levels. Because most people spend as much as 90 percent of their time indoors, this is a big concern. Poor indoor air quality is caused by burning kerosene, wood or oil; smoking tobacco products; releases from household cleaners, pesticides or building materials; building materials and furnishings (e.g., newly installed flooring, upholstery or carpet); deteriorated asbestos-containing insulation; and certain cabinetry or furniture (made of certain pressed wood products); central heating and cooling systems and humidification devices; excess moisture; outdoor sources such as radon, pesticides, and outdoor air pollution; and even certain air fresheners (2012b; EPA 2020a).

Children are especially vulnerable to air pollution because, compared to adults, they breathe at a faster rate, their metabolic rates are higher, and they consume more food and water per pound of body weight (Chatham-Stephens, Mann, Schwartz and

Landrigan 2012). Our children are growing up in a world in which environmental toxins are literally everywhere—outdoors, indoors at home, and indoors at schools. Air pollution and cigarette smoke contribute to asthma, the most common chronic disease in childhood, which has increased 160 percent in the past 15 years for children under age 5 (Chatham-Stephens et al. 2012). Just in our schools alone many children face a slew of hazardous airborne toxins, including lead (via lead-based paint, lead-contaminated dust and soil, lead-containing art supplies, and lead-lined water pipes and water coolers); pesticides (vial inhalation, ingestion and dermal contact); mercury (lab equipment, thermometers, thermostats, batteries, and fluorescent light bulbs); arsenic (usually via pest control or treating a water supply that was contaminated); outdoor pollution (factories, power plants, smelters; automobiles) indoor air pollution (via ill-maintained buildings, formaldehyde and other volatile organic materials such as scented products, paints and lacquers, rug cleaners and paint strippers); mold (via expired food products and too much moisture or inadequate ventilation); asbestos (generally restricted to buildings built prior to the 1980s); and radon (via decaying uranium in soil and water) (Chatham-Stephens et al. 2012).

It is an understatement to say that air pollution is a major environmental health threat. "Exposure to fine particles in both the ambient environment and in the household causes about seven million premature deaths each year. Ambient air pollution (AAP) alone imposes enormous costs on the global economy, amounting to more than US $5 trillion in total welfare losses in 2013" (WHO 2020a).

To put this enviromare in perspective, bear in mind the thing we most take for granted—that every time we inhale there will be oxygen—is contaminated and if we don't take steps toward preserving the quality of our air, we risk extinction along with countless other plant, marine and animal species.

2. Water Pollution. Arguably, water is the second most important need of humans (and other living species) that must be met. While we can only go without oxygen for seconds or minutes, most humans can live between three to eight (or more) days (depending on such variables as how much water-rich foods, such as fruits, juices, or vegetables, the person consumes in their regular diet and environmental conditions the person is exposed to) without water (Delaney 2020). Despite the importance of water, the United Nations (2020) claims that globally, 785 million people remain without basic drinking water services and one out of four health care facilities in the world lack basic drinking water services. WaterAid, a non-profit organization that began in 1981, paints a far gloomier picture stating that some 3 billion people, from indigenous communities in Brazil to war-shattered villages in northern Yemen, have nowhere to wash their hands with soap and clean water at home (*Associated Press* 2020b). While access to clean water is always important, it is especially important during a pandemic; after all, as we all know, we are told to wash our hands repeatedly, every single day and especially after touching anything at all that might have the coronavirus attached to it.

The longer one goes without water, the greater the potential negative effects on health. Since water constitutes a basic essential of life, logic alone would dictate that we do all that we can to keep it clean. All living things need water to thrive. We drink it and we consume it for a number of other reasons including for cooking and cleaning. Nonetheless, a great deal of our water is not clean, and humans are responsible for a great deal of water pollution. The EPA (2012d) defines water pollution as "the addition of sewage, industrial wastes or other harmful or objectionable material to water in concentrations or in sufficient quantities to result in measurable degradation of water quality." Any body of water can become polluted, regardless of its size and location, and this is due in part to atmospheric transport of pollutants, agricultural by-product runoffs, industrial harmful pesticides, oil spills, and hazardous wastes and human neglect. "Water pollution occurs when water supplies are contaminated. Sources of water pollution include industrial waste, harmful agricultural run-offs (e.g., chemical fertilizers and manure), sewage, fracking fluids, pharmaceutical products, pesticides, animal and human waste and decay, and plastics" (Delaney 2020: 122).

One of newest sources of water pollution is "fatbergs." Heed this warning from *Newsweek* in 2019, "Fatbergs are a growing scourge infesting cities around the world—some are more than 800 feet long and weigh more than four humpback whales. These gross globs, which can cause sewer systems to block up and even overflow, have been plaguing the U.S., Great Britain and Australia for the past decade, forcing governments and utilities companies to send workers down into the sewers armed with water hoses, vacuums and scrapers with the unenviable task of prying them loose" (Watling 2019). Gracious! What sort of creature is this fatberg? Will we need Godzilla to fight it? While it sounds like fatbergs are fictional, "at their core, fatbergs are the accumulation of oil and grease that's been poured down the drain, congealing around flushed nonbiological waste like tampons, condoms and—the biggest fatberg component of all—baby wipes" (Watling 2019). Baby wipes have been around since the 1960s, but they became a significant source of water pollution in the mid–2000s when they were marketed to adults as an alternative to conventional toilet paper. Researchers investigated London fatbergs and found additional debris of pens, false teeth, and watches.

Common sense and rational thought dictates that we should do all that we can to protect our precious and limited water supplies. Why then, wasn't there a public outcry in January 2020 when the Trump administration ended federal protection for many of the nation's millions of miles of streams, arroyos and wetlands? This environmental rollback could leave the waterways more vulnerable to pollution from development, industry and farms. Trump has freely admitted that he has targeted environmental and public health regulations because he values business over ecology (Knickmeyer 2020).

The earth consists mostly of water but less than one percent of the planet's water is drinkable (Glennon 2009). Due to the earth's atmosphere, water can neither escape nor enter our planet and consequently we have a fixed amount of water. However, once water is polluted it may never be drinkable again. The 1972 Clean Water Act (an updated version of the 1948 Federal Water Pollution Control Act) was designed to protect our access to clean by establishing the basic structure for regulation discharges of pollutants into the waters of the United States and regulating quality standards for surface waters (EPA 2012e).

Sewage, industrial wastes and other harmful materials dumped into water supplies is a significant cause of water pollution. Sewage is the term used for wastewater that generally contains a high concentration of decomposing organic matter such as feces, urine and laundry waste. In developed nations, sewage is broken down by treatment at sanitation facilities. Treated water can be put back into the system but the contaminated water, along with its pollutants, are generally disposed into the sea. Among the toxins flushed by residents are chemical and pharmaceutical substances and harmful viruses and bacteria dumped by ill people via tissue and bandage products. The EPA "indicates that about 1.2 trillion gallons of wastewater coming from industrial and household sources contaminates the nation's water on an annual basis" (Green and Growing 2018). In Colorado, the Gold King mine wastewater accident of 2015, wherein 3 million gallons of wastewater spilled, led to a plume of contamination that included such deadly chemicals as arsenic, cadmium, lead, and other heavy metals that seeped into nearby streams and rivers and surface ground areas (Brown 2015).

People flush all sorts of materials, accidentally or on purpose, that they shouldn't and still our sewage-related problems pale in comparison to that of developing nations. People in developing nations may lack access to sanitary conditions and clean water and find themselves exposed to untreated sewage water that contaminates the environment and cause diseases such as diarrhea. In many developing countries, wastewater is directly pumped into freshwater bodies or sea, without using any sort of treatment system. In any region of the world, when wastewater reaches marine waters, they become polluted and unsafe for human and animal use. Depending on the particular chemicals that get discharged, wastewater can severely harm aquatic life (Green and Growing 2018).

The dumping of leftover pharmaceutical products via sewer lines has become increasingly a cause of concern. Researchers from Umea University in Sweden, for example, have been studying the harmful effects of flushed leftover drugs on perch. They found that perch exposed to anti-anxiety medication have greatly modified their behaviors, including wandering from the safety of the group, leaving them more vulnerable to natural enemies (Dennis 2013). Scientists are just beginning to understand the many possible harmful consequences of improper dumping of leftover drugs and at this time, it seems that animals are adversely affected but humans not so much. "Pharmaceuticals have already been linked to behavioral and sexual mutations in fish, amphibians, and birds, according to EPA studies" (Peterson 2007).

Other sources of waste being dumped into our water supplies include old mines that leak 50 million gallons of waste (e.g., arsenic, lead and other toxic metals) daily (Brown 2019). Companies that built mines for silver, lead, gold and other "hardrock" minerals move on once they're no longer profitable, leaving behind tainted water that still leaks out of the mines or is cleaned up at taxpayer expense (Brown 2019). Manure runoff from farm animals may contaminate underground water suppliers and into streams and lakes. Manure contains bacteria and other organisms that can cause disease; some of the more familiar germs include Campylobacter, Cryptosporidium, E. coli O157:H7, and Salmonella (Wisconsin Department of Health Services 2015). In many regions of the world, local ecosystems are at risk from poop from some type of bird, including geese. Canada geese, for example, leave large quantities of feces and a single goose defecates every 20 minutes up to 1.5 pounds each day. The droppings contain carbon, nitrogen, phosphorus, and pathogens that transmit disease and parasites from geese to humans (Hower 2013). Multiple these daily environmentally harmful droppings by the total number of geese and realize that they defecate in/near shallow waters and the problem reveals itself. The CDC, in general, warns people not to get too close to chickens, ducks and geese as it may result in Salmonella infections and outbreaks (Lee 2017).

In many parts of the country, it is common for municipalities to dump salt on icy roads and sidewalks in an attempt to stop sliding. Nationwide, more than 20 million tons of salt is used annually, and the runoff of sodium chloride crystals ends up in our streams, rivers and lakes, especially in the Northeast and Midwest. Plant species can also be harmed by the salt that doesn't reach water supplies.

Steps being taken to limit the role of this enviromare include the use of reclaimed water, commonly known as wastewater or recycled water, at many municipalities, limiting our use of water (e.g., turning the water off while brushing your teeth and only turning on to rinse), and throwing trash away in proper places rather than waterways. The EPA is promoting and attempting to implement a green infrastructure, establishing guidelines for maximum water usage per day for individuals and businesses, implementing guidelines for minimizing the wasting of water, trying to find ways to minimize polluted runoff, encouraging the storing of rainstorm runoff, setting guidelines to limit vessel sewage discharges, and overseeing a range of programs contributing to the well-being of the nation's waters and watersheds (EPA 2012f).

We all had better do our part, as water is the most precious commodity found on this planet and we cannot live without it.

3. Land Pollution. Assuming there is breathable air and clean water, the next need humans must secure is food and shelter. Humans are land creatures, and we are designed to find shelter on land. We build our homes, domesticate animals, establish businesses and so on, all on land. And while it is true that a certain amount of our food comes from marine ecosystems, we grow, hunt or gather most of our food on land. *Land* is defined as the thin layer of topsoil on the earth's surface, both land and water, plus natural resources in their original state, such as mineral deposits, wildlife, fish and timber.

Land pollution involves both soils and sediments that may be inundated with a variety of chemical and microbial contaminants from any number of sources including municipal wastewater (and associated biosolids), industry, and agriculture (EPA 2012g). Chemical contaminants range from heavy metals (e.g., mercury and polychlorinated biphenyls) to emerging contaminants such as endocrine-disrupting chemicals (EDCs) (EPA 2012g). All forms of land pollution pose potential human health problems, affect ecosystems, compromise the integrity of the soil, and potentially harm nearby bodies of water. And of course, there are the accompanying economic costs to consider as well. In short, this enviromare represents another serious threat to the well-being of humans and other species.

As defined by the *Encyclopædia Britannica* (2012), *land pollution* refers to the deposition of solid or liquid waste materials on land or underground in a manner that can contaminate or degrade the soil and groundwater, threaten public health, and cause unsightly conditions and nuisances. There are a wide variety of waste materials including solid waste (or municipal waste), construction and demolition waste debris, and hazardous waste (e.g., poisonous chemicals). As for examples of unsightly conditions and nuisances that pollute our land, one merely needs to check their surrounding area for signs of land pollution as there exists all kinds of litter (e.g., cigarette butts, fast food wrappers, bottles and cans, and abandoned appliances); plastics (e.g., plastic bags); oil spills that happen inland; debris (e.g., from construction); and so on. The contamination aspect of land pollution may come about in a variety of ways including overgrazing by domesticated animals, deforestation, agricultural mismanagement, the increased use of chemical fertilizers and pesticides, erosion, urban sprawl, and strip mining. All of these activities may lead to the decay of the topsoil, resulting in land pollution (Delaney 2005).

Conserve Energy Future (2020) describes a number of causes of land pollution, some of which overlap with the sources of land contamination described above. Deforestation to create dry or barren land strips away its ability to be fertile. The use of toxic chemical fertilizers and pesticides on land used to grow crops by farmers in attempt to feed the ever-expanding human population may result in contamination and poisoning of the soil. Overpopulation also leads to a higher demand to convert land into housing and factories and businesses to meet the needs of the people. Several land spaces have been destroyed due to extraction and mining activities. Overcrowded landfills and the need for more space to dump trash created by humans (including large waste articles such as wood, metal, bricks, plastic) results in land pollution. Nuclear energy leads to radioactive waste materials that contribute to land pollution (more on this topic later in the chapter). The leftover solid waste material from sewage treatment plants is sent to landfill sites causing land pollution.

The United Nation's Intergovernmental Panel on Climate Change (IPCC) (2019) highlights the connection between land pollution and climate change by stating that better land management can contribute to tackling the problem of climate change. The key to better land management is reducing greenhouse gas emissions. As we described in Chapter 2, the earth's temperature, including land temperatures, have increased dramatically since the dawn of industrialization. The IPCC supports the governments of the world who backed the 2015 Paris Agreement to reduce CO_2 levels. Global warming has a negative effect on land and our hopes to maintain food security for the rapidly expanding population. Conversely, poor land use adds to global warming as agriculture, forestry and other types of land use account for 28 percent of human greenhouse gas emissions (IPCC 2019). "When land is degraded, it becomes less productive, restricting what can be grown and reducing the soil's ability to absorb carbon. This exacerbates climate change, while climate change in turn exacerbates land degradation in many different ways" (IPCC 2019). Planetary warming has placed a strain on the earth's land as demonstrated by extreme weather events and changing plant and animal reactions to climate change. As an example of weather impact on land consider that the intensified rainstorms we are experiencing, along with the corresponding flooding, comes the risk of soil erosion and landslides. With global warming comes the increased probability that existing drylands will experience desertification. Roughly 500 million people already live in areas that experience desertification. Climate change also brings with the increased likelihood of drought, heatwaves, and dust storms that will cause havoc with the land. Farmers will have to learn to contend with the effects of climate change as warmer temperatures and increased drought stresses plants and crops. Invasive pests and extreme weather threaten crops that do grow. Soil erosion and land degradation reduces soil fertility and brings with it the idea to add chemical supplements to the topsoil, which in turn, will cause additional problems in the near future. The IPCC (2019) offers a glimmer of hope, but it will take some serious changes to current practices: "There is real potential here through more sustainable land use, reducing over-consumption and waste of food, eliminating the clearing and burning of forests, preventing over-harvesting of fuelwood, and reducing greenhouse gas emissions, thus helping to address land related climate change issues."

Another serious form of land pollution comes as a result of the deep well injection of waste. Deep well injection is a procedure that involves pumping liquid waste through a steel casing into a porous layer of limestone or sandstone buried deep into the earth's crust by using high pressures to force the liquid into the pores and fissures of the rock, where it is to be permanently stored. Ideally, the injected waste is sufficiently buried away to avoid its seepage into pools of fresh drinking water. Injection wells are utilized as a method for disposal of treated wastewater and this is deemed a better option than direct discharge of wastewater into surface water or on land (Hosansky 2001). There have been instances (i.e., the Love Canal case and the Rocky Mountain Arsenal case), however, wherein the treated wastewater escaped its burial area and caused detrimental effects to the environment.

Perhaps the biggest concern with polluted land is the realization that land pollution inevitably equates to a scarcity of food and the scarcity of food brings with it starvation, malnutrition, hunger and global unrest (e.g., protests, civil wars and wars against nations). If we want to eat, we had better protect the land.

4. Chemical and Nuclear Pollution. We have already introduced a number of ways that chemicals can cause harm to the environment. There are over 86,000 chemicals listed on the Toxic Substances Control Act (TSCA) Inventory and it's the EPA's job to review new chemicals every year (EPA 2019a). (The EPA provides free access to the inventory online.) Chemicals have become an essential aspect of life in developing nations among industries such as agriculture, manufacturing, oil refining, waste and drinking water management, and pharmaceutical production (Kaplan 2006). However, with so many chemicals it is impossible to understand how all the possible combinations can affect the environment. It's also of little wonder that we have a great deal of chemical-related pollution. Many chemicals, hazardous chemicals in particular, have the potential to cause great harm to humans, all other living species, and the environment.

Hazardous chemicals are substances for which a facility must maintain a material safety data sheet (MSDS) or safety data sheet (SDS) under the Occupational Safety and Health Administration (OSHA) Hazard Communication Standard (specific standards for the general industry, shipyard employment, marine terminals, longshoring, and the construction industry), which lists the criteria used to identify a hazardous chemical. MSDSs and SDSs are detailed information sheets that provide data on health hazards and physical hazards of chemicals along with associated protective measures. Over one-half million products are listed on MSDSs or SDSs (EPA 2020b). All facilities that use or store hazardous chemicals must abide by the requirement to fill out safety data sheets. Facilities must also submit an annual inventory of those chemicals by March 1 of each year to their State Emergency Response Commission (SERC), Local Emergency Planning Committee (LEPC), and local fire department. The same information submitted by facilities must also be made available to the public (EPA 2020b).

Beyond their pollution-creation capability, facilities that operate with hazardous chemicals are under risk of terrorist attack. In 2016, there were over 13,000 (down from 15,000 in 2006) recognized hazardous chemical facilities in operation. The Department of Justice estimated that nearly half of these facilities had the potential to affect upwards of one thousand people in the surrounding area if they were hit by terrorists (Kaplan 2006). The EPA estimates that 123 of those plants could affect more than one million people (Kaplan 2006). Terrorist attacks then, could have potentially catastrophic consequences. "The vulnerabilities of chemical-related industries, although recognized before 11 September 2001, became dramatically more pressing that day after terrorists demonstrated the capacity for catastrophic strikes against key U.S. targets" (Lippin et al. 2016: 1307). Over the years, the DOJ has increased its level of concern over terrorist attacks from the physical, only to now include the worry of cyber threats from remote locations. This bit of information alone begins to set the tone of danger for this enviromare.

Mass-produced chemicals represent a potential large-scale problem, but small-scale chemical operations also pose a risk; in particular, illegal drug labs. Secret labs operated by private citizens that specialize in making illegal drugs such as methamphetamine (meth) are springing up in large numbers across the nation. The Drug Enforcement

Administration (DEA) (2020) keeps a list of known drug labs on a state-by-state basis in its National Clandestine Laboratory Register Data. This list contains addresses of locations where law enforcement agencies reportedly found chemicals or other items, including the presence of either clandestine drug labs or dumpsites. (This list is made available to the public at the DEA website.)

Environmentalists, among others, have promoted the idea that we need to lessen our dependency on the burning of fossil fuels. However, with the planet's ever-expanding population and continued thirst for energy, how can we reduce our dependency? To do this, we need alternative energy sources. In Chapter 2, we discussed renewable forms of energy (wind, solar, water flow, geothermal energy, natural and human-made biomass, and tidal). Combined, these sources represent just 11 percent of the total energy consumed in the U.S. An additional 8 percent of U.S. energy comes from nuclear power. As described in Chapter 2, nuclear energy is not a renewable source of energy, but it is a sustainable source. There are many people who believe that nuclear energy is the solution to our quest to reduce/eliminate fossil fuel use. In fact, as early as the 1950s, researchers predicted that nuclear energy would supply about 21 percent of the world's commercial energy; but, by 2000, it produced only 6 percent of the world's commercial energy and 19 percent of its electricity (Harper 2012). As of 2020, nuclear energy accounted for 13.2 percent of the total global energy needs.

The U.S. Energy Information Administration (EIA) (2020) reports that as of October 31, 2019, there were 58 (down from 65 a decade earlier) commercially operating nuclear power plants with 96 (down from 104 a decade earlier) nuclear reactors in 29 (down from 31 a decade earlier) U.S. states. On April 30, 2020, the Unit 2 reactor at the Indian Point Energy Center along the Hudson River (24 miles north of Manhattan, NYC) closed for good; in April 2021, Unit 3 will also close. The closing of these two units continues the trend across the United States to decommission a number of nuclear plants, primarily over a fear of radioactivity danger (Esch 2020). The Unit 2 reactor generated a quarter of the electricity used in New York City and suburban Westchester County, where the plant is located. New natural gas plants and efficiency measures are supposed to make up for this lost energy. The Indian Point plant was linked to killing fish in the Hudson River, water contamination and experienced recurrent emergency shutdowns. There were also concerns about potential terrorist attacks (Esch 2020). Of the active nuclear plants, 32 of them have two reactors and 3 plants have three reactors. The Palo Verde nuclear power plant in Arizona is the largest nuclear plant, and it has three reactors with a combined net summer electricity generating capacity of 23,937 megawatts (MW). Despite the trend to shut down nuclear power plants, two new nuclear reactors were actively under construction in 2019 (EIA 2020).

The decrease in the number of total power plants and reactors has resulted in a rise in CO_2 emissions as non-nuclear power sources create higher levels of carbon dioxide (Conca 2019). Put another way, nuclear energy has one big advantage over the burning of fossil fuels; that is, it produces a relatively low emission level of carbon dioxide into the atmosphere. Nuclear energy is derived from the atomic nucleus in one of three ways: fusion (the fusing together of atomic nuclei); fission (the breaking down of the binding forces of an atom's nucleus); or decay (the slower, natural fission process of nucleus breaking down into a more stable form) (Delaney 2008). The most common procedure for producing nuclear energy involves the fission of uranium and plutonium or thorium or the fusion of hydrogen into helium. Currently, nuclear energy is produced almost entirely from uranium within nuclear fission reactors wherein neutrons split Uranium 235 and Plutonium 239 to release a great deal of high-temperature heat energy, which in turn powers steam turbines that generate electricity (Harper 2012). The energy created by the fission of an atom of uranium produces ten million times the energy produced by the combustion of an atom of carbon from coal.

If nuclear energy offers a better alternative to burning fossil fuels, why hasn't it become our number one source of energy? The answers to this question are all related to one explanation, the concern over the safety of nuclear energy. The threat of core reactor meltdown and human operating errors (i.e., the Chernobyl accident), concerns over how to safely transport and store nuclear radioactive waste, toxic exposure, and the threat of nuclear weapons—especially in the "wrong" hands. As a result of the fear of nuclear technology in the wrong hands, the United States, among a select few other nations, monitors closely the nuclear development of nations around the world. As the old 1980s cliché states: "One nuclear bomb can ruin your whole day." This sentiment sums up the fear that world leaders have about nuclear energy in the "wrong hands." It goes without saying that nuclear warfare not only threatens humanity, but all other living species and the totality of the environment. This fact underscores the potential threat of this enviromare in assisting the forthcoming sixth mass extinction.

Beyond military concerns, the general public has reservations about the nuclear energy because of nuclear meltdowns and industrial accidents. Major nuclear meltdowns gain so much notoriety that, like hurricanes and earthquakes, their names become synonymous with devastating nightmares. For example, most people recognize the names "Three Mile Island" because of a U.S. nuclear plant in Pennsylvania that allowed radioactive gases to escape and "Chernobyl," a nuclear power plant in the former Ukrainian Soviet Socialist Republic, now the Ukraine, which experienced a complete meltdown.

While nuclear energy does not produce much of the dangerous CO_2 toxin or other greenhouse gas emissions, it does produce radioactive waste—something fossil fuels do not. From an environmental standpoint this may be the biggest issue with nuclear energy, as meltdowns and major nuclear accidents are, thankfully, relatively rare, but even smooth-running nuclear power plants come with the problem of waste disposal. No one seems really clear on how to handle radioactive waste. Can it be buried underground safely? What if it leaks and gets into the water system or reaches the topsoil? Do we have containers that will last long enough to store nuclear waste until it no longer poses an environmental threat? Again, no one really knows how to best handle the waste problem. Consider, for example, that in 2019, Congress demanded that the Department of Energy investigate an aging, cracking U.S. nuclear dump threatened by climate change and rising seas in the Marshall Islands (Rust 2019). According to law (National Defense Authorization Act), the energy agency must submit a report that includes an "assessment of how rising sea levels might affect the dome" (Rust 2019). The Runit Dome, a waste site known alternatively s the Tomb, or simply as the Dome, holds more than 3.1 million cubic feet—or 35 Olympic-size swimming pools—of U.S.-produced radioactive soil and debris, including lethal amounts of plutonium and radiation is leaking from the site. "Nowhere else has the United States saddled another country with so much of its nuclear waste, a product of its Cold War atomic testing program" (Rust 2019). The Department of Energy has acknowledged that the facility is vulnerable to rising sea levels and storm waves.

Radioactive waste is a problem because it contains chemical elements that do not have a practical purpose. Generally speaking, nuclear waste can be classified as "low-level" or "high-level" radioactive waste. Low-level nuclear waste (the majority of radioactive waste) consists of materials used in nuclear reactors (e.g., cooling pipes and radiation suits) and waste from medical procedures (Delaney 2008). High-level radioactive waste is generally material from the core of the nuclear reactor and includes spent fuel, plutonium, uranium, and other highly radioactive materials used during fission (Delaney 2008).

The concern over nuclear power plant safety is the result of many worries including a design flaw in the electric power systems of nuclear power plants. In 2016, seven electrical engineers who work for the U.S. Nuclear Regulatory Commission (NRC) took the unusual step of petitioning the NRC as private systems in hopes of compelling regulators to fix a "significantly concern" that affects all but one of the then nation's 100 nuclear plants (Knauss 2016). "The engineers say there is a design flaw in the electric power systems [and that] the flaw prevents the detection of certain disruptions on power lines connected to the plants. If a degraded power line were called into service during an emergency, the reactor's motors, pumps and valves could burn out, preventing a safe shutdown" (Knauss 2016: B3).

5. Solid Waste Pollution. Almost any item can eventually become an example of solid waste, as solid waste is often a polite way of describing garbage. A more formal and extensive way of describing solid waste is provided by the New York State Department of Environmental Conservation (2012): "Solid waste means any garbage, refuse, sludge from a wastewater treatment plant, water supply treatment plant, or air pollution control facility and other discarded materials including solid, liquid, semi-solid, or contained gaseous material, resulting from industrial, commercial, mining and agricultural operations, and from community activities, but does not include solid or dissolved materials in domestic sewage, or solid or dissolved materials in irrigation return flows or industrial discharges that are point sources subject to permit under 33 USC 1342 … or as defined by the Atomic Energy Act of 1954." Once again, solid waste is garbage! Examples include tires, seepage, scrap metal, latex paints, furniture and toys, domestic refuse, lawn and gardening clippings, discarded appliances and vehicles, uncontaminated used oil and anti-freeze, construction and demolition debris, asbestos, packing materials, empty aerosol cans, and food items. The EPA (2019) excludes a number of wastes which are not solid wastes including domestic sewage and mixtures of domestic sewage; point source discharge; irrigation return flow; radioactive waste; in-Situ mining; pulping liquors; spent sulfuric acid; reclamation in enclosed tanks; spent wood preservatives; coke by-product wastes; splash condenser dross residue; hazardous secondary materials from the petroleum refining industry; excluded scrap metal; shredded circuit boards; spent materials generated within the primary mineral processing industry from which, minerals, acids, cyanide, water, or other values are recovered by mineral processing or by beneficiation; petrochemical recovered oil from an associated organic chemical manufacturing facility; spent caustic solutions from petroleum refining liquid treating processes used as a feedstock to produce cresylic or naphthenic acid; and a number of other hazardous materials (that would go into the hazardous waste category and not solid waste).

Americans produce a great deal of solid waste, generally referred to as municipal solid waste (MSW) by the EPA. The U.S. EPA has collected data on the generation and disposal of waste in the United States for more than 40 years. The data is collected to measure the success, or failure, of waste reduction and recycling programs across the nation. In 2017, Americans generated 267.8 million tons of MSW or 4.51 pounds per person per day (EPA 2020c). Of this total amount of MSW, approximately 67 million tons were recycled, and 27 million tons were composted for a 35.2 percent recycling and composing rate. An additional 34 million tons (12.7 percent) were combusted with energy recovery and more than 139 million tons of MSW (52.1) were landfilled (EPA 2020c). Paper and paperboard products (25 percent) made up the largest percentage of MSW generated, followed by food (15.2 percent), plastics (13.2 percent), yard trimmings (13.1 percent), metals (9.4 percent), wood (6.7 percent), textiles (6.3 percent), glass (4.2 percent), and rubber and leather (3.4 percent) (EPA 2020c).

The World Bank (2018) states that high-income countries, which account for just 16 percent of the world's population, generate more than one-third (34 percent) of the world's waste while East Asia and the Pacific region is responsible for generating close to a quarter (23 percent) of all waste. By 2050, waste generation in Sub-Saharan Africa is expected to more than triple from its current level, while South Asia will more than double its waste stream. That accounts for this huge increase in waste generation, rapid urbanization and growing populations. As a result, the World Bank (2018) claims that without urgent action, the global waste will increase by 70 percent at this current rate by 2050. As an interesting illustration of omnipresence of trash found around the world, the Nepalese government decided to clean up the trash found on Mount Everest, the world's highest mountain, and removed over 24,000 pounds of garbage. Among the trash items were food wrappers, cans, bottles, and empty oxygen cylinders. And by the way, four dead bodies were also found (Lewis 2019). In a cold bit of mountain climbing reality, 300 climbers have died on Everest since 1953, but officials are unsure of how many bodies are still on the mountain (Lewis 2019).

Solid waste that is not disposed of properly becomes land pollution and a blight on communities. Trash can be seen along our highways, streets and neighbor's yards. Some individuals, especially those in rural areas, may burn a great deal of paper-related trash while regulated waste facilities may do the same. The burning of solid waste materials produces nitrogen oxides and sulfur dioxide as well as trace amounts of toxic pollutants, such as mercury compounds and dioxins (EPA 2012h). Solid waste may end up in streams, rivers, lakes and oceans (water discharges from power plants or illegal dumping), potentially harming marine life and local ecosystems and negatively affecting water quality. As the world's population continues to expand, we have more and more people producing trash. More and more landfills and treatment plants will become necessary. Large urban municipalities already have a hard enough time disposing of their trash, often shipping it hundreds or thousands of miles away. The solid waste enviromare is only going to get worse and will most assuredly continue to haunt us into the future.

6. Noise Pollution. Of all the enviromares, noise pollution is the least likely to hasten the arrival of the sixth mass extinction. Furthermore, noise pollution is primarily an urban pollutant, but large industry and the sounds emanating from farmers' machinery often intrudes on the lives of people in rural areas. Generally, it is one's quality of life that is compromised more than the environment.

Noise pollution refers to any intrusive noise that a person finds to be annoying. As defined by the EPA (2012i), noise pollution is "unwanted or disturbing sound." Sound becomes unwanted when it interferes with normal activities such as sleeping, conversation, or in any way disrupts or compromises one's quality of life. Noise pollution does not receive the same attention as air, land or water pollution in part because we cannot see it, taste it or smell it. But noise pollution does adversely affect the lives of millions of people

(EPA 2012i). Noise pollution has been attributed to stress-related illnesses, high blood pressure, speech interference, hearing loss, sleep disruption and lost productivity. Noise induced hearing loss is the most common and adverse health effect related to noise pollution (EPA 2012i).

People may be exposed to noise pollution because of where they live, work or travel. Individuals may take steps to protect themselves from the harmful effects of noise pollution by wearing devices such as ear plugs or noise reduction headphones. The commonest sources of urban noise are transport, industry and residential influence (e.g., noise from neighbors) (Kim, Ho, Brown, Oh, Park and Ryu 2012). A number of subway trains, including those in New York City, are above ground and contribute a great deal of noise pollution. As disturbing as trains can be in the daytime, hearing the trains pass by all night can have negative effects on one's health. For example, "living in the vicinity of railway tracks produces habituation to nocturnal noise on sleep architecture" (Tassi et al. 2010). What this means is, people who are constantly exposed to high levels of noise while they sleep learn to become less responsive to noises, they black it out. This can be dangerous in cases of emergencies, as such people may fail to heed warnings of danger. Scientists have also found that loud sounds are attributing to something known as "hidden hearing loss" (HHL) an infliction that is caused by exposure to loud noises and contributes to hearing loss. HHL cannot be measured with most common tests and people with this type of hearing loss often have difficulty with complex listening tasks, such as deciphering speech among background noise (Michigan Medicine—University of Michigan 2017).

Noise pollution does not affect just humans; instead, many animal species are bothered by loud noises. If you own a dog, chances are you already know how much the high pitch sounds of fireworks on Independence Day adversely affect your pet. Research conducted in Australia has found that noise pollution is a disruption in the lives of animals. Among the findings of researchers Parris and McCauley (2020): anthropogenic noise pollution affects a range of animals across multiple habitats; animals are altering their natural behaviors or relocating to avoid noisy areas; changes in animal behavior can an effect for whole ecosystems; and marine animals are affected by the noise from a range of human activities including vessel traffic, oil and gas exploration, seismic surveys and military sonar. Some specific examples of their research indicates that certain bird populations have declined because of continuous exposure to noise. Several species of birds have begun to adjust their vocal calls in an attempt to be heard above the unnatural noise. "Male great tits (*Parus major*) for example, have been noted to change the frequency of their call in order to be heard over anthropogenic noise. Female great tits prefer lower frequency calls when selecting a mate, but these frequencies are less attractive to females.... Males are therefore placed in a difficult position—sing at a lower frequency and not be heard, or sing at a higher frequency and potentially be dismissed!" (Parris and McCauley 2020). In another study, urban European robins (*Erithacus rubecula*), highly territorial birds who rely strongly on vocal communications, have adjusted the timing of their singing to compensate for acoustic pollution (Parris and McCauley 2020). The mating call of male pobblebonk frogs, which could historically be heard up to 800 meters away, can now only be heard up to 14 meters away because of urban noise pollution. This has equated to lower rates of mating and a reduction in their population. These are just a couple of examples of animals negatively affected because of noise pollution.

The problem with noise is itself nothing new—the philosopher Arthur Schopenhauer

(1788–1860) wrote long ago that "general noise-making has been a daily torment to me all my life," adding, "There are people I know, indeed very many of them, who smile at these things, because they are insensitive to argument, to ideas, to poetry and works of art, in short to intellectual impressions of every kind.... Noise, however, is the most impertinent of all interruptions, since it interrupts, indeed disrupts, even our thoughts" (Schopenhauer 1979: 234–235). But the increasing level and unceasing amount of noise in everyday life is surely having a negative impact on our ability to think clearly, which adversely affects our chance to thrive. As Simon and Garfunkel might say, we should never underestimate the sounds of silence.

7. Celestial Pollution. The final enviromare is celestial pollution—space junk. Although people over the years have jokingly (we hope they are joking) suggested that perhaps we should just send all of our junk into outer space (a very expensive project), the fact is, space is already littered with trash orbiting the Earth. Among the space junk in our orbit are rocket fragments, used-up boosters, Soviet nuclear reactors, and obsolete satellites. Sometimes larger pieces of space debris collide with one another, creating many smaller pieces of junk. For example, in 2007, a Chinese anti-satellite weapon test left space junk particles in our orbit and in 2009 two satellites crashed into one another. According to a Congressional Research Service report, "On January 11, 2007, the People's Republic of China (PRC) conducted its first successful direct-ascent anti-satellite (ASAT) weapons test, launching a ballistic missile armed with a kinetic kill vehicle (not an exploding conventional or nuclear warhead) to destroy the PRC's Fengyun–1C weather satellite at about 530 miles up in the low earth orbit (LEO) in space" (Kan 2007: 1). Showing that out-of-control, defunct satellites can do just as much damage, if not more, than space weapons, the 2009 collision between an Iridium communications satellite and the defunct-era Cosmos 2251 spacecraft expended a great deal more destructive energy than China's infamous anti-satellite missile test did in January 2007 (Marks 2009). The collision resulted in the creation of an extra 10,000 pieces of debris shards varying in size from centimeters to tennis-ball sized, a figure more than triple the number created in the ASAT test (Marks 2009).

In fact, there are tens of millions of pieces of junk are floating around in space; albeit most of these are tiny particles. But even tiny particles can cause problems as a piece smaller than a pea could easily kill an astronaut in space by tearing a hole through their suit. Relatively small particles could also cause damage space stations such as the International Space Station. Traveling at speeds on the imperial scale of 15,000–20,000 mph, even a tiny piece of debris could cause catastrophic damage to spacecraft, including satellites that relay cellular phone calls and other important voice and data communications around the globe (Orr, Male and Graham 2009; Tallis 2015). To keep watch over the potential disastrous consequences of space junk interfering with future launches and existing satellites and space stations, the U.S. Air Force commissioned the Lockheed Martin Corporation to develop a surveillance system that will provide a continuous watch over our orbit. As of 2020, Lockheed was still constructing its "Space Fence," a sophisticated system that will dramatically improve the way the U.S. Air Force identifies and tracks objects in space (Lockheed Martin 2020). The "Space Fence radars will permit the detection of much smaller microsatellites and debris than current systems. Additionally, Lockheed Martin's Space Fence design will significantly improve the timeliness with which operators can detect space events, which could present potential threats to GPS satellites or the International Space Station" (Lockheed Martin 2020).

U.S.-based Lockheed Martin is not the only company looking to keep an eye on space debris. A Japan-based venture is working to prevent space-debris collisions by clearing the earth's orbit of junk. Astroscale Holdings Inc. has developed the world's first in-orbit debris capture and removal craft that is designed to capture a targeted piece of space junk that is on path to collide with a satellite or space stations. Ideally, the craft can capture the celestial junk and send it back toward earth, burning up on re-entry into the atmosphere (Horie 2019).

The concern with space junk is not restricted to what happens in space, however. Asteroid and meteorite impacts were causes of the third and fifth mass extinctions. Space junk, although not traveling as fast or far to reach our surface as celestial-originated asteroids and meteorites, may create devastating results if or when it crashes on the Earth's surface. With all the detritus orbiting our planet, it is inevitable that pieces of space junk will crash into our fragile environment, perhaps causing extreme harm to ecosystems. Notable examples include the 2011 defunct German ROSAT satellite that was hurtling toward the atmosphere, fearing that the fragmented parts (the satellite would fall apart upon entry) would spread across the surface possibly causing harm to humans, marine life and other species (Baetz 2011). The debris landed in the Indian Ocean and no significant damage was reported. We dodged a bullet made of space junk when the ROSAT satellite fell apart and landed where it did, causing no major damage. In 2018, China's Tiangong-1 space station crashed on the planet's surface after re-entering Earth's atmosphere. Only about 10 percent of the bus-sized, 8.5-tonne spacecraft survived being burned up on re-entry. The station landed northwest of Tahiti in the South Pacific (News.com.au 2018).

We've been lucky so far that celestial pollution has not caused much harm to us, but will that remain the case in the future?

Popular Culture Section 3

Those Lovable Mutants and the Nuclear Waste That Spawned Them

There are many examples of mutant creatures in popular culture that were spawned as a result of nuclear waste and toxic dumping, but perhaps none as famous and enduring as Godzilla.

Godzilla is such a popular mutant that he has appeared in dozens of films. The first *Godzilla* (a.k.a. *Gojira*) film was released in 1954. This film is considered the granddaddy of all "monster" movies. The Godzilla monster came about as a result of a lizard being exposed to nuclear fallout and then mutating. This Japanese-made film series reflects a time in history when Japan was still coming to grips with its role in H-bomb testing in the Pacific and its being attacked by nuclear bombs. Godzilla's rampaging of Japan reflected an entire population's angst and concern with their own possible impending mutation. In Godzilla, then, the Japanese saw themselves and that is why the monster is beloved. In fact, as the number of sequels increased, Godzilla would become a hero in the film franchise because he fought other mutant monsters, including Mothra, Mechagodzilla, Gigan, and Destoroyah.

From television, *The Simpsons* provides us with another example of a lovable mutant, Blinky, the three-eyed fish. The environmental conscience of *The Simpsons* is usually articulated by Lisa Simpson. At the opposite extreme of Lisa and her desire to enlighten the masses about the countless examples of enviromares in existence is Mr. Charles Montgomery Burns. Mr. Burns is the owner and operator of the Springfield Nuclear Power Plant. Burns' primary concern rests with maximizing profits regardless of the consequences. He has little regard for the masses, let alone his workers, and he certainly has little interest in protecting the environment.

Mr. Burns and his nuclear power plant are regularly subjected to investigations by the Environmental Protection Agency (EPA) and criticism from environmental watch groups, concerned citizens, and of course, Lisa Simpson. In an early *Simpsons* episode (#17), "Two Cars in Every Garage and Three Eyes on Every Fish" (first air date November 1, 1990), Bart and Lisa Simpson are fishing at the "Old Fishing Hole" located near the Springfield Nuclear Reactor and, much to their surprise, Bart catches a three-eyed fish. An investigative reporter happens to witness this event and publishes an article in the local newspaper headlined "Fishin' Hole, or Fission Hole?" A public uproar ensues, the state governor calls for an investigation of the power plant and Mr. Burns is left to try and diffuse the situation.

Attempting to put a positive spin on mutant fish as a result of illegal nuclear waste dumping is a daunting task but Burns comes up with a clever idea. Burns decides to run for governor with an end-goal of suspending the investigation. Burns also decides to convince the public that his "little friend" Blinky (the three-eyed fish continuously blinks its 3 eyes) is not a scary mutant but instead, a lovable petlike creature. Burns also claims that Blinky represents an evolutionary advancement over other fish. To accomplish this, Burns holds a mock interview with an actor playing naturalist Charles Darwin. Burns introduces Darwin to the viewing public and asks the naturalist to explain his concept of "natural selection" and its applicability to Blinky.

> **Darwin:** Glad to, Mr. Burns. You see, every so often Mother Nature changes her animals, giving them bigger teeth, sharper claws, longer legs, or in this case, a third eye. And if these variations turn out to be an improvement, the new animals thrive and multiply and spread across the face of the earth.
>
> **Burns:** So, you're saying this fish might have an advantage over the other fish, that it may in fact be a kind of super-fish.
>
> **Darwin:** I wouldn't mind having a third eye, would you?

Burns continues his political speech emphasizing that Blinky is simply a product of natural selection and a "miracle of nature." Burns has, in effect, suggested that nuclear radiation speeds the evolutionary process (Delaney 2008). If this is the case, does that provide comfort to you? What do you think?

Summary

In this chapter, the issue of human population and the increasing demand humans place on the environment leads to the question, "Are there too many people in the world?" Such a question arises when we realize that there are more than 820 million people worldwide that face hunger and malnutrition and another 2 billion people that face insufficient nutritious food. Billions of people also lack access to clean drinking water.

Eighteenth-century social thinker Thomas Malthus, one of the earliest "alarmists" who expressed grave concerns about overpopulation, suggested that human population was growing geometrically while food production was growing arithmetically with the conclusion that there would come a time when there would be too many people for the available food supply. Although his theorem was incorrect, Malthus was on to something by claiming that many of the growing human population would go without easy accessibility to food and water.

Malthus also used the concept of the "Four Horsemen" to illustrate nature's role in keeping human population under control. The four horsemen concept comes from the Bible. Malthus argued that the forces of war, famine, pestilence and disease would keep population "in check" by killing off those who could not find needed food supplies. Years ago, Tim Delaney coined the terms the "Five Horrorists" and "enviromares" as part of his related but much different version of the four horsemen concept. Delaney argued that war, famine, pestilence and disease are not solely caused by forces of nature, but rather, are mostly human made. Thus, humans have created the forces that will lead to their ultimate demise via the sixth mass extinction. The four horsemen concept is further updated with the introduction of a fifth "rider," the enviromare. An enviromare is an environmental nightmare and each enviromare is tied to a distinct category of pollution. The five enviromares refer to the variety of types of pollution: air, water, land, chemical and nuclear, solid waste, noise and celestial.

Delaney and Madigan take the Five Horrorist concept and elaborate on the environmental aspects of this perspective. Ultimately, it is argued that humans are speeding the sixth mass extinction process through the creation of their own demise—the Five Horrorists.

Chapter 4

Humans Will Be the Cause of the Sixth Mass Extinction

As deadly as the five horrorists are, there are additional human activities contributing to the deterioration of the environment and ultimately speeding the process of the sixth mass extinction. The topics to be discussed in this chapter are extracting fossil fuels; the creation and mass abuse of plastics; food waste; harmful agricultural practices; deforestation; marine debris; electronic waste (e-waste); and medical waste. These activities contribute, either directly or indirectly, to the sixth mass extinction. We begin our discussion with a look at the extraction of fossil fuels.

Extracting Fossil Fuels

We have discussed in some detail the threat of human dependency on fossil fuels and how this destructive behavior is contributing to the sixth mass extinction. Before we can use fossil fuels they must be extracted from the ground. As we know, fossil fuels come from plant and animal matter that died millions of years ago. "Soil and sediment built over time, putting pressure on the material and forcing oxygen out. This plant matter turned into kerogen, which becomes oil as it warms up to 110 degrees Celsius. Natural gas then forms from oil at temperatures above 110 degrees Celsius" (Huebsch 2017). *Kerogen* is a fossilized mixture of insoluble organic material found in shale and other sedimentary rock that yields oil upon heating.

In the following pages we will take a quick look at the extraction process for the primary fossil fuels of coal, crude oil and natural gas.

Coal Extraction

Coal was referenced in Chapter 2 as a useful net energy yield product because domestic mining of coal is subsidized, thus minimizing costs to fossil fuel corporations. It was also stated that coal production in the U.S. is second in the world (behind only China). The amount of the subsidies the U.S. allocates for fossil fuel extraction is both frightening and disturbing; especially when we consider that the amount is ten times more than what is spent on education (Ellsmoor 2019). The International Monetary Fund (IMF) states that $5.2 trillion USD was spent globally on fossil fuel subsidies in 2017, the equivalent of over 6.5 percent of the global GDP for that year; it also represented a half-trillion dollar increase since 2015 when China ($1.4 trillion), the United States ($649 billion) and Russia ($551 billion) were the largest subsidizers (Ellsmoor 2019). The IMF report "explains that fossil fuels account for 85% of all global subsidies and that they remain largely attached to domestic policy" and this is in spite of "nations worldwide committing to a reduction in carbon emissions and implementing renewable energy through the Paris Agreement" (Ellsmoor 2019). The IMF believes that if nations reduced subsidies in a way to create efficient fossil fuel pricing in 2015, global carbon emissions would have been lowered by 28 percent and fossil fuel air pollution deaths reduced by 46 percent. To make matters worse, Mr. Trump pulled the U.S. out of the Paris Agreement June 1, 2017. The move is one of many that demonstrates Trump's priority to support the fossil fuel industry over the health of Americans and all people across the globe.

The United States is not the only country to ignore the sentiment of the Paris Climate Change accord. For example, in 2019, the Australian government gave final approval for construction to begin on a controversial coal mine to be built by Indian company Adani. "The mine, in Queensland's Galilee Basin has been the subject of years of hold-ups over environmental approvals" (*BBC News* 2019). Pro-environment protestors fear that this approval could lead to the construction of six other mines in the area. Although the Carmichael site itself is barren land, environmentalists argue that activating the site could still harm the fragile ecosystem of the nearby Great Barrier Reef (*BBC News* 2019). Writing for *Rolling Stone*, Jeff Goodell (2019) pulled no punches in his assessment of Trump and the Australia decision to approve the Carmichael mine: "Thanks to President Trump and his transparent and perverse desire to enrich his golfing buddies in the fossil fuel industry and to accelerate the climate crisis, the U.S. is the most notorious climate criminal in the world right now. But the Aussies are giving us a run for our money." Goodell connects his concern for the future of Great Barrier Reef and the Galilee Basin, an unspoiled region of Queensland, with his condemnation for the mining and burning of coal, which is heating up and acidifying the oceans and killing of coral reefs. "Australian politicians are essentially saying they are willing to sacrifice one of the great wonders of the world for a few jobs for their pals and some extra cash in their pocket" (Goodell 2019). The industrialization of Queensland coast, which is at the edge of the Barrier Reef, will equate to water pollution and more coal barges floating over the reef (Goodell 2019).

The coal extraction process itself involves either mining coal from the surface ("opencast") or underground. "Coal miners use large machines to remove coal from the earth. Many U.S. coal deposits, called *coal beds* or *seams*, are near the earth's surface, while others are underground. Modern mining methods allow coal miners to easily reach most of the nation's coal reserves and to produce about three times more coal in one hour than in 1978" (EIA 2019c). *Surface mining* is generally utilized when coal is less than 200 feet underground. Surface mining involves large machines removing the topsoil and layers of rock (known as overburden) in order to expose coal seams. *Mountaintop removal* is a variation of surface mining where the tops of mountains are dynamited and removed to access coal seams. Surface mining represents about two-thirds of U.S. coal production as it is less expensive than underground mining. Underground mining, sometimes called deep mining, is utilized when the coal is several hundred feet below the surface; some underground mines are thousands of feet deep and require a series of tunnels that may extend out from the vertical mine shafts for miles. Miners ride elevators down deep into mine shafts and travel on small trains in the tunnels to get to the coal (EIA 2019c).

After the coal is mined, it is transported to plants that clean and process coal (to eliminate rocks, dirt, ash, sulfur, and other unwanted materials). Trains transport nearly

70 percent of coal deliveries in the U.S. for at least part of the way from mines to consumers. Barges are used to transport coal on rivers and lakes and ships are used to transport coal on the Great Lakes and the oceans to consumers in the U.S. and other countries (EIA 2019c).

Ideally, once mining has been completed at a particular site the disturbed area will be covered with topsoil for planting grass and trees, but this is something the fossil fuel industry likes to promote, rather than actually follow-through on. Generally, when companies are finished mining, they simply abandoned the mines. It is estimated that there are as many as 500,000 abandoned mine in the U.S. (AbandonedMines.gov 2020). Abandoned coal mines share many of the same negative environmental consequences as any mine, especially with regard to sedimentation and nearby water supplies. "Surface runoff can carry AML-originated silt and debris down-stream, which is often contamination by metals and acid, eventually leading to stream clogging. Sedimentation results in the blockage of the stream and can cause flooding of road and/or residences and pose a danger to the public. Sedimentation may also cause adverse impacts on fish" (AbandonedMines.gov 2020). Highly acidic water rich in metals is a serious problem at many abandoned mines and poses significant risks to surface water and ground water making them corrosive and unable to support many forms of aquatic life and vegetation; humans may also be affected by consuming water and fish tissue with a metal content (AbandonedMines.gov 2020).

There exists a running counter (much like the Worldometers referenced in Chapters 2 and 3), the World Counts, that keeps a running total on the "Tons of Coal Mined" globally (by the year, month, week or day). It is fascinating to watch how quickly the numbers climb. As an idea, in a 12-hour period on April 25, 2020, nearly 10 million tons of coal was mined, and year to date (April 25, 2020) the total was already nearly 2.5 billion tons (The World Counts 2020a). This site claims that coal provides 40 percent of world's electricity and it produces 39 percent of the global carbon dioxide emissions. A number of the negative effects of coal mining itself are also provided: destruction of landscapes and habitats (the surface is stripped of earth and rocks, and mountain tops blasted away); deforestation and erosion (as part of the process to clear the way for a coal mine, trees are cut down or burned, plants uprooted and the topsoil scraped away); ground water contamination (minerals from the disturbed earth can seep into ground water and contaminate water ways with chemicals that are hazardous to our health); chemical, air and dust pollution (e.g., mercury, arsenic, fluorine and selenium); methane in the atmosphere (methane is less prevalent in the atmosphere compared to carbon dioxide, but it is 20 times more powerful as a greenhouse gas); coal fires (underground mines can burn for centuries); health hazards (coal dust inhalation can cause black lung disease, and residents in nearby towns can experience cardiopulmonary disease, hypertension, COPD, and kidney disease); and displacement of communities (the negative consequences previously described can lead to people moving away from their homes) (The World Counts 2020a).

Clearly, the relatively inexpensive costs of coal (thanks to tax-payer subsidies) are off-set by the high costs to the health and welfare of humans, animals, and plant life.

Crude Oil Extraction

Crude oil is a gooey liquid fossil fuel compromised of hydrocarbons and other organic compounds such as oxygen, sulfur, nitrogen and small amounts of metal.

"Petroleum products are produced from the processing of crude oil and other liquids at petroleum refineries, from the extraction of liquid hydrocarbons at natural gas processing plants, and from the production of finished petroleum products at blending facilities. Petroleum is a broad category that includes both crude oil and petroleum products. The terms oil and petroleum are sometimes used interchangeably" (EIA 2019d). The EIA keeps data on U.S. oil refineries, their location, and ownership. It is typical for oil production to be measured by the barrel. The U.S. petroleum refineries produce about 19 to 20 gallons of motor gasoline and 11 to 12 gallons of ultra-low sulfur distillate fuel oil (most of which is sold as diesel fuel and in several states as heating oil) from one 42-gallon barrel of crude oil (EIA 2019d).

Geologists search for oil via such methods as satellite imagery, gravity meters and magnetometers; once a steady stream of oil is found, underground drilling can begin (OilPrice.com 2009). The most common method of crude oil extraction is drilling via onshore oil derricks and offshore oil rigs. Drilling itself is a simple process with a standard industrial method developed to provide maximum efficiency. The first step is obvious, a rig is set up above the site of the oil and drilling begins. "Once a steady flow has been identified at a particular depth beneath the ground a perforating gun is lowered into the well. A perforating gun has explosive charges within it that allow for oil to flow through holes in the casing. Once the casing is properly perforated a tube is run into the hole allowing the oil and gas to flow up the well. To seal the tubing a device called a packer is run along outside of the tube" (OilPrice.com 2009). This allows for the final step, the placement of a structure that is shaped like a pine tree which allows oil workers to control the flow of oil from the well. There are numerous environmental concerns with crude oil drilling including disruptions of wildlife habitats (oil spills can be deadly to animals); air and water pollution that is harmful to human communities; dangerous emissions that contribute to climate change; oil and gas development infrastructure that ruins pristine landscapes; fossil fuel extraction that is bad for tourism; and the light pollution caused by the oil derricks creates light pollution that impacts wildlife and the wilderness (The Wilderness Society 2019).

Crude can also be extracted from desert oil sands. The drilling process is a little different with oil sands, also known as "tar sands," as oil sands are typically a mix of sand and clay mixed with water to form a crude oil known as bitumen. Oil from bitumen is a very heavy liquid or sticky black solid with a low melting temperature (King 2020). "The method used to extract bitumen from an oil sand depends upon how deeply the oil sand is buried. If the oil sand is deeply buried, wells must be drilled to extract the bitumen. If the oil sand is close to the surface, it will be mined and hauled to a processing plant for extraction" (King 2020). There are a number of environmental concerns with oil sands mining and processing including greenhouse gas emissions and degradation of local water quality. In the U.S. the water concerns are especially important because the known oil sands and oil shale deposits are located in arid areas of Utah. Several barrels of water are required for each barrel of oil produced (King 2020).

With offshore extraction, oil companies must first find the crude and then built a platform in order to place the drilling equipment atop. The platforms must be strong enough to deal with inclement weather, especially in such places as the North Sea. Off-shore oil rigs are gigantic structures as the drilling piping must reach the depths beneath the water. The environmental problems are consistent with other forms of fossil fuel extractions beginning with the release of carbon emissions into the atmosphere and extending to oil spills (see Chapter 2) that will cause a huge environmental nightmare to oceanic sea creatures and plant life.

Natural Gas Extraction

Although natural gas is a fossil fuel, it represents a cleaner energy source than crude oil or coal. As a result, natural gas is in high demand. On the plus-side, natural gas is rather plentiful in the United States and in many other parts of the world. On the down-side, natural gas deposits are commonly located in rock formations thousands of feet below the earth's surface. And this is where the controversy over natural gas extraction comes in to play. Natural gas extraction, also known as hydraulic fracturing, and most commonly known as hydrofracking, involves a controversial method of drilling.

According to the EPA (2012j) hydraulic fracturing is a well stimulation (drilling) process that creates fractures within a reservoir that contains oil or natural gas and allows for maximum extraction. A hydraulic fracture is formed when a fracking fluid is pumped down the well at pressures that exceed the rock strength, causing open fractures to form in the rock and the potentiality of deadly toxins being released. Once a rock formation has been breached, the natural gas is released and, ideally, caught within well system. The fracking fluid consists of a mixture of water, sand and chemicals. If a sufficient amount of sand grains is trapped in the fractures, the fractures will remain propped open after the pressure of the fracking fluid is reduced. This process increases the permeability of the shale which allows much greater flow of gas back up the wellbore (Nersesian 2012).

Environmentalists, medical practitioners, and residents of communities near fracturing sites have expressed numerous concerns as a result of fracking. Among these concerns are undersurface disturbances (e.g., earthquakes); land surface disturbances; the high volume of water used during the course of the fracking process; exposure to deadly chemicals; the potentiality of underground and standing water contamination; and the overall cost-benefits analysis of hydraulic fracturing.

The concern with undersurface disturbances is the result of critics who claim that fracking not only involves the release of methane emissions, the pollution of water and air, but also the number of earthquakes sometimes set off by the practice or the disposal of related waste. When the U.K. government placed a moratorium on new fracking permits in November 2019 it acted out of concern for associated quakes (Wethe 2019). Joining the U.K. with moratoriums on fracking are Romania, Denmark, Ireland, South Africa and the Czech Republic; four countries have outright banned fracking: France, Bulgaria, Germany, and Scotland. Four U.S. states have also banned fracking: Vermont, New York, Maryland, and Washington.

Because hydraulic fracturing blasts water, sand, and hundreds of unidentified chemicals at high pressure down a well to get underground to crack rocks known as shale so that the oil or gas trapped inside can escape the process creates tiny tremors. These tremors would be mostly inconsequential, that is, unless the shale is situated along a geological fault, which then increases the odds of an earthquake occurrence (Wethe 2019). In addition to the drilling process potentially leading to an earthquake is the disposal of dirty water that comes out of the well along with oil or gas. "When this so-called produced water is injected back into the ground via another well, it can cause a pressure change that triggers an earthquake if geological conditions are right" (Wethe 2019). Data is a little sketchy on the true number of earthquakes caused by fracking in the U.S. and around the world, but the state of Oklahoma decided to find out if quake risks could be mitigated. Regulators there directed oil and natural gas producers to close some wells and reduce injection volumes in others. As a result, the number of Oklahoma earthquakes registering a magnitude of 3.0 or greater declined for three straight years.

The NYSDEC (2011) reports that the water, sand and chemical mixture that makes up fracking fluid is 98 percent water, and the other 2 percent is sand and chemicals. This leads to the next two issues, where will water come from for drilling and how dangerous is this chemical mixture? Fracking involves the use of, on average, over 5 million gallons of water at each well. The NYSDEC (2011) reports that the average water use per well in New York is approximately 3.6 million gallons. Across the United States we are talking about the deliberate use, or waste, of billions of gallons of fresh water. Drilling companies may acquire this water from existing underground sources or purchase it from suppliers. A number of people wonder whether this is an efficient use of our limited clean water supply.

Environmentalists and residents of communities that host drilling sites are also concerned with the chemicals that are being pumped underground and threaten drinking water. According to the NYSDEC, an agency that reviewed the fracking records of gas drilling companies, found that at least 260 types of chemicals, many of them toxic (e.g., benzene), have been used during the fracking process (Mouawad and Krauss 2009). These chemicals tend to remain in the ground once the fracking has been completed, raising the fears about long-term contamination (Mouawad and Krauss 2009).

In Chapter 5 of their Supplemental Generic Environment Impact Statement, "Natural Gas Development Activities and High-Volume Hydraulic Fracturing," the NYSDEC lists all of the known chemicals used in fracking. The list is extensive and clinical in description. As a sampling of their findings, we find that petroleum distillate products, such as kerosene, petroleum naphtha, petroleum base oil, and aromatic hydrocarbon among others are used. Among the potential negative health effects associated with this category of additive chemicals are adverse effects on the gastrointestinal and nervous systems and skin irritation, burning and peeling. Aromatic hydrocarbons, such as benzene, toluene, ethylbenzene and xylenes (or "BTEX") are used, and these chemicals are connected to harmful effects on the nervous system, liver, kidneys and blood-cell-forming tissues. Benzene has been linked to an increased risk of leukemia (NYSDEC 2011). The description of chemicals used in fracking and their potential adverse health effects on humans goes on and on. These same chemicals can harm other living species as well.

For many people, it is the threat (or the reality) of water contamination that remains as the greatest concern among those against fracking. As previously stated, billions of gallons of fresh water are used during the hydraulic fracturing process; but the problem with fracking and its adverse effects on fresh water supplies does not end there. Evidence of underground and surface water contamination abounds.

The term "fracking" is like a new kind of "F-word." The very mention of hydraulic fracturing makes some people cringe. In a capitalistic society it is easy to understand why gas and oil companies want to continue fracking, it's very profitable, and that's the very nature of capitalism, make as much money as possible and damn the consequences. Furthermore, the United States, like every other nation, has energy needs. We need some source of energy to fuel our growing appetites for essentials and non-essentials. We need to reduce our dependency on fossil fuels, and natural gas is a fossil fuel. Still, it does burn cleaner than coal or oil, so the movement toward natural gas could be viewed as commendable. A number of people are in favor of fracking because it represents a growth in the local community employment sector. Understandably, people who need jobs and are qualified to work in the fracking-related field find fracking wells beneficial. Tax revenue is a welcome relief for many communities as well. The benefits of the provision of jobs and an increased tax revenue base that hydrofracking can provide to economically depressed communities were among the focused topics of the 2012 film *Promised Land* (see Popular Culture Section 4 at the end of this chapter).

Plastics

In many cases, recent college graduates are unsure about the direction of their life course. This is more likely to happen to college students who did not start to look for work while in school. After graduation they have a party and then wonder, now what? If only someone would offer me words of wisdom, some graduates ponder. This is not a new feeling for college graduates and such anomic experiences have been articulated in popular culture. For example, in the 1967 classic film *The Graduate*, starring Dustin Hoffman as recent college graduate Ben Braddock, we have a character that is talented but aimless. In the film, Braddock has a conversation with a businessman named Mr. McGuire (Walter Brooke) who tells him, "I want to say one word to you to you. Just one word. Are you listening?" McGuire asks. "Yes, I am," replies Ben. "Plastics!" McGuire exclaims. Not sure what to make of this, Ben responds, "Exactly how do you mean?" McGuire states, "There's a great future in plastics. Think about it. Will you think about it?" McGuire was urging Ben Braddock to chart a course in the sale of plastic materials, pretty good advice considering it was offered more than a half century ago and the plastics-related consumer products boom had yet to occur.

It was the 1960s that witnessed the production boom in plastics. All sorts of products seemed to benefit from the switch to plastics. Anyone alive in the 1960s, or earlier, remembers products like gallons of milk sold in glass containers. You spill a gallon of milk and the glass container would break, and you would have to clean up milk and shards of glass—a project daunting enough to cry over. With plastic containers, you drop a gallon of milk and (if the lid stays on) you have no messy clean up and "no need to cry over spilt milk!"

At initial glance, plastics appear to be one of humanity's greatest convenience creations. Comedian George Carlin describes the creation of plastics as *the* purpose of human existence. As an observational humorist, Carlin, to put it generously, had a rather unique way of looking at the world and one thing is for certain, he was not an environmentalist. His rather cavalier attitude about the environment includes the subject of plastics. His classic plastics comedic routine (available for viewing via any Internet search engine by typing "Carlin and plastics") includes the notion that plastics may not be biodegradable but counters with the notion that the planet Earth will simply incorporate plastic into a new paradigm: the earth plus plastic.

Carlin states that the earth does not share our prejudice towards plastics because the ingredients for plastic come from the earth. Carlin goes on to suggest that the earth probably sees plastic as just another one of its children. Environmentalists would not describe plastics as a child of the earth, but rather as an invasive species. Carlin goes on to suggest (kiddingly, one would hope) that because the earth did not create plastics, or did not know how to create plastic, it allowed humans to spawn. In other words, the planet needed us to make plastic, and therefore, the answer to our age-old egocentric philosophical question as to "Why are we here?" is to create plastic.

So there you have it, the reason humans are on this planet is to create plastic. Now that we have plastics, what are we going to do with them? What is the planet going to do with all the plastic items we have created and pollute the landscape, waterways, and landfills with? Because humans created plastics and are the only species to benefit from them, it would be a good idea for humans to, at the very least, create biodegradable plastics. Otherwise, we risk some day being so overrun with plastics that are environment will suffer direly.

George Carlin joked that plastics represent the very reason humans exist on this planet. Most people view plastics as a modern convenience that has made everyday life simpler. Environmentalists warn that the creation of plastics and our inability to find a way to effectively incorporate biodegradable plastics into consumer use represents, essentially, a contributing factor toward a compromised environment.

Although we are not always consciously aware of it, plastics play an important role in many aspects of our daily lives. Plastics are used to manufacture everyday products such as beverage containers, toys, furniture, cleaning supply containers, and medical devices. There are two major categories of plastics: thermosets (or thermosetting plastics) and thermoplastics. A thermoset solidifies or "sets" irreversibly (meaning, once cooled and hardened, these plastics retain their shapes) when heated and is used for its durability and strength and is therefore used primarily in automobiles and construction applications, but also in adhesives, inks and coatings (EPA 2012k). A thermoplastic softens when exposed to heat and returns to its original condition at room temperature; they can be easily shaped and molded into products such as milk jugs, floor coverings, credit cards, and carpet fibers (EPA 2012k). Examples of thermoplastics would include polyethylene (PE), polypropylene (PP), and polyvinyl chloride (PVC) (Freudenrich 2020).

Carlin had described plastics as one of Earth's children and while this statement is absurd, the over-reliance and use of plastics has led to so much planetary pollution we could say that Earth has become a "plastic planet" (Chatterjee and Sharma 2019). Chatterjee and Sharma (2019) argue that the 20th century was the revolution of the plastic industry and that the 21st century represents "the time to face its consequences" (p. 54). In 2017, plastics generation was 35.4 million tons in the United States, representing 13.2 percent of municipal solid waste (MSW) generation (EPA 2019b). In 1960, plastics were less than 1 percent of the MSW (EPA 2012k). In 2019, the average American created 250 pounds of plastic waste, much of that coming from packaging (NPR 2019). The EPA (2019b) estimates that the recycling rate in 2017 was just 8.4 percent. So, what do we do with all this plastic waste? For decades, the U.S. and other industrialized nations have shipped most of their plastic waste overseas—primarily to China, where cheap labor and voracious factories dismantled the crap and turned it into new plastic goods. But China banned all plastic waste imports a few years ago amid concerns that emissions from processing were harming the environment. Many scrap dealers then rerouted their cargo to smaller recyclers in nearby Southeast Asian countries, which were suddenly overwhelmed by mounting foreign refuse (Bengali 2018). "Malaysia became the top destination for U.S. plastic waste, importing more than 192,000 metric tons in the first 10 months of 2018—a 132% jump from the year before, according to federal government data. Thailand took in more than four times as much American plastic as it did in 2017, Taiwan nearly twice as much" (Bengali 2018: A3). But now, these nations, understandably, no longer want to be a dumping ground for other nations' trash. Malaysia's Prime Minister Mahathir Mohamad announced in 2018 that in an effort to protect the environment, pledged to eliminate all single-use plastics by 2030 (Bengali 2018). And that is exactly what all nations need to do if there is any hope of eliminating the harm caused by using and discarding plastics.

Despite the fact that some plastics can be recycled, the existence of a growing market of biodegradable plastics, and a growing market for some recycled plastic resins, such as PET and HDPE, it is clear that plastics continue to have a negative impact on the environment. The U.S. National Park Service (2003) reveals the time it takes for certain items to decompose in the environment; among its listed items are newspaper, 6 weeks; plastic bag, 10–20 years; cigarette butt, 1–5 years; tin cans, 50 years; foam plastic cups, 50 years; foam plastic buoy, 80 years; disposable diapers, 450 years; and plastic beverage bottles, 450 years.

Once plastics enter the environment, it breaks down into smaller and smaller pieces but never goes away; instead, these tiny particles become microplastics. Microplastics are any plastic particle that has a diameter of less than 5 mm (Kiprop 2018). Microplastics are classified as either primary or secondary with primary microplastics produced that way for external human use and secondary microplastics as those that occur as a result of the breakdown of large plastic debris. A large collection of secondary microplastics is found at the Great Pacific Garbage Patch in the Pacific Ocean (Kiprop 2018). (Note: The Great Pacific Garbage Patch is discussed later in this chapter with marine debris.) Among the sources of microplastics are tires, textile, paint ropes, and waste treatment (Kiprop 2018). In California alone, research findings indicate that there are more than 7 trillion pieces of microplastics in California (Xia 2019). Microplastics are dangerous because, essentially, we cannot get rid of them and humans and animals alike end up ingesting them. Research indicates that humans, on average, consume a credit card's size worth of microplastics each week (Xia 2019). While some animals may consume microplastics, others will have the tiny particles embedded into their tissue. Barnacles, crabs, and a variety of fish have ingested microplastics and suffer such consequences as liver toxicity (Sahagun 2014). Chatterjee and Sharma (2019) conclude that that microplastics are very harmful to an array of marine habitants like corals, planktons, fish, seabirds, and marine mammals and state, "The adverse effects of microplastics pollution in the marine environment spans from molecular level of organisms to its physiological actions and include poor health of organisms" (p. 61). As an interesting spin on the microplastics food chain, Karmenu Vella, the European commissioner for environment, maritime affairs and fisheries states, "When we have a situation where one year you can bring your fish home in a plastic bag, and the next year you are bringing that bag home in a fish, we have to work hard and work fast" (*The Post-Standard* 2018).

Chances are, you use plastic bottled water, or whenever you are in a public place you see many people with plastic bottled water. There are water fountains almost everywhere, in schools, libraries, public buildings, and so on, so why do people with easy access to water carry bottled water with them as if their daily lives are depended upon it? In some cases, the answer is tied to the realization that the quality of water is poor in many drinking fountains and in other cases, as a result of COVID-19, water fountains are being removed from many public facilities over health concerns with people sharing such a communal source of water. (People concerned about public water fountains in the Western world would be alarmed to see communal sources of water in most developing nations.) Many people have poor water quality coming from the water taps in their homes. As a result, people often turn to water packaged within plastic containers. The vast majority of single-use plastic water bottles are made out of polyethylene terephthalate, better known as PET plastic (Gleick 2009). PET material is valued in manufacturing because the material is both strong and light. PET is a thermoplastic polymer that can be either opaque or transparent, depending on the exact material composition. But PET products are made from petroleum hydrocarbons and therefore, bad for the environment. As a result, many people are using plastic-free reusable water bottlers including Hydro Flask (stainless steel and BPA free, and vacuum sealed and insulated); lifefactory glass bottle; Klean Kanteen; and Cayman Insulated water bottles (Wells 2019).

Many people are switching to plastic-free containers of water but there is another related issue of concern here, namely, what is the source of water? Most people do not go to a stream or lake for fresh water; so, they still have to attain, or purchase, water from some source to fill these containers. That would imply the use of a fountain or water tap. For that reason, and many others, most people still purchase plastic bottles of water. Purchasing plastic containers of water has a different associated problem, that most people are oblivious to, and that is, the private ownership of water. It is interesting to note that water from home taps and drinking fountains contain trace amounts of microplastics but water from plastic bottles contain far more microplastics. The WHO found that 90 percent of the world's most popular bottled water brands contain microplastics, an average of 325 plastic particles for every liter of plastic bottled water (Readfearn 2018).

There are few times when people in countries like the United States need to rely on bottled water to meet their daily needs. And yet people have become consumers of privatized water companies almost to the point of addiction. If one believed in conspiracy theories, he or she might come to the conclusion that privatized water companies are attempting to control the nation's water supply. Let's first identify the three largest American bottled water companies: Nestlé, Coke, and Pepsi. Nestle operates in the United States under multiple brands, including Poland Springs, Arrowhead, Ozarka, and Ice Mountain; Coke owns Dasani; and Pepsi owns Aquafina (United Nations Water 2010).

The sale of bottled water began modestly enough as a trend in the 1970s when Perrier introduced bottled sparkling water to urban professionals via small green glass bottles. Immediately, bottled water became associated with conspicuous consumption, an unnecessary purchase designed to simply show off wealth. In 1989, PET plastic bottled water was introduced, and the industry took off. At that point the two major carbonated beverage industries, Coke and Pepsi, joined in the craze. The controversy, or conspiracy, takes shape when one asks, where do these companies get all their water? Generally, "they buy small, cheap plots of land in small communities, install a water pump to access the community's underground supply, and pump to their heart's content with little-to-no-overhead, taxes, regulation or accountability" (United Nations Water 2010). The laws governing water usage in the U.S. make this all possible, as underground water in most states is governed by a law established in the late 1800s called "absolute dominion," which essentially translates to "he who has the biggest pump gets the most water" (United Nations Water 2010).

Small communities all across the United States have been subject to this process and they are often oblivious to what is happening right beneath their very own feet. Privatized water companies not only sell the water they've pumped from communities as bottled water to eager consumers, but they also sell the water back to the communities they took the water from (United Nations Water 2010). Privatized water companies make billions of dollars in profits while gaining control of the nation's water supplies because people have been duped into paying for bottled water rather than drinking it, without paying for the bottle, from a tap. And, in case you did not know it, most bottled water is nothing more than tap water with a fancy name and label. Aquafina, for example, simply means finest water, and yet it is simply regular water that is available nearly everywhere in the U.S. That conspiracist might ask, "How much longer will free water be available?" Privatized water companies around the globe are taking control of water supplies, and the United States is one such nation where this is occurring.

Privatized water issues aside, our primary concern with plastic bottled water here rests with its environmental impact. It takes about 1.75 kg of petroleum in terms of energy and raw materials to make 1 kg of high-density polyethylene (Kumar, Panda, and Singh 2011). For every ton of PET plastic produced, including transportation energy to move resin to fill location, there is an energy requirement of 100,000 megajoules (Gleick 2009). In 2007, the United States had to produce one million tons of PET just to meet consumer demand for bottled water; and the global level, three times that amount was needed (Gleick 2009). There is also an energy requirement and corresponding cost to treat the water at a bottling plant. Treatment methods vary but may include some or all combination of ultraviolet radiation, micro or ultrafiltration, reverse osmosis, and ozonation (Gleick 2009). Additional bottled water costs include the total energy required to clean, fill, seal, label, and package bottled water, the total transportation energy (which is dependent on two major factors—the distance from the bottling plant to the market and the mode of transportation), and the salaries of people involved in the entire bottled water industry. It is estimated that the total energy costs can range from 5.6 mega joules to 10.2 megajoules per bottle. A barrel of oil contains approximately 6,000 megajoules, thus the estimated energy cost required to meet the demand for bottled water was equivalent to 32 to 54 million barrels of oil, or a third of a percent of total U.S. energy consumption (Gleick 2009).

The Environmental Working Group (2012) provides us with a good review of bottled water consumption and its adverse environmental impact. Consider that every 27 hours, Americans consume enough bottled water to circle the entire equator with plastic bottles stacked end to end; in just two weeks, those bottles would stretch to the moon; the federal government does not mandate that bottled water be any safer than tap water, in fact, bottled water is less regulated; close to half of all bottled water is sourced from municipal tap water; it takes an estimated 2,000 times more energy to produce bottled water than to produce the equivalent amount of tap water; and bottled water production and transportation for the U.S. market produces as much carbon dioxide as 2 million cars. Furthermore, while plastic water bottles can be recycled, the majority are not, and as a result, plastic water bottles are the fastest growing form of municipal solid waste in the United States—each year more than 4 billion pounds of PET plastic bottles end up in landfills or as roadside litter (The Environmental Working Group 2012). And bear in mind, most plastic never actually degrades; it just breaks down into smaller and smaller pieces (microplastics) that continue to cause harm for the environment.

The up-to-450-year process of decomposing of plastics; the overuse of plastic materials for consumer goods, including paper bags and wrapping and bottled water; the use of petroleum to produce plastic goods; and the realization that privatized water companies are gaining control over the available fresh water supplies should be enough to make people think twice about using plastics. George Carlin was not correct, humans were not put on earth to create plastics but now that we have, we have to find a way to wean ourselves free of plastics. In short, the use of plastics is definitely contributing to the sixth mass extinction.

Food Waste

As described in Chapter 3, there are people in affluent nations that literally throw away perfectly good food and yet more than 820 million people face hunger worldwide (in 2019) and an additional 2 billion people who do not have access to safe, nutritious and sufficient food. However, as shall see below, it is not just people at home wasting food; instead, there are many industries that contribute to food waste. Hunger and malnourishment are contributors to the sixth mass extinction, but food waste is an ally.

Food waste includes uneaten portions of meals and trimmings from food preparation activities in kitchens, restaurants, fast food chains and cafeterias (Miller 2012). The causes of food waste are plentiful and include individuals who waste food; farm waste; agricultural waste; marine waste; industrial waste; retail stores; and restaurants and institutions. Estimates of food waste caused by humans vary slightly depending on the source but seem to average to about one pound of food per person, per day. The United States Department of Agriculture (USDA) for example, found that American consumers waste about one pound of food per day, or 225–290 pounds per year. "This means that roughly 20% of all food put on the plates of Americans is trashed every year, or enough to feed 2 billion extra people annually" (Troitino 2018). The EPA calculates that each American discards a little less than a pound of food per day. The Garbage Project at the University of Arizona estimates each person wastes about 1.3 pounds of food per day while the U.S. Department of Agriculture places the figure at slightly higher than 1.3 pounds (Miller 2012). In 2017, it was estimated that food waste represented about 15.2 percent of municipal solid waste (EPA 2020c).

People waste food for any number of reasons: they prepared too much food at home and simply discard the leftovers; parents allow children not to eat prepared food and then make them eat their own food; ordering too much food at a restaurant and declining to take the leftovers home with them (via a "doggie bag"), or perhaps they are traveling and have no means of preserving the prepared foods; throwing away parts of foods that people don't like (e.g., bread crust); overconsumption of food (eating more than is required for daily nourishment, which may lead to being overweight or obese); dining at a food buffet and taking an abundance of every type of food just to sample a bite or two; and so on.

An even greater amount of food is wasted at the source of most of our food supply—farms. According to a study from the University of Arizona, 40 to 50 percent of all food ready for harvest never gets eaten (*Food Production Daily* 2004). In 2020, it was estimated that 20 billion pounds of produce alone is lost on farms each year (Food Print 2020). Food loss at the farm may occur because of severe weather (e.g., an early freeze that kills fruits and vegetables before harvest), disease that destroys crops, and predation such as squirrels that devastate an apple orchard or deer, rabbits, and other species that eat crops. Cattle and sheep, among other meat sources, may become unsuitable for human consumption before slaughter because of severe weather (e.g., herds of cattle that die during a snowstorm, volcano or hurricane), disease, and, in a few instances, predation. Farmers may also plant more than consumers demand to hedge against pests and weather. Market conditions may also impact decisions by farmers, for example, if the price of a food item (e.g., produce) on the market is lower than the cost of transportation and labor, farmers will leave their crops un-harvested—a practice known as dumping. "During the COVID-19 pandemic of 2020, for example, farmers lost a major portion of their business due to restaurant and schoolroom closures.... This led them to the painful decision to plow over edible drops and dump up to 3.7 million gallons of milk per day onto fields rather than go through the additional cost of harvesting and processing products they could not sell" (Food Print 2020).

In addition to the farming, there are other industries responsible for food waste. The Food and Agriculture Organization of the United Nations (FAO) estimated that 8 percent of the fish caught in the world's marine fisheries is discarded—about 78.3 million tons per year. Other studies estimate that 40 to 60 percent of fish caught by European trawlers in the North Sea are discarded at sea and in the U.S. between 16 and 32 percent of bycatch are thrown away by American commercial fishing boats (Food Print 2020). The process of discarding "throws the ocean's ecosystem off balance by increasing food for scavengers and killing large numbers of target and non-target fish species" (Food Print 2020). Food loss occurs at manufacturing facilities when edible portions, such as skin, fat, crusts, and peels are trimmed from food; although some of this is recovered and used for other purposes (e.g., animal feed). Still, an estimated two billion pounds of food are wasted in the food processing or manufacturing stage (FAO 2005). The transportation and distribution industries are responsible for food waste especially in developing nations where access to adequate and reliable refrigeration, infrastructure, and transportation can equate to food waste when food spoils.

The retail industry is also responsible for food waste. Grocery stores may overstock certain items, especially perishable items (e.g., milk, vegetables) only to throw the food away after the expiration date. Bakeries in food stores may throw away donuts, bagels, muffins and other baked goods after they are day old fearing they are too stale to sell. An amazing five billion eggs are thrown out each year because the Department of Agriculture will not guarantee the carton's freshness if just one egg in a dozen cracks (Blomberg 2011). Chances are, if you buy eggs, or if you have ever watched someone buy eggs in the produce section of a store, you open the carton and check to see if the eggs are intact. How many other food items do people open up to make sure none of them are cracked? Imagine if people did this with cookies, potato chips and other packaged food items. Restaurants and institutions (e.g., schools, hotels, and hospitals) throw away a great deal of food as well, with estimates ranging between 7 to 11 billion pounds per year. "Approximately 4 to 10 percent of food purchased by restaurants is wasted before reaching the consumer" (Food Print 2020). Restaurants waste food by serving oversized portions, inflexibility of chain store management and extensive menu choices. Diners leave, on average, 17 percent of their meals uneaten (Food Print 2020). As an interesting and horrifying tidbit of information, with restaurants across the United States (and many other parts of the world) closed during the COVID-19 pandemic, rats that relied on food waste at eating locations were forced to look elsewhere for their meals. In Baltimore, for example, exterminators reported that the calls for rats in people's homes doubled during the coronavirus pandemic. One Baltimore exterminator said, "They're climbing. Typically, they're coming in through an old dryer vent that's not well sealed or well covered" (*Baltimore Sun* 2020).

There are a number of ways we can reduce food waste. As individuals, we can be mindful of the food we purchase and prepare to eat. We should prepare only what we plan on eating and if there is food left over, save it and eat it as leftovers. If we go to restaurants, we should order only the food we plan to eat or bring home in a doggie bag. We should be very mindful of our behaviors at a buffet-style meal. We can encourage businesses to modify their buying and selling habits so that food waste is minimized. Consumers and retailers alike need to realize that a misshaped food item does not equate to a defect in taste. We need to accept that some foods are misshaped, bruised, or "ugly" and get over it.

Individuals and businesses alike can start backyard compost bins, worm composting bins, or any other variation of composting. Composting is an efficient way to turn food scraps into something reusable and of environmental value. Compost is created by combining organic wastes (e.g., yard trimmings, food wastes, cow manure) in proper ratios into piles, rows, or vessels; adding bulking agents (e.g., wood chips) as necessary to accelerate the breakdown of organic material to fully stabilize and mature through a curing process (EPA 2011d). Composting reduces landfill space dedicated to food waste, reduces methane emissions, regenerates poor soils, suppresses plant disease and pests, reduces or eliminates the need for chemical fertilizers and promotes higher yields of agricultural crops (EPA 2011d).

In 2011, the EPA introduced its "Food Recovery Challenge" in hopes of reducing food waste in the United States. The "Challenge" is designed for consumers, grocers, universities, stadiums, and other venues to rethink business as usual and learn to purchase leaner stocks of fresh food and find different ways to divert food waste away from landfills (EPA 2012l). The Challenge involves a three-step process: assessment of current food-related practices; second, set a goal for the amount of food waste to be reduced; and lastly, track efforts across the three food diversion action areas—prevention, donation, and composting (Zanolli 2012). Within its first year of introduction, the Challenge has 77 participants, among them MGM resorts, New Season Market, North Texas Food Bank, Middlebury College and the University of Texas at Arlington (Zanolli 2012).

Of particular interest is the Food Recovery Hierarchy shaped liked an inverse pyramid established by the EPA's Food Recovery Challenge. At the top of the inverse pyramid, which represents the most preferred food waste reduction scheme, is "source reduction," in other words, reduce waste at the source (e.g., farms). Descending down the pyramid to the least preferred option are "feed hungry people," "feed animals," "industrial uses," "composting," and "incineration or landfill" dumping (EPA 2012l).

Waste diversion efforts are being made by many. For example, in 2016 Starbucks announced a new initiative to donate unsold food to charity. The then-7,600 Starbuck stores in the U.S. would give away surplus ready-to-eat meals to food banks as part of a new program, FoodShare (Ruiz-Grossman 2016). We mentioned how buffets are a source of food waste and no place compares to Las Vegas for lavish buffets. Catholic Charities CEO Deacon Tom Roberts states, "There's probably more food consumed in Las Vegas per capita than any other place in the world. The numbers, frankly, are astounding" (*CBS News* 2018). Since 2007, much of the leftover food from MGM properties has been used to feed pigs because serving it to humans involved complicated health law restrictions. However, in 2018 MGM Resorts started a flash frozen process wherein uneaten food at banquets could be immediately stored in a high-tech refrigeration unit and donated to places like Catholic Charities. Italy started a program that provided incentives to businesses to end food waste and instead donate unsold food to charities rather than throw it away (McCarthy 2016). Italy hopes to save up to 1 billion tons of excess food per year; it recovered 500 million tons in 2015. In 2016, France became the first country in the world to ban supermarkets from throwing away or destroying unsold food, forcing them instead to donate it to charities and food banks (Chrisafis 2026). Most European supermarkets have adopted an "ugly foods" social movement by selling produce with superficial blemishes. In England, the food chain Waitrose sells misshapen, bruised and ugly produce and has made a lucrative profit by doing so. Shoppers like the cheaper prices associated with such "flawed" food; and fruits are used in jam-making and vegetables are cooked right away. In the past strict EU standards forbade such a practice and all bruised, misshaped or ugly foods were to be thrown away, juiced or shipped off to feed animals (Poulter 2011). Most American supermarket chains do not sell ugly foods but in 2016, Whole Foods Market announced that it would not trash blemished foods (Park 2016).

One other way to reduce food, and this option is not for everyone, is "freeganism," otherwise known as "dumpster dining." Freeganism refers to the anti-consumerist movement wherein people reclaim and eat food that has been discarded. Freegans, the people who participate in dumpster dining, are the scavengers of the developed world who live off consumer waste in an effort to minimize their own spending and to deny their support of corporations (More 2011). Freegans have embraced a subcultural perspective on values; they do not value how much something is worth, but rather, they value using and consuming goods that have now become reused.

Freegans include "dumpster divers," and they may also be members of Food Not Bombs. Food Not Bombs is an international group that encourages members to collect surplus food before it goes into the bins and then give it to street people (Edwards 2006). Many freegans go to grocery stores and collect all the food that is past date and pulled from shelves to be discarded but is still edible and safe to eat and distribute the food to the hungry. Freegans avoid food that is produced by industrial agriculture because of the chemicals and waste associated with corporate farms (Edwards 2006). Freegans also promote a vegan or vegetarian lifestyle.

There are many other entities and movements attempting to save and donate food, too many to identify all of them here. Each of us can play a part in the attempt to save food. Anna Brones (2017) provides a list of 101 ways to use food waste as a means of illustration (see: https://www.pastemagazine.com/food/zero-waste/101-ways-to-use-food-waste/).

The bottom line is, food waste is bad for the environment and in a number of ways. According to the Natural Resources Defense Council (NRDC), "Getting food from the farm to our fork eats up 10 percent of the total U.S. energy budget, uses 50 percent of U.S. land, and swallows 80 percent of all freshwater consumed in the United States. Yet, 40 percent of food in the United States goes uneaten. This not only means Americans are throwing out the equivalent of $164 billion each year, but also that the uneaten food ends up rotting in landfills … where it accounts for a large portion of U.S. methane emissions." A great deal of food waste that is diverted from landfills can be turned into compost. Compost is an organic material that can be used a natural fertilizer, which in turn can create healthier soil and reduce the need for synthetic fertilizers (EPA 2011d). Wasted food can also be converted into an energy source. While rotting food causes harmful methane emissions, controlled anaerobic digestion turns food waste produced methane into a valuable energy source when it is captured and harnessed for energy use (EPA 2011d). Globally, much of the world's resources are used to produce food (40 percent of its land, 70 percent of its freshwater, and 30 percent of its energy) and therefore, every piece of food that is thrown away represents wasted resources (e.g., unnecessary use of chemicals, energy and land). In a study conducted by the FAO, it is estimated that the direct and indirect costs (from impacts such as agricultural greenhouse gas emissions and erosion) add up to $2.6 trillion worldwide annually (EatForTheEarth 2014).

Harmful Agricultural Practices

Our previous discussion on food waste helps to serve as a transition to our look at harmful agricultural practices that contribute to the overall compromising of planet's environment. The connection between wasting food and harmful agricultural practices is both related to the realization that many people are ignorant about the origins of their food supplies. Such an observation is not new, in fact. Aldo Leopold (1949) stated, "There are two spiritual dangers in not owning a farm. One is the danger of supposing that breakfast comes from the grocery, and the other that heat comes from the furnace." This observation by Leopold was part of his great concern that the planet's wilderness was disappearing and that people have become disconnected from environmental reality. And remember, he made this observation more than 70 years ago.

There are people, especially young urbanites, who have little or no idea about the origin of their food. This basic lack of understanding contributes to the mentality that it is okay to throw away food and a *c'est la vie* attitude about how food is grown and produced. Because so most people live in urban areas away from agricultural farms, it is understandable that so many simply take for granted that food of sorts is available at the grocery store. However, before food reaches the grocery store there is a long process and this process often involves harmful agricultural practices that lead to a negative environmental impact.

The food production process is further complicated by the ever-expanding world's population and the corresponding increasing demand for food. Thus, we must somehow find the balance between feeding billions of people while preserving the environment's natural resources. Addressing the demand on increased food supplies has been a concern of scientists for nearly a century now. For example, in the 1920s vitamin A and vitamin D were discovered as valuable tools that allowed animals to grow strong without benefit of sunlight or exercise (McDowell 2006). This meant that farmers could maintain healthy cows indoors year-round and produce a higher yield product (via the cow's milk or use as beef). Before long, it was discovered that animals living in close proximity without outdoor exercise led to widespread disease among the confined animals. This led to the 1940s introduction of antibiotics as a feed supplement to keep livestock healthy (Castanon 2007). Following World War II, affluence and the start of the U.S. baby boom led to demands of luxury items, including the incorporation of meat in the daily meals of many people. The demand for meat led to the demise of family farms and the dominance of factory farms.

Furthermore, the attempt to produce the largest yield of food crops led to the development of genetically modified planted crops and genetically modified foods. Genetically modified foods are made up of organisms, giving us the term "genetically modified organisms," or GMO. These organisms alter the DNA of food products. Food modified genetically by organisms is supposed to represent an improvement in food production as it will lead to more consistent products and resist pests, pesticides and fertilizer. However, according to the National Center for Biotechnology Information (NCBI) there are a number of potential risks associated with GMO foods, including genetic alterations that can cause environmental harm as it's possible that modified organisms could be inbred with natural organisms, leading to the possible extinction of the original organism (Lallanilla 2019b).

Monsanto, a multinational agricultural biotechnology corporation based in the United States, is the world's largest producer of genetically engineered (GE) seeds, accounting for over 90 percent of the GE sees planted. Monsanto is responsible for the creation of some truly deadly pesticides, such as Roundup, in an attempt to protect its GE crops. Roundup, a brand name of a systemic, broad-spectrum glyphosate-based herbicide (acquired by Bayer in 2018) essentially kills everything it comes in contact with other than Monsanto's GE crops. Such a product represents an environmental nightmare and yet it is still sold in stores. Understandably, environmentalists and many citizens are concerned about the precise genetic make-up of the altered food being served to consumers. According to internal documents, some of which date back for more than a decade, Monsanto and German chemical giant BASF were aware for years that their plan to introduce GE agricultural seed and chemical system would probably lead to damage on many U.S. farms (Gillam 2020). In some internal emails, employees appear to joke about sharing "voodoo science" and hoping to stay "out of jail" (Gillam 2020). Monsanto and BASF designed the "Roundup system" as a means for farmers to spray glyphosate herbicides over the top of certain crops that Monsanto genetically engineered to survive being sprayed with pesticides. Eventually, weeds became tolerant of such herbicides (Gillam 2020). Monsanto and BASF have since designed a new industrial system built around a different herbicide called dicamba. Despite a number of lawsuits against the two giant agricultural companies, they have remained steadfast that their products are safe "when used correctly" (Gillam 2020).

The most vocal environmentalists to address harmful agricultural practices are vegetarians or vegans who promote eliminating meat (and in many cases, dairy) products from our diets. Their arguments are sound. Consider the "Laws of Thermodynamics" and its application to eating. It takes more energy to eat higher up on the food chain than it does at the lower end (Puskar-Pasewicz 2010). This idea can be applied two different ways, first, a simpler creature such as a grasshopper expends less energy to eat grass than a bird expends to eat a grasshopper, and it takes even more energy for a predator to eat a bird (Puskar-Pasewicz 2010). Secondly, it takes less energy, or fewer resources, (e.g., fuel, land, and water) to produce lower levels of food such as grains than it does to produce meat from livestock (Bittman 2009). For example, to produce 1 calorie of human-edible corn it takes 2.2 calories of fossil fuels, but to produce 1 calorie of human-edible meat it takes 40 calories of fossil fuels (Bittman 2009). In addition to the amount of energy and resources it takes to produce meat, we feed 40 percent of the world's grains and cereals to livestock that could otherwise be used to feed humans in a more sustainable fashion (Stuart 2009).

Despite the concerns of environmentalists, vegetarians and vegans, the world demand for meat was more than 320 million tons in 2017 (Ritchie and Roser 2019). Some data to ponder regarding meat consumption: the demand for meat has large environmental impacts including increasing greenhouse gas emissions (methane from livestock accounts for an estimated 14.5 percent of greenhouse emissions), agricultural land, and freshwater use. The world is consuming four times the quantity of meat it did fifty years ago; pigmeat is the most popular meat globally, but the production of poultry is increasing most rapidly; 80 billion animals are slaughtered each year for meat; meat consumption increases as the world gets richer; the world produces around 800 million tons of milk each year; and beef and lamb have much larger environmental impact than pigmeat and poultry (Ritchie and Roser 2019). The world also demands a great deal of cereal as well, 2.7 million tones in 2019 (FAO 2020a).

Agricultural practices are designed to maximize food production but often neglect the environment. We have an agricultural system designed where we must plant, maintain, fertilize, harvest, and process antibiotic enhanced products through the use of machinery that relies on fossil fuels that are sent to processing plants where they are eventually delivered to stores in trains and trucks. Place aside for a moment the concerns over meat consumption in favor of a look at meat processing and ponder, what happens when meat processing plants close, and thus curtail the availability of meat for consumers at supermarkets? Such was the case during the early stages of the COVID-19 pandemic when workers were subjected to unsafe working conditions (e.g., required to wear a mask and yet no masks were provided to workers and continuous shifts that did not allow for a thorough cleaning of meat-packing facilities). During the month of April, thousands of workers contracted the disease and dozens died. Such a large number of employees at meat-processing plants called in sick that some plants had to close, leading to shortages of beef, pork and chicken. Let's bear in mind that Tyson Foods and its two top rivals—JBS and Cargill Inc—control about two-thirds of American beef processing plants in the U.S. and its easy to see how a handful of plant closures can dramatically impact the industry. The pork and chicken industries are also dominated by a couple of large corporations. By the beginning of August 2020 there were over 16,000 cases of COVID-19 at 239 meat processing plants spread across 23 U.S. states. There were also 24 plant closures because of the coronavirus outbreaks among employees, who are jammed together on processing lines. Those closures were enough to wipe out 25 percent of pork-processing capacity and 10 percent for beef (Skerritt 2020).

Meanwhile, beef, pig, and poultry farmers, much like their dairy farmer counterparts, were left with a huge amount of "product" but no place to send it. This reality is a direct result of consolidation in the meat-processing industry and not a problem with supply. Acting like an uncaring billionaire, John Tyson, the chairman of Tyson Foods, suggested (in late April) that meat plants should keep running despite the outbreaks. (John Tyson did not volunteer to work on the assembly-line to lead by example.) Another billionaire, President Trump signed an order (he invoked the Defense Production Act) for the facilities to stay open amid the pandemic. The 1950 Defense Production Act was passed in response to production needs during the Korean War. Trump declared that "it is important that processors of beef, pork, and poultry ('meat and poultry') in the food supply chain continue operating and fulfilling orders to ensure a continued supply of protein for Americans" (Jackson 2020). Trump went so far as to promise plant owners that they would be shielded from legal liability if they are sued by employees who contract coronavirus while on the job (Jackson 2020). Such action of the part of Trump is not a surprise as he routinely represents the needs of the rich and powerful over those of everyday workers. To be fair, though, Americans and countless other across the globe love to eat meat and the meat industry is a major source of employment. Nonetheless, in this particular instance consumers should answer this question: "Are the lives of workers at meat-processing plants expendable just so we can have an ample supply of meat at grocery stores?" Furthermore, as we shall learn later in this chapter with our discussion on marine debris, Tyson Foods and other agricultural businesses located near the Mississippi River are directly responsible for turning the Mexican Gulf Coast into a dead zone because of their toxic chemical dumping into the river that runs to the coast.

Trump's actions led to sharp criticism from unions and activists who want to protect workers and the environment. Marc Perrone, president of the United Food and Commercial Workers International Union, the nation's largest meatpacking union, said the government must put the safety of the workers first (Jackson 2020). Environmentalists, and again, especially those who are vegan and vegetarian countered that there are plenty of food sources for protein including, beans, cottage cheese, sauerkraut, kefir, mushrooms, lentils, a wide variety of fish (e.g., shrimp, halibut, and tuna), avocados, almonds, oats, milk, quinoa, Yogurt, tofu, and eggs. (Vegans would not support dairy as an option.)

Environmental activists have long pointed out that the beef and dairy industries are not an efficient way of utilizing valued commodities such as land, water, workers, grains and money and putting them into a system that makes our food more expensive and less sustainable (Fairlie 2010). Furthermore, because we are using a great amount of energy to eat higher on the food chain, we are taking natural resources from the earth that are not needed. We are using more fossil fuels, cutting trees and contributing to global warming through our agricultural practice of producing meat. It is estimated that livestock have contributed up to 18 percent of all greenhouse gas emissions (Food and Agriculture Organization 2006). The grazing of livestock occupies 26 percent of the Earth's terrestrial surface, while feed crop production requires about a third of all arable land (Food and Agriculture Organization 2006).

Livestock do not simply graze; instead, current agricultural practices allow for livestock to overgraze. Overgrazing refers to herbivory (animal consumption of plants) that extract an unsustainable yield of floral biomass from an ecosystem by either wild or domesticated animals (Hogan 2010). Generally, it is domesticated animals (cattle and sheep) overgrazing that causes the greatest interest among environmentalists. Overgrazing causes 34 percent of the world's soil degradation, putting it ahead of deforestation, cropland agriculture, over-exploration and industrialism (Withgott and Brennan 2007). Overgrazing contributed to the "Dust Bowl" that occurred in the Great Plains Region in the 1930s (Withgott and Brennan 2007). Among the causes of overgrazing are excessive animal density, lack of rotation in grazing, and grazing at times inappropriate for flora production (Hogan 2010). In brief, overgrazing occurs when vegetation is consumed before it can be replaced. Overgrazing not only harms the topsoil; the soil underneath the surface becomes more vulnerable to erosion, which will make it difficult for vegetation to regrow, leading to more erosion. The nonnative weeds which livestock will not typically consume will be able to out-compete the native vegetation, causing further environmental problems (Withgott and Brennan 2007).

Deforestation

Trees, and especially forests, are essential for life. They help to cool and moisten our air and they fill it with oxygen to help sustain life. Predictably, most of the largest (geographically speaking) countries of the world have the most forested areas. The top ten countries with the largest forest area are (forested area in square kilometers appears inside the parentheses): Russia (8,149,3000); Canada (4,916,438); Brazil (4,776,980); United States (3,100,950); China (2,083,210); Australia (1,470,832); The Demographic Republic of the Congo (1,172,704); Argentina (945,336); Indonesia (884,950); and India (802,088) (Cary 2019). Globally, forests cover less than 30 percent of the Earth's land area.

Forests perform a number of ecological functions including regulating the water cycle, preventing soil erosion, regulating river flow and maintaining wind patterns. "Forests act as a source of food, medicine and fuel for more than a billion people. In addition to helping to respond to climate change and protect soils and water, forests hold more than three-quarters of the world's terrestrial biodiversity, provide many products and services that contribute to socio-economic development and are particularly important for hundreds of millions of people in rural areas, including many of the world's poorest" (FAO 2020b). With the continuing human population explosion comes a greater demand for land and how to best utilize it. For some people, cutting down forests and clearing the land for farms and homes is considered a viable solution; for others, the very idea of deforestation represents an environmental nightmare that will end badly, very badly.

Deforestation refers to the clearing, or permanent removal, of the Earth's forests on a massive scale, almost always resulting in damage to the quality of the land, causing soil erosion, poor water quality, reduced food security, impaired flood protection, and an even greater number of people moving to urban areas (Smith 2012). Deforestation releases carbon stored in trees and soil. The greatest driver of deforestation is agriculture, especially providing land for beef cattle. Livestock, as we know, need a lot of land for grazing, and as previously stated, livestock overgraze. As research has shown, if humans reduced their meat consumption the demand for land would decrease and along with it, greenhouse gas emissions (Rosen and Phillips 2019). At the current rate of deforestation, the world's rainforests would completely vanish within one hundred years; then again, humans would die off before we reach that point, as the lack of oxygen and increased CO_2 counts would lead to our premature extinction. Humans would not be the only species to die off as a result of continued deforestation, as many living creatures reside in the rainforest; and, of course, like humans, all living things on earth are depended upon oxygen to survive.

Rainforests are of particular concern when deforestation is discussed as an environmental topic. "The term 'rainforest' refers to a specific type of ecosystem that is dominated by forest land and receives a higher than average amount of rainfall every year. The rainforest system is considered one of the oldest types of ecosystem in existence today.... In fact, over half of all living plant and animal species in the world today inhabit rainforest ecosystems" (World Atlas 2020e). The tropical rainforests have great value to all living creatures as they "take in vast quantities of carbon dioxide (a poisonous gas which mammals exhale) and through the process of photosynthesis, converts it into clean, breathable air. In fact, the tropical rainforests are the single greatest terrestrial source of air that we breathe" (SavetheRainforest.org 2011a). Researchers estimate that the world is home to 5–6.8 million square miles of tropical rainforest which means that only 2.5 percent of the surface of the earth is covered by rainforests, or approximately 8 percent of the land on earth (World Atlas 2020e). And yet, nearly half the medicinal compounds we use every day come from plants endemic to the tropical rainforest (SavetheRainforest.org 2011a). The largest rainforests are found in the Amazon Basin in South America, in Western African countries near the equator, and in such South Pacific countries as Indonesia and the Philippines. Areas of Central America, Mexico, Hawaii, and other islands of the Pacific and Caribbean are also home to tropical rainforests. The World Atlas (2020e) claims that 53 percent of the rainforests found around the globe are located in the Americas; 27 percent in Africa; and the remaining 20 percent in Asia and Oceania.

With the importance of the rainforest to all living creatures, including humans, why would the planet's rainforests be in danger of deforestation? There are four main causes of deforestation: agriculture; logging; nature (e.g., wildfires); and overpopulation, urban sprawl and industrial use. As we can ascertain by these categories (other than nature), financial gain and human self-centeredness are the underlying reasons for deforestation. Agricultural practices represent the biggest driver of deforestation (National Geographic 2012). Farmers, both small and corporate, cut forests to provide more room for planting crops or grazing livestock. Family farmers may clear a few acres of forest to feed their families by cutting down trees and burning them in a process known as "slash and burn" agriculture (National Geographic 2012a). The burning of the forest releases nitrogen and fertilizers held in by the trees, allowing the fertilizers to go back into the soil again as ash. This agricultural method is effective for a few years until the soil has lost all of its nutrients needed to help crops grow. There are thousands of fires associated with slash and burn agriculture occurring each day in the Amazon rainforest (Tennesen 2008). Although we can appreciate the need for farmers to provide for their own families and produce additional food products for others in the short run, the slash and burn technique of deforestation will cause far more serious problems in the long run.

Generally considered the biggest culprit in deforestation is Brazil and its treatment of the Amazon rainforest. The Amazon rainforest extends across a huge part of South America with approximately 60 percent of it located in Brazil (followed by Peru with 13 percent and Colombia with 10 percent and smaller amounts in neighboring nations). The Amazon rainforest is considered so important to the earth that it is referred to as the "lungs of our planet" and the "Earth's air-conditioner." The Amazon rainforest is responsible for generating 20 percent of our planet's oxygen. "Deforestation between August 2018 and July 2019 reached 3,769 square miles, a 30% increase over the previous year, Brazil's National Institute for Space Research said. The area of deforestation is the largest recorded since 2008 and is about the size of the U.S. states of Delaware and Rhode Island combined" (*Associated Press* 2019a). Citing data provided by the Brazilian National Institute of Space Research (INPE), the FAO, and MapBiomas, Rhett Butler (2020) created a table to demonstrate the number of square kilometers of the Brazilian rainforest lost since 1970. Pre-1970 there were an estimated 4.1 million square kilometers of rainforest, in 2018 that figure dropped to 3.4 million square kilometers, equating to a loss of just over 700,000 square kilometers. The Brazilian rainforest is being destroyed for a number of reasons but primarily for agriculture and logging. In 2019, fore than 70,000 fires—most set by farmers and ranchers to clear land—have dramatically accelerated the pace of deforestation in Brazil. "Cattle ranching remains the biggest driver, responsible for 80% of deforestation, according to the Yale School of Forestry and Environmental Studies" (Hyde 2019: A4).

Logging operations, which provide the world's wood and paper products, are also responsible for the cutting of countless trees annually. The logging industry was one of the earliest and most profitable industries in the United States. The Adirondacks forest was among the earliest in the United States to be exploited for profit, with nearly 150 lumber camps employing over 7,000 people during the early 1900s (McMartin 1994). Logging has been a profitable industry primarily because of our need for lumber to build homes, businesses and other buildings and the conversion of trees into paper and paper-related products. Logging is still a profitable industry in the United States with a market size of $1.5 billion in 2020 and accounting for 88,951 employees (IBIS World 2020). During the period 2015–2020, the industry had a negative 0.9 percent growth (IBIS World 2020).

The United States, of course, is not the only nation involved in deforestation, as there are other nations with far more extensive rainforests. Peru has a lucrative logging industry and no wonder considering that nearly two-thirds of the country is covered by the dense jungle of the Amazon rainforest. However, in addition to causing great ecological harm, the World Bank estimates that 80 percent of Peruvian timber export stems from illegal logging (the Peruvian government requires loggers to show that the timber came from designated areas where logging is permitted) (*BBC News* 2014). Peru also engages in selective logging, a process that involves cutting down a select number of trees in a given area while leaving the others intact (Kozloff 2010). Selective logging is sometimes considered a sustainable alternative to clear-cutting logging, which involves cutting a large swath of forest, leaving little behind except wood debris and a denuded landscape (Shwartz 2005). Although selective logging sounds like a sustainable option to clear-cutting, it is actually very harmful to the remaining forest as for every tree that is cut down, thirty others are harmed in the process (Kozloff 2010). Trees surrounding the ones selectively cut are pulled down in avalanche form by the vines of the falling trees. Selective logging also requires the creation of numerous roads which are built by giant tractors and skidders that rip apart the soil and forest floor (Kozloff 2010).

Brazil has also begun selective logging, which is estimated to be responsible for an additional 20 percent of tree lost over the next two decades (Walker 2013). If this figure comes to fruition, the forest's ecology will begin to unravel. When intact, the Amazon rainforest produces half of its own rainfall through the moisture it releases into the atmosphere. Without that rain, the rest of the rainforest will die off. In the present, because trees are being wantonly burned to create open land (for farms, grazing and urban sprawl) in the frontier states of Para, Mato Grosso, Acre, and Rondonia, Brazil, has become one of the world's largest emitters of greenhouse gases (Walker 2013). Furthermore, the events set in motion by Brazilian logging are almost always more destructive than logging itself, as once the trees are extracted and the loggers have moved on, the roads serve as conduits for an explosive mix of squatters, speculators, ranchers, farmers, and, invariably, hired gunmen (Walker 2013). Squatter rights lead to land thievery and fraudulent land titles. In 2018, villagers in a remote area of Brazil's Mato Grosso state were gunned down when they refused to let a timber exporter log on their land. The logging company sent in hit men to eliminate the villagers who protested (Langlois 2018).

The third cause of deforestation is nature itself, usually in the form of wildfires. Wildfires may be caused by lightning strikes or when the wind blows across dry kindling wood creating sparks which trigger fires. Higher temperatures in the rainforest will make the land drier and may also contribute to wildfires. When a forest fire occurs, large areas will burn quickly. The wildfires not only harm the trees, but other living forms of vegetation and animal species. Humans may also contribute to forest fires via agricultural practices, sparks from the engines of equipment used in farming and logging, and human carelessness like throwing lit cigarettes on the ground or not putting out campfires.

Overpopulation, urban sprawl and industry are contributing forces to deforestation. The ever-expanding population places a high demand on our natural resources and wood products rank near the top in our demands on nature. Although the vast majority of people live in urban areas, the expansion of our population leads to a corresponding expansion or sprawl of urban boundaries. Industries beyond farming and logging have demands that include the prerequisite need for space to conduct business. Agricultural practices such as overgrazing prevents the growth of young trees which assures that future generations will be without rainforests producing the amount of oxygen our planet needs to sustain life. Add to this the realization that an environment that is struggling to sustain certainly is not thriving.

Deforestation results in the loss of biodiversity; loss of pharmaceutical products; canopy gaps; soil erosion; flooding; release of greenhouse gases; rising global temperatures; an alteration in rainfall patterns; and the release of carbon dioxide into the atmosphere that contributes to global warming. The reduction of the rainforest also means that there are fewer trees to absorb greenhouse gases already in existence and further fuels global warming. Eventually, we reach the point where deforestation results in less oxygen being produced, more CO_2 gases released into the atmosphere, and climate change. This pattern contributes to changes in rainfall patterns where droughts will become more common in some areas and heavy rains occur in other areas.

When tracts of the rainforest are destroyed the biosphere as a whole has been tampered with. Animal and plant species may become extinct and layers of vegetation will be destroyed, including the top layer, where the canopies of the trees are located and are home to bird species. When trees are removed, the forest is denied portions of its canopy, which blocks out the sun's rays during the day and holds in heat at night. This canopy gap leads to more extreme temperature swings that can be harmful to plants and animals (National Geographic 2012a). Canopy gaps allow for the growth of weeds that interferes with plant growth. Because canopy gaps allow the sun to shine on species that need shade, many of these species will have to flee the rainforest (Fimbel, Grajal and Robinson 2001). Selective logging is responsible for a great deal of recent canopy gaps. When trees are cut, the soil risks erosion deterioration during periods of rain and dry soils become susceptible to more pests and diseases. Dry soil also contributes to global warming, as dry soil in direct sunlight lacks the ability to cool down, making the surrounding air hotter; and this heat is released in atmosphere (Tennesen 2008).

The rainforest provides humans with a number of plants that are used for pharmaceutical purposes and it is likely that we have only begun to scratch the surface of the potential benefits of rainforest vegetation. A number of our medications are reliant on tropical plant sources, including certain types of surgery that require bark from specific South American plants (Tennesen 2009). Rainforest plants also provide us with latexes, gums, dyes, waxes and oils used in many products that are associated with medical procedures and consumer products. In brief, rainforests contain known and undoubtedly unknown gifts to the medical world.

Clearly, the rainforest specifically and forests in general are critical for the overall well-being of humans and countless animal and plant species. Continued deforestation will help to speed the sixth ME process.

Marine Debris

Marine debris refers to habitants that are contaminated with man-made items of debris and can be found from the poles to the equator and from shorelines, estuaries and the sea surface to the depths of the ocean, making this problem a global issue (Thompson, La Belle, Bouwman and Neretin 2011). The California Coastal Commission (CCC) (2019) describes marine debris as "global pollution problem that impacts human health and safety, endangers wildlife and aquatic habitats, and costs local and national economies millions in wasted resources and lost revenues." *Marine debris* is generally defined as any object discarded, disposed of, or abandoned by humans that enters the coastal or marine environment (NOAA 2007). This debris may be either deliberately or unintentionally littered into a river, waterway, lake or river. Types of marine debris include plastics; glass, metal, Styrofoam, and rubber; derelict fishing gear (DFG); and derelict vessels. As we have already discussed, plastics are an environmental nightmare because they are not biodegradable. As difficult as it is to clean land surfaces of plastics debris, it is an even more daunting challenge in our water. Plastic marine debris includes various domestic and industrial products such as bags, cups, bottles, balloons, strapping bands, plastic sheeting, hard hats and resin pellets and discarded fishing gear (NOAA 2007). A majority of marine debris is plastics, and along with their unsightliness, plastics debris includes harmful chemicals that cause a loss of biodiversity, increases acidification and contributes to the rise in the sea level (Thompson et al. 2011).

Glass, metals, Styrofoam and rubber are all items that break down into smaller pieces, but they generally do not biodegrade, meaning that just like microplastics, once these items are in the water, they will stay there and continue to cause harm to marine life (NOAA 2007).

Derelict fishing gear includes such items as fishing nets and lines, and other recreational or commercial fishing equipment that has been lost, abandoned, or discarded in the marine environment (NOAA 2007). Modern fishing gear is generally made of synthetic materials and metal, and therefore, can take a long time to biodegrade. Thousands of derelict vessels litter ports, waterways and estuaries, creating a threat to navigation, recreation, and the environment (NOAA 2007). These vessels will begin to fall apart and when they do, they will dump their fuel and any number of other toxins or poisons that may have been carried onboard.

There are ocean- and land-based sources of marine debris. Ocean-based sources are items that were dumped, swept, or blown off vessels and stationary platforms at sea and include fishing vessels (primarily gear from recreational and commercial fishing vessels); stationary platforms (e.g., plastic drill pipe thread protectors, hard hats, large gallon storage drums); and cargo ships and other vessels (any range of items from the small, sneakers, televisions and plastic toys, to the large, entire containers). Land-based sources are debris generated from land that may have been blown, swept or washed out to sea. Humans may also directly deposit debris in waterways as well. Land-based marine debris includes littering, dumping and poor waste management practice (intentional or unintentional disposal of waste); storm water discharges (e.g., street litter, like cigarette butts and filters, medical items like syringes, food packaging items and so on); and extreme natural events (e.g., such as the case following major storms like Hurricane Sandy) (NOAA 2007). According to the CCC just 20 percent of marine debris items originate from ocean-based sources (e.g., commercial fishing vessels, cargo ships, or pleasure cruise ships), the remaining 80 percent is from land-based sources, like litter (from pedestrians, motorists, beach visitors), industrial discharges (often in the form of plastic pellets and powders), and garbage management (ill-fitting trash can lids, etc.).

Unbeknownst to many people is the fact that the Mexican Gulf Coast region that extends along the shores of Texas, Louisiana, Mississippi, and Alabama is the largest "dead zone" in the U.S. and that this dead zone can extend to nearly 9,000 square miles of ocean. NOAA estimates that the dead zone costs U.S. seafood and tourism industries $82 million a year (Nature Conservancy 2020). The dead zone has existed for decades and its size varies annually due to weather conditions (e.g., wind patterns and the number of hurricanes in the region). The source of the dead zone is the containments runoff from the Mississippi River into the Gulf of Mexico. The runoff includes massive amounts of nutrients—particularly nitrogen and phosphorous—from lawns, sewage treatment plants, and farmland, but especially from Big Agriculture. The single largest source of this contamination is Tyson Foods, the largest meat producer in the United States (Canfield 2017). "As an idea of how toxic U.S. agriculture has become to the environment, last year alone [2016] 1.15 million metric tons of nitrogen pollutants from Midwest agricultural runoff flowed into the Gulf of Mexico from the Mississippi River.... This is roughly 170 percent more pollution than was caused by the 2010 BP oil spill" (Canfield 2017). In this scenario, we have harmful agricultural practices combining with marine debris and the consequences are an environment nightmare, a nightmare that could be avoided.

Marine debris harms wildlife in a number of ways, including entanglement; ingestion; and disruption of habitat (CCC 2019). Common items like fishing line, strapping bands, and six-pack rings can hamper the mobility of marine animals, cause injury, and interfere with eating, breathing or swimming, all of which can have fatal consequences (CCC 2019). Birds, fish, and mammals often ingest pieces of plastic mistaking it for food. The CCC (2019) reports that 245 different species have been found to ingest marine debris. Marine debris acts as an invasive species to many marine life habitats causing great disruption. This disruption of habitat extends all the way to the bottom of the ocean. This fact underscores the reality that marine debris is not simply a surface problem (CCC 2019).

Nonetheless, it is at the surface of waterways and especially of oceans that we see the true devastation of marine debris. Ocean marine debris is subject to ocean currents and wind. Large amounts of marine debris follow the same ocean and wind currents getting caught in accumulation zones, or "garbage patches." The size of these patches constantly varies depending on shifts in the currents and winds. There is, however, a notable example of garbage patch in the North Pacific Ocean known as the "Great Pacific Garbage Patch," or the "Garbage Island" for short, that is found in the North Pacific Ocean. The Garbage Island was referenced in a 2011 episode of *How I Met Your Mother* ("Garbage Island") wherein upon learning of the garbage island the "Marshall" character becomes so distraught about the ocean pollution that he can't have sex with his wife Lily. Marshall proclaims, "Guys, I'm going environ-mental."

The Great Pacific Garbage Patch is like a floating landfill of trash and marine debris found in the Pacific Ocean north of Hawaii and larger than the size of Texas. The Great Pacific Garbage Patch is really a loose collection of islands of garbage patches in close proximity that sometimes link together into large pieces and other times floats apart into smaller islands (many of which are larger than the Hawaiian Islands). The "Great Pacific Garbage Patch" is located at a natural collecting point at the center of a set of revolving currents called the North Pacific Gyre. Ocean gyres are large, circular rotations of water that are caused by the Coriolis effect—an effect whereby a mass moving in a rotating system experiences a force (like jet streams and western boundary currents) acting perpendicular to the direction of motion and to the axis of rotation. The middle of the gyre is more of a meteorological phenomenon than an actual place: a consistent high-pressure zone north of the Hawaiian Islands that, combined with the extremely weak currents, helps keep the ocean surface as placid as lake water (CNN World 2010). The garbage island stretches for hundreds of miles forming a nebulous, floating junk yard on the high seas (Mother Nature Network 2010). The "garbage patch" label is actually more fitting as there are actually thousands of smaller floating islands of trash. Thus, the mass is not the size of Texas but the stretched out "galaxy" of trash islands covers the amount of water the size of Texas or France. About 80 percent of the debris in the Great Pacific Garage Patch comes from land, much of which is plastic bags, bottles and various consumer products. The rest of the material consists of free-floating fishing nets (10 percent of the trash), items dropped by recreational boaters and from offshore oil rigs, and containers dropped by large cargo ships (MNN 2010).

As we have mentioned, there are multiple garbage patches floating in the oceans. Another such trash island has appeared in the Arctic Circle, off the coast of Norway near the Lofoten Islands. Norway's Lofoten Islands are "dreamscape of jagged mountains and glassy water that is so beautiful, Disney used the polar paradise as its muse when creating their blockbuster animation *Frozen*" (Vigliotti 2017). There are no inhabitants on these islands and yet everywhere on the shoreline and in the ocean are plastic bottles, especially water bottles, and plastic bags. It's estimated that there are around 300 billion pieces of plastic from all over the world, but primarily from the Americas and Europe, floating in this once pristine ecosystem (Vigliotti 2017). It appears that the Arctic Circle has become the world's new dumping ground (Vigliotti 2017).

The floating islands of trash are harmful to the environment in general but also to a great number of specific sea creatures such as sea turtles, seabirds and seals, and marine vegetation life such as coral reefs (Howell, Bograd, Morishige, Seki and Polovina 2012). Aquatic life, including many of the variations of sea life that humans eat, may ingest dangerous chemicals, such as Mercury (H_g). "Mercury (H_g) and its derivative compounds have been parts of widespread pollutants of the aquatic environment. Since H_g is absorbed by fish and passed up the food chain to other fish-eating species, it does not only affect aquatic ecosystems but also humans through bioaccumulation" (Morcillo, Esteban and Cuesta 2017: 386).

Because most of the trash in garbage islands isn't biodegradable (a great deal of the garbage is plastics) these islands are destined to float forever. This also means that the ocean will be compromised forever as well. If "The Great Pacific Garbage Patch" continues to grow, we may have to recognize it as our eighth continent. Tragically, it might some day be the only continent to survive the sixth mass extinction.

E-Waste

Technology, often a wonderful thing, is ever-present and ever-changing in society. Technology has been applied to many facets in human life and is generally viewed in a positive manner. But, as we know, and as demonstrated throughout the text so far, technology brings with it a great deal of harm, especially to the environment. We have discussed a variety of ways—the introduction of industrialization, methods of burning of fossil fuels to create energy, and the creation if nuclear power—that technology contributes to pollution. Through technology, a number of devices, especially in communications and entertainment, have been created. These objects become obsolete in a short period of time. A product that once served our needs has been updated with new features and suddenly we "must" have this new product. Such is the way of life in consumerism—buy, buy, buy! So, we purchase new phones, new electronic devices, and new everything. But what happens to our old, "outdated" electronic devices? Most people simply throw them away and when the masses join corporate entities by throwing away huge numbers of electronic items, we have the beginning of electronic-waste, or e-waste. There are an estimated 400 million units of consumer electronics that are "retired" each year, many of which are not properly disposed of via collection and recycling centers (Wright 2020).

E-waste is a popular, informal name for electronic products that have been deemed obsolete because they are outdated and are simply thrown away. In most instances, the products deemed as waste could be reused, refurbished, or recycled, but instead, they are looked upon as trash and unceremoniously thrown away. Electronic devices that are thrown away represent more than just solid waste materials; they are an environmental hazard because of their toxins. Electronic devices consist of any combination of toxins including lead, mercury, nickel, cadmium, chromium, polyvinyl chloride, and brominated flame retardants (EPA 2012m; Mulvaney and Robbins 2011; Environmental Recyclers International 2015). "Delving into the many chemical and heavy metal materials that make up most electronic devices, it seems that toxins and electronics go hand in hand. Though many manufactures of electronic devices are finding innovative ways to work around toxins found in e-waste, there is still a long journey ahead before realizing completely toxin-free devices" (Electronic Recyclers International 2015).

So, in terms of metric tons, how much e-waste is there? According to Statista (2020), there was 49.8 million metric tons of e-waste produced globally in 2018. The majority of e-waste generated around the world is from small and medium-sized electronic equipment (e.g., televisions, computers, printers, scanners, mice, keyboards, and cell phones). "Technological advancements and growing consumer demand have defined the era in which electronics have become a prominent part of the waste stream. The global quantity of electronic waste in 2014 was mainly comprised of 12.8 million metric tons of small equipment, 11.8 million metric tons of large equipment and 7 million metric tons of temperature exchange equipment (including cooling and freezing equipment)" (Statista 2020). The amount of e-waste is expected to grow 4 to 5 percent year to year. Most e-waste is generated in Asia (16 million metric tons of e-waste produced each year, 2014–2016) while the Americas and Europe both produce around 11.6 million metric tons (Statista 2020).

E-waste toxins pose risks to humans and other living species. Mercury, for example, a toxin found in fluorescent lamps, printed circuit boards, laptops and LCD screen backlights can adversely affect human health, marine life, and vegetation. Mercury can be spread from dumping sites when it seeps into groundwater or becomes airborne, accumulates in rain clouds and is spread in raindrops. Because of its threat to the environment, the EPA (2012m) recommends that unwanted electronic devices should be disposed of in properly managed electronics waste management facilities; reused or donated (e.g., to a charity organization); or, preferably, recycled for its valuable resources (such as precious metals, copper, and engineered plastics).

When e-waste is collected for managed waste disposal it is often exported to foreign facilities that specialize in tearing apart electronic devices for recyclable parts such as circuit boards. As shown in a very revealing 2010 *60 Minutes* episode that began as a 2008 CBS News story, a great deal of American e-waste ends up in Hong Kong. On November 9, 2008, CBS News first reported on a place now known as one of the most toxic places on Earth—Guiyu, China (*CBS News* 2010). Guiyu is a place that government officials and gangsters want to keep hidden from the outside world. It's a town where you can't breathe the air or drink the water and where children are born with lead poisoning. Children are also exposed to the dioxin-laden ash as the smoke billows around Guiyu, and finally settles on the area. *CBS News* updated its story in August 2009, and then *60 Minutes* in January 2010 provided us with photographic evidence of the environmental horrors of managed e-waste operations.

The *60 Minutes* exposé began its trail of toxic e-waste in Englewood, Colorado, at a managed recycling company called Executive Recycling that proudly proclaimed on its website, "Your e-waste is recycled properly, right here in the U.S.—not simply dumped on somebody else" (*CBS News* 2010). Instead, *60 Minutes* learned that Executive was sending container after container of electronic waste to Victoria Harbor, Hong Kong, a violation of U.S. and Chinese law. The *60 Minutes* crew headed to Hong Kong and followed the trail of containers to southern China—"a sort of Chernobyl of electronic waste"—the town of Guiyu (*CBS News* 2010). Guiyu hosts poor workers (60,000 e-workers in 2005) who tear apart the electronic devices in their bare hands to separate parts that can be used from the unusable parts that are discarded (usually burned, but also dumped in the water supplies). The crew had a couple of run-ins with local law enforcement officials but were able to film images of enviromare-level horrors including open, uncontrolled burning of plastics; disgustingly dirty water that cannot be used by humans (water has to be trucked in); air so dirty you can see the smog hover over the town; and very poor living conditions (*CBS News* 2010).

The World Health Organization (WHO) (2020b) states that children are especially vulnerable at foreign facilities that recycle e-waste. This is because of their proportionally low weight (compared to adults) and the fact that, because they are still growing so too are their functional systems (e.g., central nervous, immune, reproductive and digestive). Many children at these facilities are exposed to e-waste-derived chemicals their daily life due to unsafe recycling activities that are often conducted at their home—either by family members or by the children themselves (WHO 2020b). "Children may also be exposed through dump sites located close to their homes, schools, and play areas" (WHO 2020b). The economically-poor are the most likely to suffer from chemical e-waste toxins as they are the people most willing to engage in recycling of e-waste. "However, primitive recycling techniques such as burning cables for retaining inherent copper expose both adult and child workers as well as their families to a range of hazardous substances" (WHO 2020b).

Instead of throwing away their electronic devices, some people may choose to donate their used electronics to any number of worthy organizations. Some donations may be tax deductible, if that serves as an incentive to donate and reuse. As a tip, one should be sure that past data and memory cards are erased or replaced before turning over cell phones, computers and the like. A number of conscientious people may want to recycle their electronic devices, but they are not sure how to do it. And while some states mandate that people must recycle their electronic devices, New York State has come up with a consumer-friendly approach to e-cycling that makes it easy to recycle and comply with a law. In 2011, the "Electronic Equipment Recycling and Reuse Act" went into effect in New York State requiring electronics manufacturers to provide free and convenient recycling of their products to consumers. Thus, the consumer merely has to return his or her e-waste product to a designated location. It's not quite like New York State's bottle recycling law wherein one can return recyclables to any retail location that sells the product, but a number of collection options exist for New Yorkers and the number of facilities is expected to increase as people become more compliant with the law. Consumer Reports (2018) states that there are three smart and responsible tactics for getting rid of unwanted laptops, phone and so forth: bring them to the recycler, many nonprofit organizations offer such a service; donate the devices, if they still work; and third, take them to a tech firm.

Medical Waste

Medical waste is a growing concern and it certainly merits recognition as a harmful influence on human health and the health of the environment. Significant events since the 1980s, including the COVID-19 pandemic, have brought to the forefront the concern over medical waste.

What is medical waste? According to Medical Pro Disposal (2018), "medical waste is any kind of waste that contains infectious material (or material that's *potentially* infectious)." This is a less than satisfactory definition although Medical Pro Disposal (2018) adds, "This definition includes waste generated by healthcare facilities like physician's offices, hospitals, dental practices, laboratories, medical research facilities, and veterinary clinics." This important information that should have been incorporated into their definition of medical waste. The EPA (2017b) defines medical waste as "a subset of wastes generated at health care facilities, such as hospitals, physicians' offices, dental practices, blood banks, and veterinary hospitals/clinics, such as medical research facilities and laboratories." The EPA (2017b) adds that medical waste is generally contaminated by blood, body fluids, or other potentially infectious materials. Medical waste examples include anything that is soaked in blood (gloves, gauze, gowns, etc.); human or animal tissues created during procedures; cultures of infectious diseases/agents; any waste produced in rooms of patients with communicable diseases; and discarded vaccines (Medical Pro Disposal 2018). Many people create medical waste in their own homes during home care procedures, for example, diabetics will have insulin injection devices and blood sugar test strips and people with any sort of injury that includes blood discharges has also created medical waste. During the COVID-19 pandemic we all became aware of prime examples of medical waste such as face masks and gloves. Other items used and discharged in large numbers during the pandemic include booties, bed linens, cups, plates, towels, packaging and disposable medical equipment.

The WHO (2018) identifies different types of medical (health-care) waste: infectious waste (waste contaminated with blood and other bodily fluids, cultures and stocks of infectious agents from laboratory work, or waste from patients with infections); pathological waste (human tissues, organs or fluids, body parts and contaminated animal carcasses); sharps waste (syringes, needles, disposable scalpels, blades, etc.); chemical waste (solvents and reagents used for laboratory preparations, disinfectants, sterilants; heavy metals; and batteries); pharmaceutical waste (expired, unused, and contaminated drugs and vaccines); cyctotoxic waste (wastes containing substances with genotoxic properties); radioactive waste (such as radioactive diagnostic material or radiotherapeutic materials); and non-hazardous or general waste (waste that does not pose any particular biological, chemical, radioactive, or physical hazard, such as paper from charts and leftover food from a care facility).

The EPA (2017b) explains that concern for the potential health hazards of medical wastes grew in earnest in the 1980s after medical wastes were washing up on several east coast beaches. (For those of us alive back then, we remember hearing about syringes and other medical waste washing ashore on thousands of New York and New Jersey beaches). In the years since then, there are semi-regular news stories of medical wastes, especially hypodermic needles, syringes, and vials of blood washing ashore. The 1980s concern led to the passage of the 1988 Medical Waste Tracking Act, which alarmingly expired in 1991 (EPA 2017b). Other federal agencies (i.e., CDC, OSHA and FDA), however, have regulations regarding medical waste. By the end of 2019, the world became aware of another reason to become consciously aware of medical waste, the novel coronavirus. By February it was clear that the pandemic would cause great harm in the United States. A month later and people were scrambling to find face masks and safety gloves. We also became aware that some people were too busy to be bothered by such things as throwing away their disposal gloves on the parking lots outside grocery stores—a form of medical waste. Disposing of this medical waste litter was now pushed onto the responsibility of store employees.

Because hazardous medical waste contains harmful toxins or chemicals, it poses an environmental threat. Proper disposal of these materials is important. Home residents must learn how to properly dispose of their medial waste, generally by collecting items and putting them in containers and taken to facilities that will dispose of them. On-site medical facilities (e.g., hospitals) have procedures for properly disposing of medical waste (anyone who has been hospitalized has an idea of how extensive these procedures may be). Ultimately, medical waste is taken to industrial facilities where they may be incinerated, autoclaving (steam sterilization), microwaving; chemical (generally reserved for waste that's chemical in nature); and biological (the use of enzymes to neutralize hazardous, infectious organisms) (Medical Pro Disposal 2018). With all the procedures other than incineration, the sterilized medical waste is taken to landfills where it will become solid waste.

With the introduction of new global diseases and pandemics such as HIV/AIDS, H1N1, SARS, MERS, COVID-19, and whatever lurks next, it is clear that medical waste will become an even bigger concern for humanity and the environment.

Popular Culture Section 4

The Pros and Cons of Hydrofracking Presented in *Promised Land*

The extraction of natural gas from underground pools has been a part of the American landscape for many years, but only recently has a public outcry against this industrial practice become a source of a growing social movement and a topic of interest to the world of popular culture. The 2010 documentary film *Gasland*, which was nominated for an Academy Award for "Best Documentary," represents one of the earliest examples of fracking as a subject of interest in the popular domain. Even the 2012 film *Promised Land*, which has fracking as a topic, was originally conceived in 2011 by co-screenplay writers Matt Damon and John Krasinski as a film about wind power (Vancheri 2012). Damon and Krasinski changed the topic from wind power to fracking in response to the rapid accession of fracking as a topic of concern among the populace. Furthermore, the rural Pennsylvania filming location of *Promised Land*, an area near fracking extraction sites, had become subject to accusations of fracking-related accidents that had a negative environmental impact, so this too contributed to the shift in focus of the film to hydrofracking. There is, however, an interesting plot shift at the end of the film that will alarm viewers who underestimate the power and greed of large corporations, and it is this power of large corporations that is the real target of *Promised Land*.

The film is primarily centered on the lives of two Global Crosspower Solutions (GCS) employees, Steve Butler (Matt Damon) and Sue Thompson (Frances McDormand). GCS is a multi-billion-dollar energy company that specializes in extracting natural gas from pockets buried deep underground. At the start of the film, Steve Butler meets with top executives of Global who are considering him for a vice-president's position within the company. Butler has an excellent reputation in the company as a field representative who meets with landowners to sign mineral rights over to the company at cheap rates and in quick fashion.

The scene shifts to a rural farming town in Pennsylvania. Butler and Thompson display their years of experience in the field and we quickly discover why they are so successful convincing landowners to give up their mineral rights. Upon arrival into town, the two Global field reps shop at a local store. They pick out clothing that the locals would wear. They stay at a cheap hotel and chit chat with local merchants. They go to the local bar in town and mingle with the customers. Thompson partakes in singing on karaoke night to show that she does not take herself too seriously and to demonstrate a willingness to make fun of herself, qualities that generally endear one to others. The pair also rent a cheap truck instead of a fancy car as they drive around town to meet with landowners. They do all of this in an attempt to be perceived as regular people instead of powerful and elitist representatives of a huge, otherwise faceless corporation.

As they visit landowners in the countryside it is clear that both Butler and Thompson have mastered their salesman's persona equipped with a down-to-earth folksiness and a great deal of smooth-double-talking charm. They also rely on subtle scare tactics, a longtime proven effective method of business in many industries (e.g., insurance). Butler attempts to relate to his contacts by telling them he came from a small town in Iowa that fell apart once the local Caterpillar assembly plant closed. He emphasizes the devastating aspects of what happens to a town with no revenue source. He promotes GCS as the solution. Most people quickly fall for his scare tactics and take the first offer he makes; and that is why he always comes in under budget (compared to the higher limit that GCS has authorized him to offer). As a parent herself, Thompson talks with the younger landowners who have school-age children. She highlights the importance of a good education, especially in today's economic market with few manufacturing jobs available and with most professions requiring a college degree. Thompson warns parents that without revenue the schools deteriorate, and the quality of their children's education is compromised, which means their futures are compromised. Thompson explains that Global will pump millions of dollars into the community, which leads to better schools and a greater chance of getting into college.

The tactics of Butler and Thompson are a key focus of interest of this film in general, and of this "Popular Culture Box" in this book specifically. After all, we know why energy companies are in favor of fracking; the question is, why would anyone else be in favor of fracking? Unless that is, they benefit financially. And there we have it, the people who take the pro-fracking side of the debate point to the provision of jobs and the increased revenue the fracking industry can provide communities. People who are financially hurting will take almost any job when offered, so when a corporation such as Global comes knocking and offers cash and employment it is understandable why so many people find such an offer attractive.

Although Butler and Thompson have secured mineral rights from more than half of the local landowners, the town must still vote on whether to allow fracking. This is a common scenario in towns across the United States. What may also be common, but documentation far more difficult to secure, are informal meetings like the one Butler has with a local politician. The local politician promises Butler he has enough pull in town to assure the vote goes favorably for Global. Butler quickly realizes the politician wants a bribe and he offers him Global's standard bribe payoff figure, a dollar amount that equates to .01 percent of the total value of the town's pool of gas. The politician demands more money, but Butler quickly puts him in his place with a quasi-threat and firm reminder of Global's financial and legal power. The politician buckles, accepts the small bribe and promises to sway the sentiment of the vote in the next morning's town meeting.

On the night before the vote, Butler gets very drunk at a local bar and oversleeps. He arrives for the meeting with a hangover. He is overconfident the vote will go his way. Up to this point in time, in every other town, the vote always had. The local politician gives a talk and attempts to convince the residents that the vote is merely routine and suggests that they quickly vote in favor of Global's desire to frack their land. Suddenly a roadblock to a quick passage emerges. Local science teacher Frank Yates (Hal Holbrook) presents a number of reasons why the residents should vote against hydro-fracking; he is concerned primarily about the safety of the procedure to the local environment, especially the water, land, and all living creatures. Butler condescendingly tries to put Yates in his place, and while off his game (because of his hangover and being unprepared for a challenge) agrees to give the townspeople a couple of weeks to consider their vote. Thompson privately expresses her disappointment with the way Butler handled the situation.

Butler speaks with his Global boss on the phone and tries to assure him that everything is fine. His boss is worried and informs Butler that Frank Yates is actually a retired internationally renowned geologist. To make matters worse, an environmentalist, Dustin Noble of Athena Environmental Group, shows up in town the next day. Normally Butler and Thompson are not confronted by environmentalist opposition, but paralleling real life scenarios, the public concern about the safety of hydro-fracking is relatively new. Noble is seemingly one step ahead of Butler and Thompson and he knows all the same PR strategies and techniques they use. Noble shows townspeople a photo of dead cows from his farm in Nebraska and states that the cause of their death was fracking-related. Noble also claims that his hometown was ruined by fracking. It looks like Butler and Thompson may be facing their first defeat.

Just prior to the big vote, Butler receives a package from Global that proves everything Noble said was lies. Butler is stoked because he knows with Noble about to be discredited, the town will surely vote in favor of Global. Butler confronts Noble as he is about to leave town. He wants to gloat over his victory and Noble's stupidity (lying to the townspeople). Instead, Butler is left to examine his own conscience when Noble reveals that he too works for Global and that Global sent him there to purposely gain the sympathy of the townspeople knowing full well that once it was revealed his "evidence" was fraudulent the town would vote in favor of Global. Noble tells Butler this is why Global always wins. They play both sides; a sure way to win any game. The problem is, hydro-fracking is not a game, and the end results may be quite dire.

Promised Land has been criticized by gas-drilling companies as an anti-fracking film that favors oil companies because it was partly funded by a state-owned media company in the United Arab Emirates (Berlinger 2012). However, Butler makes the point that gas is a far better source of energy than oil or coal because it burns cleaner. He also points out that unless people drastically reduce their dependence and consumption of fossil fuels, there will be continued drilling of gas, oil and coal to meet the needs of consumers.

Butler's sobering commentary on contemporary society is certainly relevant to the real world, as most people and nearly every business is completely dependent upon energy provided by fossil fuels. The environmental costs of fracking are very high, perhaps too high to justify, but at the rate we consume energy via fossil fuels, something has to give.

Summary

In this chapter we looked at additional ways, beyond overpopulation and the Five Horrorists that were discussed in Chapter 3, in which humans are contributing to the inevitable sixth mass extinction. Our discussion began with the topic of extracting fossil fuels including coal, oil and natural gas extraction. Fossil fuel extraction comes at great cost to the health and welfare of humans, animals, and plant life.

The second topic involved an examination of our use of plastics, including thermoplastics and plastic bottles. The realization that once plastics enter our environment and break down into smaller and smaller pieces known as microplastics which never disappear underscores the harm of plastics. Landfills and waterways are filled with pieces of plastic; in fact, there are so many plastic products in the ocean that garbage islands, such as the Great Pacific Garbage Patch in the Pacific Ocean have become a permanent fixture.

Food waste was the next topic discussed. The causes of food waste are plentiful and include individuals who waste food; farm waste; agricultural waste; industrial waste (e.g., marine fisheries, manufacturing, transportation, and distribution centers); retail stores (e.g., grocery stores that throw away food after the expiration date); and restaurants and institutions (e.g., schools, hotels, and hospitals). Some attempts to overcome food waste include, composting, freeganism, donating food that is still edible but was about to be thrown away, and selling "ugly food."

Harmful agricultural practices were the fourth topic reviewed and this includes the use of genetically modified foods made up of organisms and a diet that includes huge sums of meat consumption and the corresponding farmland dedicated to domesticated farm animals. Environmental activists have long pointed out that the beef and dairy industries are not an efficient way of utilizing valued commodities such as land, water, workers, grains and money and putting them into a system that makes our food more expensive and less sustainable.

Perhaps the most dangerous of human behaviors contributing to accelerated rate of the sixth ME is deforestation. As we all know, trees produce oxygen, and without forests, especially rainforests, there will not be enough oxygen created to sustain human and other species' lives. Deforestation also results in the release of carbon dioxide into the atmosphere and contributes to global warming. There are four main causes of deforestation: agriculture; logging; nature (e.g., wildfires); and overpopulation, urban sprawl and industrial use.

Other topics discussed include marine debris, electronic waste (e-waste), and medical waste. Marine debris is a type of global pollution problem that impacts human health and safety and endangers wildlife and aquatic habitants. E-waste is a popular, informal name for electronic products that have been deemed obsolete because they are outdated and are simply thrown away. In most instances, the products deemed as waste could be reused, refurbished, or recycled, but instead, they are looked upon as trash and unceremoniously thrown away. Medical waste includes all waste generated at health care facilities, such as hospitals, physicians' offices, dental practices, blood banks, veterinary hospitals/clinics, and by individuals conducting their own health care regiments but do not properly discard of their waste.

Chapter 5

Nature and Human Skepticism Will Be the Cause of the Sixth Mass Extinction

There are those who question the role of humanity as a contributor to the next mass extinction. Some point out that nature has been responsible for past five mass extinctions and believe that nature remains as the primary cause for any future ones. Others may acknowledge that humans are causing harm to the environment but not to the point of thereby causing inevitable extinction. And still others cling to the belief that there is no scientific proof that climate change exists, nor do they believe that human behavior has any significant adverse impact on the environment. In this chapter, we will look at the role of nature as the potential cause of the sixth mass extinction and examine the skepticism held by some toward the role of humans in the sixth mass extinction.

Nature's Role in Causing Harm to the Environment

A great deal of evidence has been presented throughout the text to highlight the negative effect of humans on the environment, and yet for as long as our planet has existed, nature itself, both worldly and outer-worldly, has been responsible for compromising the earth's environment in the past. In fact, the first five mass extinctions were all the result of forces of nature originating on the planet or bombarding our planet from outer space. (This would have to be the case, of course, as humans had not evolved yet.) The first mass extinction (the late–Ordovician) was caused by glaciation; the second ME (late–Devonian) was caused by glaciation and global cooling; the third ME (end–Permian) by asteroids and volcanoes; the fourth ME (end–Triassic) by asteroids, volcanoes, and climate change; and the fifth mass extinction (end–Cretaceous) by asteroids and volcanoes. The significance of this, argue skeptics of human responsibility for the planet's compromised environment, rests with the argument that it will be nature that causes the next mass extinction—if there even is an impending next mass extinction (which some skeptics doubt).

Presently, there is ample evidence for nature's adverse role in the environment, including volcano eruptions; lightning strikes; wildfires; storms, including superstorms, hurricanes, droughts/water shortages, floods/tsunamis, snowstorms, and tornados; the influx of invasive species from one ecosystem to another; vapors emitted into the air from sulfur springs and the decay of dead vegetation and animal species; and outer-worldly forces.

Volcanic Eruptions

Volcanoes have been shaping the earth's formation since the very beginning of our planet's history. Volcanoes give rise to islands and land masses. Generally known to have a mountain-like shape, a volcano consists of a deep magma chamber where magma—a mixture of molten or semi-molten rock—accumulates, pipes that lead to surface vents, and the vents through which lava is emitted during a volcanic eruption (Lenntech 2011). Volcanoes that have not erupted for some time are "dormant" and volcanoes that have not erupted even in the distant past are called "extinct" (Lenntech 2011). Volcanic activity and eruption are usually triggered by alterations of tectonic plates, resulting in landslides or earthquakes (Lenntech 2011).

While volcanic eruptions are generally viewed in a negative environmental light, they also have positive effects on the environment. When a volcano erupts it emits a great deal of ash into the atmosphere and in the short-term this ash is generally very harmful to the environment. But in the long term, the ash layer, which contains many useful minerals, stimulates a very fertile soil because of its nutrients (*BBC News* 2012; Hall and DeCamp 2012). The dramatic scenery created by volcanic eruptions is often a tourist draw and thus good for the local economy; that is, until such eruptions become deadly. Volcanoes also represent a potentially valuable resource for energy extraction, known as geothermal energy, as the heat that rises from the earth's crust can be converted to energy to generate electricity. Such an energy source is highly valued because it is a clean energy source, and the resources are nearly inexhaustible (*BBC News* 2012; Hall and DeCamp 2012).

Volcanic eruptions are better known for the harm they commit, and they represent the prime forces behind the third and fourth mass extinctions and a contributing force to the fifth ME. Volcanic eruptions often force people living near volcanoes to abandon their homes and land, sometimes forever, as their lives and livelihoods are threatened. In some cases, entire towns and cities are buried by ash, mud and lava that spill as a result of an eruption. While people near an erupting volcano may lose their lives, those living farther away are also threatened by potential destruction via a compromised atmosphere, crop devastation, disruptions to transportation systems (e.g., interruptions of air travel) and electrical grids, and flooding. Volcanoes pose a potential direct threat to almost a half a billion people (Zuskin et al. 2007). "There are about 1,500 potentially active volcanoes worldwide, aside from the continuous belts of volcanoes on the ocean floor at spreading centers like the Mid-Atlantic Ridge. About 500 of these 1,500 volcanoes have erupted in historical time. Many of those are located along the Pacific Rim in what is known as the 'Ring of Fire'" (United States Geological Survey 2020c). The "Ring of Fire" is a zone of frequent earthquakes and volcanic eruptions that arcs oceanic trenches partly encircling the Pacific Basin. "On average, there are about 50–70 volcanoes that erupt every year. Some of them erupt multiple times, while others have one eruption. They typical number of individual eruptions per year is more in the range of about 60–80" (Volcano Discovery 2020).

As a case study example of a volcanic eruption located within the Pacific Rim's "Ring of Fire" that was preceded by earthquake we can look at the Mt. St. Helen's eruption on

May 18, 1980. Located in the southwestern part of Washington State, an earthquake shook Mt. St. Helen's and the mountain's north face collapsed in one of the largest debris avalanches ever recorded. "The slide uncorked the volcano, baring magma that exploded with 500 times the force of the Hiroshima bomb in the most destructive eruption in U.S. history. The cataclysmic chain of events killed 57 people and thousands of animals, took out 250 homes, 47 bridges and 185 miles of highway, clogged rivers with sediment, flooded valleys and blocked the Columbia River shipping channel. The landslide also remade Spirit Lake—once a beloved recreation spot at the volcano's base—raising the lake bed by 200 feet and dumping debris that functions as a huge dam holding back 73 billion gallons of water" (Read 2020: A1). Forty years later and there is concerned that a breach is possible, although not necessarily imminent. The results would be disastrous as a massive surge of water, mud and debris could inundate cities below and disable four Columbia River ports (Read 2020). To the point of the positive aspects of volcanic eruption, hotels, restaurants, and gift shops have flourished in the surrounding area because of tourists who flock to the area to purchase T-shirts and jars of volcanic ash (Read 2020).

The biggest environmental concern with regard to volcanic eruptions rests with the wide variety of toxins that are spilled into the air. Volcanic eruptions consist primarily of water vapors, but they also contain carbon dioxide, sulphur dioxide, hydrogen sulphide, hydrogen chloride, hydrogen fluoride, carbon monoxide and volatile metal chlorides (Lenntech 2011). According to the United States Geological Survey (USGS) (2010), the volcanic gases that pose the greatest potential hazard to people and the environment are sulfur dioxide, carbon dioxide, and hydrogen fluoride. Sulfur dioxide gas can lead to acid rain pollution downwind from a volcano. Globally, large explosive eruptions that inject a tremendous volume of sulfur aerosols into the stratosphere can lead to lower surface temperatures and promote the depletion of the Earth's ozone layer. (The ozone layer is generally considered to be the stratosphere which shields life on Earth from most UV-B and UV-C, the most harmful varieties of ultraviolet radiation.) Because carbon dioxide is heavier than air, the gas may flow into in low-lying areas and collect in the soil. To provide a glimpse to the potential harmful effects of volcano eruptions, volcanoes release more than 130 million tons of CO_2 into the atmosphere every year (USGS 2010).

According to the EPA (2011e), volcanic eruptions usually release into the atmosphere millions of tons of ash that include hazardous particle matter. Most of these particles are quite small, less than 10 micrometers in diameter, but pose great harm to humans because they can pass through the nose and throat and get deep into the lungs. The most frequent health problems that occur among people exposed to volcanic ash are respiratory disorders, stress and irritations of eyes and skin (EPA 2011e).

Volcanic eruptions also create clouds with dangerous toxins and when mixed with atmospheric moisture, acid rain results. Acid rain can slow down the growth of plant life downwind, cause water supplies to be contaminated with lead, and cause great harm for people with heart or respiratory ailments such as asthma (Hall and DeCamp 2012). On the big island of Hawaii, chemical reactions form two types of toxic clouds called vog and laze. Vog contains sulfur dioxide, sulfuric acid and a number of other chemicals that react to the oxygen in the air to form pollution clouds (vogs) (Hall and DeCamp 2012). When the trade winds are calm or moderate, the vog stays on the east side of the island when the trade winds pick up. It then shifts to the western coast of the big island where the tourists are located. The vog is toxic and considered highly hazardous to humans (Hall and DeCamp 2012). Laze is created when lava enters the ocean, creating another type of chemical reaction resulting in white-plume clouds. Laze clouds can drop rainwater, a type of acid rain, on people and causes irritation to the throat, lungs, eyes and nose (Hall and DeCamp 2012).

In sum, volcanic eruptions represent a force of nature capable of causing mild to extreme harm to the environment and can indirectly or directly cause the sixth ME. See the end of the chapter for a description of some of the most devastating volcanoes to occur during the human era and for a discussion of deadly volcanic eruptions depicted in popular films.

Lightning Strikes

Essentially a high amplitude direct-current pulse with a well-defined waveform, lightning strikes are a natural phenomenon with potentially devastating effects to humans and all other living things, non-living objects (e.g., homes and automobiles), and the environment in general. The earth's environment is under constant attack by lightning strikes. Across the planet there are about 100 lightning strikes per second, or 8,640,000 strikes per day (Strike Alert 2010). The average lightning strike is six miles long. Lightning reaches 50,000 degrees Fahrenheit, four to five times as hot as the sun's surface. Lightning kills more people on an annual basis than tornadoes, hurricanes, or winter storms and is second only to flash floods in the annual number of deaths caused by storm-related hazards in the United States (Strike Alert 2010). Until recently, death caused by lightning strikes was listed as the second most common cause of storm-related deaths in the United States (behind flash floods) because nearly half of all lightning strikes go unreported (Cooper and Kulkarni 2012). But NOAA data supports that lightning strikes are the leading storm-related cause of death (Cooper and Kulkarni 2012).

In the United States, lightning causes approximately 150 to 300 fatalities per year and causes serious injuries to an estimated 1,000 to 1,500 persons (Alyan et al. 2006). There is a 1 in 5000 chance that each of us will get hit once by lightning (National Geographic 2012b). Victims of lightning strikes have a 25 percent to 32 percent chance of fatality, with the majority of deaths due to immediate cardiac arrest (Alyan 2006). The National Geographic (2012b) reports, however, that 9 out of 10 victims of a lightning strike will survive.

Each lightning strike can heat the air around it to five times hotter than the surface of the sun (National Geographic 2012b). When lightning strikes a home, people inside are usually safe because homes are grounded by rods or other forms of protection. However, if an occupant touches running water or is using a landline phone there is risk of shock by conducted electricity (National Geographic 2012b). A lightning strike on a tree vaporizes the water inside creating steam that may blow the tree apart (National Geographic 2012b). One of the safest places for a human during lightning storms is inside an automobile because the tires and metal frames conduct electricity and can carry a charge to the ground (National Geographic 2012b).

The greatest environmental harm from lightning strikes rests with the realization that their extreme heat can cause structural damage to homes and other buildings, which have to be replaced at energy costs (e.g., costs of raw materials, water to fight the fire, and electrical power interruption); cause damage to nature (e.g., the destruction of trees and other forms of vegetation) and cause wildfires. The devastating California fires of August 2020 were mostly caused by lightning strikes which burned more than 1.4 million acres of the state's scenic coastal mountains, killed seven people, destroyed more than 2,100 building, and made air unhealthy across the northern Bay Area (Serna 2020). While lightning strikes are quite common (as we already explained) the 2020 "summer heat wave magnified by climate change combined with tropical moisture and storm energy [that created] thunderstorms" resulted in lightning strikes pelting "a region with a history of difficult firefighting" in an area that has experienced numerous fires in the recent past (Serna 2020: A11).

The impact which lightning can have on living things can be extremely detrimental, and in some cases the ill effects can be long-lasting. According to the National Lightning Safety Institute, "For centuries there have been documented records from reliable individuals reporting unusual behavior (anxiety, restlessness, and irritability) associated with some pets and livestock prior to thunderstorm activity. This behavior has been observed in animals for as much as an hour or more before the first sound of thunder is heard in the distance. It is speculated that some animals are reacting to hearing long-wave sound energy below the 20-Hz level from an approaching thunderstorm" (Lightingsafety.com 2013).

Wildfires

Wildfires are often caused by lightning strikes. According to Craig Clements, a professor at San Jose State's Fire Weather Research Laboratory, "the destiny of all wildfires is shaped by a fire behavior triangle—fuel, weather, and topography" (Serna 2020: A11). Wildfires have occurred naturally on this planet since long before humans arrived, still occur now, and will continue to occur for as long as there is something to burn and a source to stimulate a spark. Wildfires can cause damage nearly anywhere on any land mass and are a natural cause of deforestation.

As a former resident of Los Angeles and Las Vegas, one of the two authors of this text (Delaney) can vividly recall the role of deserts and a shift in wind patterns as a cause of natural wildfires. For example, in Southern California there is a phenomenon known to all local residents called the "Santa Anna winds"—a reference to a shift in the trade winds that blow westward. This shift in the wind pattern blows winds of 40 mph or higher from the east across the Mojave and Sonoran deserts, picking up grains of sand that travel in the wind sparking dried up vegetation, much like scraping two rocks together to generate a spark in order to start a campfire (something any scout or good outdoors person can do). Wildfires are the common result, along with hotter than normal coastal Southern California and northern Baja California temperatures. For nature, this is a way of clearing away brush to keep other forms of vegetation life strong. For humans who have decided to build homes in natural occurring wildfire areas, the result is a loss in homes, other structures, livestock and a great deal of resources (e.g., water, energy to fuel and operate fire-fighting equipment) that are allocated to put out these natural occurrences.

It is common knowledge that smoke (more specifically toxic smoke inhalation) kills more people than fire itself during any type of fire. Fire uses up the oxygen you need and produces smoke and poisonous gases that kill. Breathing just a small amount of smoke and the corresponding toxic gases can make you drowsy, disoriented and short of breath. The odorless, colorless fumes can lull people into a deep sleep before the flames even reach them (FEMA 2012). It is not surprising to learn, then, that smoke from wildfires causes a great deal of harm as well.

In the summer of 2008, a lightning strike started a wildfire in eastern North Carolina that burned for weeks, covering nearby communities in smoke (EPA 2011f). This particular wildfire created a lot of smoke because it was fueled by burning peat, decayed vegetable matter found in swampy areas. The EPA studied the results of this fire and found that there was 37 percent increase in emergency room visits for people with symptoms of heart failure during a three-day period of dense smoke exposure and the following five days, and an increase in visits for respiratory problems (EPA 2011f).

Tracking the most harmful wildfires is a difficult challenge especially in light of the increased threat from climate change. As we discussed in Chapter 2, the western part of the U.S. has to contend with warmer, drier conditions, increased drought, and a longer fire season that increases the risk of wildfires. It seems like every year in states such as California, Oregon, Washington, Arizona, Colorado and New Mexico, we hear of a wildfire described as the "most destructive" in history. Aided by high winds, wildfires quickly tear across rugged, brushy forests and grow to thousands of acres within hours. Flames often eat through secluded communities and become raging infernos. Life or death evacuations are often characterized by pure terror, confusion, turmoil, and bravery. We described the brutal 2019–20 Australian wildfire season in Chapter 2 where bushfires spread unabated throughout a large portion of the island nation.

Many people, animals, and plant life often die during massive wildfires. Wildfires may have adverse effects on large tracts of vegetation, forests, marine life, and drinking water supplies. Jay O'Laughlin has conducted research on the ecological impact of wildfire on fish and wildlife habitats. He argues that fish populations "respond almost immediately to sediment from a disturbance event, either wildfire or logging, by reduced reproduction" (O'Laughlin 2005: 66). Interestingly, O'Laughlin (2005) concludes that while sediment produced by a disturbance will have short-term adverse effects on the reproduction of fish, within a couple of decades they will benefit by a higher reproduction rate than what had existed pre-disturbance. As for the forests of western United States, O'Laughlin (2005) simply states, "fire is inevitable" (p. 67). In other words, we cannot be surprised by wildfire occurrences in certain areas; if we want to save on energy resources and human lives, people should simply avoid living in areas where wildfires are known to occur. This is the same principle that is applied, but also ignored, in advising people not to live on cliffs (e.g., Malibu, California) where erosion and earthquakes are known to occur. The resulting mudslides cost people their homes and consumes valuable energy resources.

The destructive 2012 Waldo Canyon wildfire in Colorado provides us with an example of another environmental problem associated with wildfires—they can adversely affect drinking water supplies. Donna Kaluzniak (2012) explains, "As wildfires burn, not only do they strip the landscape of soil stabilizing trees and vegetation, they leave a layer of ash and debris. When subsequent rainfall washes these solids into nearby waterways, drinking water sources can be contaminated." The Texas Commission on Environmental Quality (2008) explains that wildfires' impact on the physical, chemical, and biological structure of aquatic ecosystems is dependent upon the fire size, intensity, severity, the proximity to surface water bodies, and the timing of fire in relation to rain events. "The majority of water quality impacts from wildfires are not apparent until after a heavy rainfall that causes soil erosion and subsequent runoff into nearby lakes, rivers, and streams. Burned ground causes more of the rainfall to quickly run off into streams, rivers, and reservoirs" (Texas Commission on Environmental Quality 2008). Following erosion, the characteristics of any future stormwater runoff will be altered and will have adverse effects on downstream water quality. Furthermore, if an area is exposed to multiple wildfires over a period of time, the soil chemistry is forever altered and all subsequent water events (e.g., rain, stormwater runoffs) will cause harm to downstream water quality.

The impact of wildfire on the chemical composition of surface of water generally reveals elevated nutrient loads, particularly phosphorus and nitrogen. "Increased nutrients may cause long-term impacts such as increased plant growth, algae blooms, and oxygen depletion" (Texas Commission on Environmental Quality 2008).

To fight wildfires, firefighters do not rely solely on water; instead, they are likely to use chemical retardants. If marine life comes in direct contact with retardants it will die because of the ammonia nitrogen in most retardants is very toxic to fish. Retardants may reach marine ecosystems as runoff (Texas Commission on Environmental Quality 2008).

Storms

Volcanic eruptions, lightning strikes and wildfires represent just the beginning of nature's attack on the environment. Nature also produces superstorms; hurricanes; droughts and water shortages; wildfires; floods, tidal waves, and tsunamis; snowstorms and avalanches; tornadoes and earthquakes; and so on. Because of nature's often negative impact on the environment, many of the doubters of human-influenced climate change argue that any variation in the climate is the result of a larger, natural, uncontrollable pattern that may extend hundreds or thousands of years. In other words, if there is such a thing as global warming occurring, it's not because of human influence but simply nature following its course. Such a belief is not without merit; after all, mass extinction periods (from beginning to end) can take as much as one million years. So, who's to say for sure that the next mass extinction will happen anytime soon or will be caused by human behaviors? The answer—scientists!

The Environmental Defense Fund (EDF) takes the same position as nearly all scientists around the world that climate change is primarily the result of human behavior. Furthermore, the EDF argues that human-made global warming negatively impacts nature to the point where additional forms of extreme weather can be expected in the coming years. "Studies show that global warming will increase the frequency or intensity of many kinds of extreme weather. While we can't attribute a particular heat wave or hurricane to global warming, the trends are clear: Global warming loads the atmospheric dice to roll 'heat wave' or 'intense storm' more often" (EDF 2011). The lesson to be learned here is—messing with nature will cause more trouble. Many years ago, Chiffon margarine had a popular commercial on TV that promoted a margarine so rich and creamy that it fooled the "Mother Nature" character into thinking it was real butter. Upon learning that she had been duped, she stood up and angrily proclaimed, "It's not nice to fool Mother Nature." She then waved her arms and thunder rolled and lightning struck.

The EDF warns that as global warming boosts ocean temperatures, one likely result will be more destructive storms such as 2005's Hurricane Katrina, the costliest in U.S. history. Although placing dollar figures to storms should be a distant second consideration to the overall harmful effects of nature's storms on the environment, there is something about placing a dollar value on natural resources, because we live in a society that embraces economic considerations. The EDF (2011) emphasizes these potential problems due to global warming:

- Fiercer hurricanes more likely—Global warming heats our oceans. Since hurricanes derive their power from warm waters, scientists expect more ferocious hurricanes.
- Wildfires on the rise—Hot, dry conditions are perfect for sparking wildfires. Scientists are starting to trace the link between our warmer climate and the recent increase in wildfires.
- More droughts expected—Higher temperatures cause water to evaporate faster, leading to dried-out soil that increases water shortages and puts crops and livestock at risk.
- Higher risk of downpours and floods—A warmer world will deliver heavier rainfalls and more flooding. Some of the flooding will likely be as disastrous as what occurred in 2005 in New Orleans.
- Bigger odds for deadly heat waves—A warmer climate means more hot weather. Scientists expect global warming to increase the chance of "killer" heat waves.

This is a mere sampling of the potential danger nature presents to humanity as a result of global warming (all of which have proven to be true). And this is in addition to the other problems nature makes us confront. Earthquakes and tsunamis are two more things to consider. And what happens when the events happen together? We found out the answer to this scenario in March 2011 when Japan was dealt an astounding 9.0 (Richter scale) earthquake followed by an almost unbelievable tsunami. The death toll exceeded 10,000 people (*BBC News* 2011). The tsunami killed human, animal, plant life, and marine; destroyed buildings and natural habitants; and caused major damage to the Fukushima Daiichi nuclear power plant.

There have been a number of Superstorms since Sandy. During the summer of 2017 there were three monster hurricanes that swept in from the Atlantic one after another, shattering storm records and killing hundreds of people. "First, Harvey brought catastrophic rain and flooding to Houston, causing $125 billion in damage. Less than two weeks later, Irma lashed the Caribbean with 180 mile per hour winds—and left the island of Barbuda uninhabitable. Hot on Irma's heels, Maria intensified from a Category 1 to a Category 5 hurricane in just 30 hours, then ravaged Puerto Rico and left millions of people without power" (PBS 2018). As the planet continues to warm, we can expect a continuation of superstorms. At the time that Irma hit landfall on Barbuda (September 6, 2017), it was the most powerful Atlantic hurricane in recorded history. "Irma held 7 trillion watts of energy. That's twice as much as all bombs used in World War II. Its force was so powerful that earthquake seismometers recorded it. It generated the most accumulated cyclone energy in a 24-hour period" (Amadeo 2019).

Superstorms are connected to climate change. There were three ways that climate change impacted Irma. First, rising sea levels worsened storm surges and flooding. Second, South Florida's average August 2017 temperature was four-tenths of a degree above normal and warmer air holds more moisture and when the warm air releases the moisture, the water falls in torrents creating greater rainfall during a hurricane. Third, global warming slows weather patterns allowing hurricanes to hover over an area longer. Since 1949, storms have slowed down by 10 percent (Amadeo 2019).

In 2018, Hurricane Florence became the sixth named storm and third hurricane to hit the U.S. Florence produced extensive wind damage along the North Carolina coast and caused widespread power outages, produced storm surge of 9 to 13 feet and devastating rainfall of 20 to 30 inches, which produced catastrophic and life-threatening flooding (National Weather Service 2018).

Beyond the descriptions of water surges, rainfall, and power outages associated with storms is the realization that America's aging infrastructure is not equipped to handle superstorms. The American Society of Civil Engineers states that America's infrastructure (e.g., systems for clean water, energy, roads, dams, and levees) in failing, earning a D-plus grade in 2017. Superstorms, in particular, are expected to overwhelm the U.S. infrastructure, forcing civil engineers to rethink how they approach their work. Ben Schafer, a structural engineer and director of the Cold-formed Steel Research Consortium at Johns Hopkins University, states, "These types of extreme events [superstorms], they're not stressors on little things we can solve—they attack our basic infrastructure. As a society, we need to start planning for constant renewal and stop acting like it's such a surprise when these extreme weather events happen" (Crulckshank 2018).

Coastal regions are especially vulnerable to superstorms and a failing infrastructure. However, there are scientists who predict that areas not historically prone to superstorms must also take warming. Stacey Shackford of the Cornell College of Agriculture and Life Sciences wrote an article for the Syracuse *Post-Standard* on November 28, 2011, that was published under the headline "Are Hell and High Water Headed Our Way Again?" In an attempt to dramatize the severity of storms that will rock New York City in the future, the article begins, "In 2080, will New York City residents take a submarine to work instead of the subway?" (Shackford 2011: C-3). Shackford and other Cornell scientists conclude that during the 21st century, New York City and New York State will regularly become besieged by huge tropical storms that will cause flooding; compromise drinking water supplies; and cause coastal erosion and damage; and compromise agriculture, energy, transportation, telecommunications and public health (Shackford 2011). Less than a year after this article was published, Superstorm Sandy hit the Northeast and did everything the Cornell scientists predicted. If all their predictions come true, the state of New York and New York City in particular, will face "hell and high water."

If the conservative public and politicians do not wake up to the reality of climate change and its link to superstorms, then perhaps the idea of a crumbling infrastructure will serve as a wake-up call. The evidence is too overwhelming to ignore.

Invasive Species

Perhaps you have heard of the growing threat invasive species present to local ecosystems. It seems as though during the past couple of decades the number of stories and academic articles on the subject have increased dramatically. There is good reason to be concerned about invasive species, another threat from nature itself against the global environment.

So, what is an invasive species? NOAA (2020) defines invasive species as "animals or plants from another region of the world that don't belong in their new environment." The New York City Department of Environmental Protection (NYCDEP) (2012) offers a better definition by including microorganisms as an example of an invasive species. "Invasive species are non-native plants, animals, and microorganisms that have been transported to new locations where they proliferate and spread." An important aspect of invasive species is that once they are introduced into their new environments, they become established and expand their range (Vilcinskas 2015). The NYCDEP (2012) reports that invasive species are capable of causing immense and often irreversible harm to the environment or human health both in the short-term and also over the long haul. Furthermore, only the total loss of biodiversity represents a greater threat to endangered and threatened species as invasive species presents (NYCDEP 2012). In the United States, the cost of invasive species damage to agriculture, forestry, fisheries and property exceeds $120 billion annually (NYCDEP 2012). In 2014, the costs of invasive species were estimated at $138 billion per year and perhaps more importantly is the recognition that 42 percent of the species on the Threatened or Endangered species lists are at risk primarily because of non-indigenous species (Energy Skeptic 2014). The economic costs of invasive species have only soared since this data was accumulated.

How are invasive species introduced to new ecosystems? There are a few answers to that question. (Note that invasive species are primarily spread by human activities.) Some invasive species are introduced to an environment purposely as a means toward financial or economic growth (McNeely 2000). Ecologist Jeffrey A. McNeely (2000) reasons that exotic or alien species hold a certain charm for people because they are new and distinct. Sometimes it is believed that a new species will carry out functions that native species could not carry out as effectively.

While some invasive species are introduced to native ecosystems on purpose for financial gain, in other instances they are purposively introduced by explorers and travelers of different lands who bring with them, or take home with them, native animal and plant species. For example, a New York City pharmacist and huge fan of Shakespeare named Eugene Schieffelin became inspired to populate the United States with many of the birds mentioned in Shakespeare's works (Bright 1998). Some accounts claim he was trying to introduce all the bird species mentioned in Shakespeare to the United States. In 1890, he released 60 starlings into New York City's Central Park; he did the same with another 60 birds in 1891. It is believed that he also introduced the starlings to control the pests that had been annoying him in New York City. Schieffelin was not the first New Yorker with the goal of introducing invasive species to the U.S., as the American Acclimatization Society (AAS) was found in the city in 1871 with the goal of introducing European flora and fauna into North America for both economic and cultural reasons (Bright 1998). A November 15, 1877, *New York Times* article describes Schieffelin as the chair of the AAS and claims that the organization had released 50 sparrows into Central Park in 1864 and 50 skylarks in 1874. The article described how well the birds adapted to their new environment and also mentioned the organization's goal to introduce fish to various waterways in the Northeast (*The New York Times* 1877).

Invasive species are also introduced to native ecosystems by accident and as a result of such factors as global markets and international trade, and boaters who fail to properly clean their boats, especially by not scrubbing the bottoms of their boats free of microorganisms. At many boat launches wash stations are provided to encourage, or mandate, that boaters wash the external parts of their boats; if all boaters do this, the threat of accidental invasive species is reduced. Other examples of the accidental introduction of invasive species to an ecosystem includes seeds that can be carried in potted plants or even in clothing, and the transportation of insects via wooden packing crates—insects often live undetected in wooden packing crates (Keller and Perrings 2011).

The introduction of a foreign species into an existing environment is almost guaranteed to lead to conflict, as native and invasive species must compete with one another for the available scarce resources, including food, breeding sites, and sunlight. This inter-species competition for food, breeding sites and sunlight is referred to as "niche usurpation" (Lindell 1996; Van Driesche and Van Driesche 2000). The new species is looked upon as an invading species by the native species; consequently, this leads to a very competitive situation wherein one species is likely to survive at the expense of the other. The introduction of invasive species to a native ecosystem creates a natural selection laboratory. The species that is the most efficient in securing sufficient energy will live while the weaker species will die off. The invasive species has the capability of thriving in the new ecosystem because they have no natural competition or predators that are able to serve as a checks and balances to hinder the invading species from taking over. As the invasive species spreads unabated throughout the new ecosystem the resources intended for local species become depleted. The continued introduction of invasive species to local ecosystems is one of the main reasons as to why there is a decline in species richness around the world (Mollot, Pantel, and Romanuk 2017).

Regardless of how invasive species are introduced to an ecosystem, the number of invasive species is climbing and so too are the corresponding problems associated with invasive species. We will limit our discussion to a few examples of invasive species on land and those found in aquatic environments.

Pigs, sheep, and ants: what do these land species have in common? You guessed it; they are examples of invasive species. Jackson Landers has traveled the United States to hunt and eat invasive species. He has hunted and eaten everything from iguanas to nutria and feral hogs, and as he describes in *Eating Aliens* (2012), wild pigs are among the most invasive species in the United States. The problem with pigs, according to Landers (2012), is that they dig constantly, eroding the earth. They are omnivores and will almost certainly cause the extinction of native species if left alone, argues Landers (2012). It doesn't end with pigs either. According to Landers (2012), North America is under attack by a wide range of invasive animals including black spiny-tailed iguanas in Florida, Asian carp in Missouri and Virginia, nutria in Louisiana, European green crabs in Connecticut, and other alien species throughout the U.S. From this perspective, it is clear that niche usurpation is in full effect across North America. How should humans handle this problem? Landers offers a unique suggestion: we should eat them! He argues that invasive species must be kept in check or they will destroy numerous native species; humans can help by hunting and killing the invasive invaders.

The Santa Cruz Island, off the coast of California, has more than its share of invasive species, including feral pigs, sheep, and fennel (a plant). The number of feral pigs varies inversely to the number of sheep, but as grazers, they both present environmental problems. It is believed that pigs were first introduced to Santa Cruz Island by a sea captain who turned a few pigs loose on the island around 1854 in order to ensure a supply of pork for passing sailors (Van Driesche and Driesche 2000). The pig population grew unchecked for several decades and they ran rampant on the island, destroying vegetation and a competing with other animal species. Eventually a bounty was placed on all feral pigs and their population dwindled. This paved the way for the sheep that were introduced to island to take over as the invasive species that grazed so much vegetation entire hillsides were subject to erosion because of the lack of roots to hold the soil in place. Attention was then placed on reducing the sheep population, which would allow for the return of the pig population. Today, the National Park Service (NPS) of Channel Islands National Park identifies non-native feral pigs and non-native fennel (an invasive weed that we will discuss shortly) as the most significant disturbances to the island's sensitive resources—native plant communities, rare plant species, and archeological sites (NPS 2012). The NPS (2012) reports the Santa Barbara Island song sparrow, the crimson flower and monkey flower of Santa Cruz along with eight other rare and unique plant species are extinct because of pigs and fennel and warn that 10 island animal species including the island fox are on the brink of extinction.

There are a number of invasive land plant species, including the Santa Cruz fennel. For some people, fennel is an herb; but on Santa Cruz Island, it is a nuisance. Fennel was introduced to the island in the 1850s as a garden herb and did not become a problem until the number of grazing cattle and sheep were put under control (Van Driesche and Van Driesche 2000). Fennel spread across the island without sheep or cattle as a grazing counterbalance. As with other invasive plant species, fennels take the nutrients and resources from an ecosystem that would otherwise be consumed or used by native species.

Known as social insects, ants are a good indicator of how other species, including humans, react to environmental changes, claims North Carolina state biologist Rob Dunn. Dunn and his colleagues have discovered that some of the least known variations of ants have begun to flourish at the expense of native ant species (and other species) because of global warming (Henderson 2011). The environmental concerns center around the realization that native ants favor us with ecological services such as dispensing seeds. Non-natives might carry diseases and become pests; this is especially true with fire ants that are flourishing in the warmth. A number of species are moving northward to avoid the heat of the southern states. These animal species pose a major threat to native species. A number of endangered species that already struggle to survive in cooler areas and on mountaintops are in grave danger of extinction (Henderson 2011).

There are many other insects that have proven to be deadly invaders of local ecosystems. We begin with the emerald ash borer (Agrilus planipennis or EAB) known as the killer of trees. The emerald ash borer has destroyed tens of millions of trees in 30 states of U.S., including all the ash trees in Detroit (Mason 2013; USDA 2020). Native to Asia, it believed that this insect arrived in the United States hidden in wood packing materials. They were first discovered in Michigan in 2002 (USDA 2020). By the summer of 2013 the emerald ash borer had threatened all of the ash trees in Central New York. Much like hydrilla, the ash borer is often spread by humans. When wood is taken from county to county or state to state, the ash borer can infest a forest where it hadn't been seen before (*The Citizen* 2012c). The invasive beetle creatures threaten the American institution of baseball too. Louisville Slugger baseball bats, the preferred bats for professional baseball in the United States for over five generations, are made from ash trees and it is feared that soon all ash trees, including those used for baseball bats, will be gone (Dahler 2013). In New York State, the DEC is trying to control the borer by setting up quarantine areas across the state to prevent the bug's spread through firewood (Coin 2018).

Another wood-boring insect is the Asian longhorn beetle (Anoplophora glabripennis). This insect was first discovered in New York in 1996 and arrived accidentally in cargo from Asia (USDA 2020b). This wood-boring pest feeds on a variety of hardwoods including maple, birch, elm, ash, poplar, horse chestnut, and willow (Coin 2018). Chances are you are more aware of this next insect that the previously discussed wood borers: the Marmorated stink bug (Halyomorpha halys). The stink bug is an invasive pest found throughout most of the United States. This species is also native to Asia and was introduced into the United States in the mid–1990s, possibly stowing away in a shipping container (EPA 2017c). The reason most of us are aware of this insect is because they find their way inside our homes and other buildings and if you disturb them or squish them to kill them, they emit a very unpleasant odor, a stinky smell, from scent glands on their abdomen (EPA 2017c). While they are nasty to look at, they do not bite people or animals, nor do they damage buildings. The reason they are so harmful to the environment is because they feed on a large number of high-value crops and ornamental plants. In New York State, stink bugs are especially fond of the ample apple orchards and across the U.S. the stink bugs destroy crops to the tune of nearly one billion dollars a year (Coin 2018).

A couple of other destructive insects with origins from Asia reached the public's attention in 2020—the lanternfly and the hemlock woolly adelgid. The lanternfly feeds on more that 70 plant species including tree-of-heaven, and plants and crops that are critical to New York's agricultural economy such as maple trees, apple trees, grapevine, and hops. First discovered in Pennsylvania in 2014, this invasive species has since been found in New Jersey, Maryland, Delaware, West Virginia, and Virginia (*The Citizen* 2020). The hemlock woolly adelgid (HWA) is an aphid-like insect that attacks North American hemlocks. HWA are very small (1.5 mm) and often hard to see, but they are easily identified by the white woolly masses they form on the underside of branches at the base of the needles (NYSDEC 2020). As climate change contributes to more mild winters, experts anticipate an increase in the number of HWA and their continued spread across the country.

The discussion of these insects represents a mere sampling of invasive insect species and it would be impossible to document them all here. However, amidst the COVID-19 outbreak in the spring of 2020 came word of a newly documented flying invasive insect to hit the United States, "murder hornets." Known as the world's largest hornet, a 2-inch killer with an appetite for honeybees, murder hornets were first found in Washington State. The giant Asian insect, with a sting that can be fatal to some humans, started to emerge from its winter hibernation in May, although it was first sighted in the U.S. in December 2019. The "murder hornets" that are native to Japan, where they are known to kill people, have a stinger that can penetrate through most beekeeper suits and deliver nearly seven times the amount of venom of a honeybee and sting multiple times (*Associated Press* 2020a). The "murder hornet" was the last thing Americans wanted to hear about as they fought the devastation caused by COVID-19.

We turn our attention now to invasive species in aquatic environments. The number of invasive species in aquatic environments is expansive; but like our discussion of land invasive species, we will provide just a few examples, beginning with the Nile perch. In 1954, the Nile perch was introduced to Lake Victoria, Africa, to counteract the drastic drop in native fish stocks caused by over-fishing. The perch contributed to the extinction of more than 200 endemic fish species through predation and competition for food (Invasive Species Specialist Group 2004). The Nile perch is oilier than local fish, so more trees were felled to fuel fires to dry the catch. The subsequent erosion and run-off contributed to increased nutrient levels, opening the lake up to invasions by algae and in particular to water hyacinth (Eichhornia crassipes). The hyacinth grew quickly and thickly, it floated along the top of the lake blocking out the sun and led to oxygen depletion in the lake, which resulted in the death of more fish and plant species (ISSG 2004).

In New York State, the legislature has deemed the state's water supply and recreational sectors to be particularly vulnerable to the detrimental effects of invasive species. The NYCDEP (2012) is especially concerned about the quality of the state's water supply because New York City's water supply is directly correlated to the health and integrity of the watershed lands, forests, reservoirs, and water supply infrastructure. The introduction of invasive species "can lead to profound changes to natural ecological communities as well as the fundamental infrastructure that sustain water quality and quantity" (NYCDEP 2012). In an attempt to assure the sustainability of the New York's water supply, the state's DEP has developed a proactive stance designed to prevent the introduction of new invasive species to water supply lands, reservoirs, and streams through public education, employee awareness training, early detection and rapid response of new invasive species, controlling the current invasive species population, and rehabilitating and restoring impaired ecosystems (NYCDEP 2012).

Many of Wildlife Forever's top 10 invasive species threats to hunting and fishing have been identified in New York's waters, including Central New York. Zebra mussels, water chestnuts, and hydrilla are the prime invading species in the many Central New York lakes and rivers. The hydrilla is a huge concern, as it can quickly grow up to 25 feet in length, creating dense beds that choke fish spawning areas and limit boating access. Another problem with hydrilla is how easily it can be spread from one lake to another from tiny pieces caught on boats and trailers (*The Citizen* 2012c). The spread of hydrilla from lake to lake is an example of inadvertent or accidental introduction of a foreign species to a new ecosystem. In light with the state's DEP recommendations, boaters are warned and advised to thoroughly clean their boats before moving them.

Despite the attempts of New York and other states that neighbor the Great Lakes, these sources of fresh water have become increasingly invaded by foreign species. The problem began in earnest with the opening of the St. Lawrence Seaway in 1959 as this passageway caused a great environmental hazard for the Great Lakes as ships were able to travel from the Atlantic Ocean throughout the lakes region. It may have represented an economic boom for the shipping industry, but it is also, at least in part, was responsible for the introduction of invasive species as ballast water from these ships transported non-native species from other continents (Atlin 2014). NOAA (2020d) reports the Great Lakes ecosystem has been severely damaged by more than 180 invasive and non-native species. These species include the zebra mussel, quagga mussel, round goby, sea lamprey, and alewife, and they quickly reproduce and spread, ultimately degrading habitat, out-competing native species, and short-circuiting food webs. As an elaboration of the harm these invasive species can cause we will look at one example, the round goby (neogobius melanostomus). The goby species impacts the ecosystem because they eat the eggs of native game fish such as trout as a specific example and because they have a negative impact on aquatic biodiversity in general (Burkett and June 2015). It will be nearly impossible to eliminate this species because they are found throughout the Lakes.

On the west coast of the United States, California has become inundated with hundreds of millions of voracious purple sea urchins that feed on towering underwater kelp forests sending the delicate marine ecosystems off the shore into such disarray that other critical species are starving to death (*Associated Press* 2019b). It is estimated that in Northern California, 90 percent of the giant bull kelp forests have been devoured by the urchins, perhaps never to return. The sea urchins have moved northward where one count found 350 million purple sea urchins on one Oregon reef alone—more than a 10,000 percent increase since 2015. The explosion of purple sea urchins is the latest symptom of a Pacific Northwest marine ecosystem that's out of whack due to the warmer-than-usual waters in the Pacific Ocean (*Associated Press* 2019b).

There is financial gain to be made from invasive species when consumers find non-native animal species to be cute, furry, scary or cool-looking, but then are grown tired of and released to the wild. Invasive plants may work the same way. Some people just like the way a certain plant looks or smells, or find the colors aesthetically pleasing. A number of plants appear on banned lists (from local, state, or federal jurisdictions) because they are known to spread outside of intended confined areas. Plants spread when seeds or pollen are blown downwind. The Norway maple, for example, is an invasive plant that pushes out native maples, including black and sugar maples. The Norway maple is number one on New York State's 2012 banned invasive list (Eisenstadt 2012). The Norway maple is a favorite of developers because it is cheap, looks pretty and grows fast. However, it reproduces quickly and chokes out other native trees. The top items on the banned plant list of New York State are multiflora rose (looks pretty, but spreads quickly, creates thickets, and makes it hard for native plants to grow); barberry (a cheap purple hedge, but grows quickly and takes over forest underbrush and emits a chemical that chokes off nearby plants); bush honeysuckle (looks and smells pretty, but gives off a substance that makes nearby soil poisonous); burning bush (an inexpensive shrub, cheap to grow, but the seeds spread easily and fields can be overrun by burning bush); and the sycamore maple (grows easily and quickly, but grows thick and takes over surrounding native trees and vegetation) (Eisenstadt 2012).

Landscapers and gardeners are responsible for purposively introducing invasive plants into local ecosystems that are not equipped to handle their negative environmental effects. Consider, for example, "The poster child for invasive plants, kudzu—aka 'the vine that ate the South'—was brought here from Japan in 1876 to decorated the Japanese Pavilion at the U.S. Centennial Exposition in Pennsylvania. The fast-growing vine was widely planted to prevent soil erosion and as a high-protein livestock forage crop. Kudzu escaped cultivation and today smothers millions of acres of land in the southeast, engulfing trees, building and anything else in its path" (DeJohn 2019). Common landscape invasives found throughout the U.S., by category, include ground covers (bugleweed, crown vetch, English ivy, and periwinkle); shrubs (Japanese barberry, Russian olive, autumn olive, burning bush, privet, bush honeysuckle, and multifloral rose); trees (Norway maple, tree of heaven, mimosa, empress or princess tree); perennials (non-native bamboo goutweed, pampas grass, cogongrass, purple loosestrife, and Chinese silvergrass); and vines (porcelain berry, Oriental bittersweet, English ivy, Japanese honeysuckle, and Chinese and Japanese wisteria) (DeJohn 2019).

Invasive plants also find their way into foreign countries in the same manner as aquatic invasives, via the ballast water from ships. As an example, an aquatic weed is creeping across the Great Lakes regions that grows very fast and is very hard to kill, the starry stonewort (*Associated Press* 2017). The starry stonewort is a bushy, bright green macro-algae which forms dense surface mats in lakes and first appeared in North America in 1978 in the St. Lawrence River in New York State. Researchers believed it arrived in ballast water from ships entering the Great Lakes. Initially, this invasive weed was of little concern, that is until, it spread onto Michigan's Lower Peninsula (circa 2008), where it has infested more than 200 inland lakes, and then parts of western New York, Wisconsin (in 2014) and Minnesota (in 2015). It has also reached some lakes in Indiana, Pennsylvania, Vermont, and Canada (*Associated Press* 2017).

Invasive plant species are exerting a serious threat to biological diversity in many regions of the world beyond the United States, of course. As an example, researchers have found 166 documented plant invasives in the Nepalese Himalayas, distributed along huge altitude ranges. Not surprisingly, the areas with the most invasive species were those that a relatively high human population densities, transportation routes, and visiting tourists (Bhattarai, Maren, and Subedi 2014). The implication is clear; humans contribute greatly to the spread of invasive species regardless of geographic location.

By definition, invasives also include microorganisms. Once again, humans are a prime contributor to the spread of this type of invasives as well. Invasive microorganisms can be grouped into three general categories: parasites (an organism that lives in or on another species) or pathogens (organisms that cause disease); mutualists (interactions between two species) or symbionts (two species that benefit one another); and saprotrophs (an organism that fees on or derives nourishment from decaying organic matter) or decomposers (organisms such as fungus that decomposes organic material). Because of their size, it is generally much harder to detect invasive microorganisms. However, in order to better understand and manage invasive species, especially if human-induced environmental change exposes more native species to conditions that promote invasive characteristics—particularly among microorganisms that are difficult to detect—it is important to pursue such knowledge (Taylor and Bothwell 2014). Solomon and associates (2019), for example, have studied the microbiota of Drosophila suzukii and its influences on the larval development of Drosophila melanogaster. Among other things, they have found that for the Drosophila species, microbes are known to affect nutrition, immunity, and a range of other traits. The Drosophila melanogaster "has emerged as a useful model for studying ecological interactions that shape the assembly of microbial communities" (Solomon et al. 2019). Among the useful bits of information found by Solomon and associates is the realization that "Drosophila suzukii is an agricultural pest that infests soft and stone fruit. Since its recent arrival in North America, this invasive species has spread rapidly causing significant economic damage due to crop loss."

So, what are the solutions to eliminating invasive species? Some would argue that once they are here, there is nothing we can do about it. Others would suggest that action can be taken. For example, we can heed Jackson Landers' recommendation that we hunt down invasive animals and eat them. Another idea is promoted by Wildlife Forever (2012), a nonprofit conservation arm of the North American Hunting Club and North American Fishing Club, with a membership of 1.3 million, which promotes the idea of trying to hunt and kill all of the invasive species, a not so practical solution, especially when we reconsider all of the examples of invasive species described here. Another possible solution is to create a genetically manipulated pest management strategy wherein genetically engineered DNA sequences of a specific species in introduced along side the invading species. With this proposed solution, Harvey-Samuel, Ant, and Alphey (2017) are working on a genetic pest management strategy that exploits a pest's mate-seeking behavior by introducing genetically modified versions of invasive insects to mate with non-modified DNA pests; the caveat is that the modified insects are sterile. "Since wild females mating a released sterile male have few or no viable offspring, mass-release of radiation-sterilized pests reduces the reproductive potential of the target population and over time can suppress or even eradicate it" (Harvey-Samuel et al. 2017: 1684). And yet another realization leads us to a possible solution we are not likely to pursue, and that entails the realization that *humans* are the invasive species (Bekoff and Bexell 2010). Humans are, on the one hand, lucky to be the species the rose to the top of the food chain; on the other hand, we are the greatest threat to the environment. Bekoff and Bexell do not suggest we eliminate ourselves, but strongly encourage a revolution in thought, heart, and action, a paradigm shift, if you will, wherein we put animals, habitats, and the future of humanity as the top priority in order to save the environment and the planet.

Vapors from Sulfur Springs and the Decay of Dead Species

As if the various topics discussed above were not enough to make one aware of how tenuous life on earth can be, one more disturbing factor should be added to the mix: the dangers of deadly fumes, often undetectable until too late. The air we breathe is something all too often taken for granted. The necessity for clean air is all too evident when human-caused pollutants create difficulties in breathing and, in extreme situations, loss of life. But even if there were no human-made factories belching foul odors into the skies, life would still be fraught with peril, as naturally occurring smells can be equally deadly.

One form of this type of danger comes from vapors emitted from deep inside the earth. Hydrogen sulfide is an extremely toxic chemical compound found within coal pits, gas wells, and sulfur springs. Usually characterized by a "rotten egg" smell, it is a product of sulfur-containing organic matter, which occurs usually under low oxygen conditions. It is also sometimes known as "sewer gas." Hydrogen sulfide is a chemical asphyxiate, meaning it can cause suffocation, nausea, vomiting, and in severe cases death. Fortunately, the odor itself is so strong that it is usually detectable even in minute traces, but prolonged exposure can cause "olfactory fatigue," meaning that reliance upon smell alone is often insufficient.

Another way that nature can cause harm is through the putrefaction of dead flesh. Decomposition creates not only horrific smells but, in many cases, the emission of the substances "cadaverine" and "putrescine," both of which can be extremely toxic.

Outer-Worldly Forces

As we learned in Chapter 1, the leading contributing cause of past mass extinctions is not of this earth; instead, it was an outer-worldly source, asteroids. Consequently, there is no reason to doubt that the demise of humans and at least an additional 75 percent of all species could become extinct as a result of asteroids or meteors bombarding our planet. This sobering reality can lead some to conclude that if this is the case, why should we worry about putting forth efforts such as reducing our carbon footprint and trying to control pollution if we could all perish in a seemingly blink of an eye? The obvious immediate answer to that question is the fact that it could be millions of years before such an outer-worldly event could occur and therefore we need to protect our environment for as long as we can while our species is still on this planet.

From the above paragraph it should be clear what we mean by the term "outer-worldly" and that is, not of this earth. Outer-worldly forces include asteroids, meteoroids, and comets and also species from different planets that might travel to earth for nefarious reasons. The authors are not writing a book about life outside this planet so we will not spend time dwelling on the possibility of life on other planets, we will treat it as a given that somewhere there is some form of life; and most likely, intelligent life forms. It would be very naive and ignorant for anyone to think that in all the universes and infinite number of alien worlds that life only exists on this planet. Some people view the "outer-worldly" concept as pertaining toward that which is imagined or spiritual, but that is not our contention of its meaning. The term "*other*-worldly" would be used to refer to the spiritual world, or the world of imagination. We are sticking with the idea that outer-worldly relates to asteroids and meteoroids, as there is not only scientific evidence of their existence, any one of us can look to the sky and see such evidence darting across deep space. And, of course, asteroid and meteor strikes hitting our planet are already well-documented.

In the "Popular Culture Section 1" description of a potential asteroid apocalypse we already clarified a number of important details regarding outer-worldly events. For example, we pointed out the distinction between a meteoroid and meteorite (a meteoroid that has entered the Earth's atmosphere) and stated how a meteorite the size of 5 to 10 miles in width could destroy the planet but that even smaller ones could cause a great deal of destruction. Asteroids, on the other hand, are much larger than meteoroids, they are fragments formed billions of years ago that have broken free of their orbit and hurdle throughout space ready to crash into some other entity found in space, such as the planet Earth. We also pointed out a couple of examples of near misses of asteroids hitting Earth and an example of a meteor that landed in Russia in 2013. The very fact that scientists and a few world leaders are preparing a defense system to protect us from asteroids and meteoroids is a testament to the threat of outer-worldly forces. So why then, aren't we more concerned about the danger outer-worldly forces represent? Perhaps it's because we have more immediate concerns here on earth. Consider this March 2020 political reality: "A poll of likely New Hampshire Democratic voters taken on the eve of their primary found that 95% disapproved of the job President Trump has been doing. Not at all shocking. But when asked which of the following outcomes they would prefer on election day this [2020] November, 'Donald Trump wins reelection,' or 'a giant meteor strikes the Earth, extinguishing all human life,' 62% picked the meteor" (Morson and Schapiro 2020: A11).

There are plenty of people who are concerned about the dangers of asteroids and as such, they have created an annual Asteroid Day, celebrated on June 30 annually. Asteroid Day, the official United Nations' day of global awareness and education about asteroids, was co-founded by astrophysicist and famed musician Dr. Brian May of the rock group Queen; Apollo 9 astronaut Rusty Schweickart; filmmaker Grigorij Richters; and B612 President Danica Remy to promote awareness and provide knowledge to the general public about the importance of asteroids in our solar system, and the role they play in our solar system today. "Asteroid Day has evolved to include the participation of major national space agencies, ESA, JAXA, Roscosmos, ISRO and NASA, and prominent scientists, astronomers, educators, and media worldwide. In 2016, the United Nations declared Asteroid Day an official day of education, initiated in part by a need to better understand the role of asteroids, following the 15 February 2013 meteor impact in Chelyabinsk, Russia" (AsteroidDay.org 2020).

Because asteroids, meteoroids, and comets are not human made, they fall under the domain of forces of nature and like all the other forces of nature that can cause harm to life forms on our planet, we cannot control them, we can merely prepare for their inevitability.

Human Skepticism and the Sixth Mass Extinction

As described to this point in this chapter, there are many elements of nature that have an adverse effect on the environment, including volcanic eruptions, lightning strikes, wildfires, storms, invasive species, vapors from sulphur and the decay of dead species, and outer-worldly forces. Combine these forces of nature with the negative impact that humans have on the environment (e.g., a reliance on burning fossil fuels, war, famine, pestilence, disease, enviromares, the development of plastics, food waste, harmful agricultural practices, deforestation, marine debris, e-waste, and medical waste) and we can see how humanity and nature have tagged team against the environment causing it to spiral deep into the sixth mass extinction era.

We know that nature and not humans caused the first five mass extinctions. This train of thought leads a number of people to be skeptical about the idea that we are currently in the sixth mass extinction era and that this ME is largely caused by human behavior. There are skeptics because of conservative religious, political, and business doctrines; the possession of an unenlightened and anti-science bias; being brainwashed by the conservative mainstream and far right radical media; an over-reliance in the ability of science and technology to conquer any threat to human existence; and the inability to fathom such an occurrence as a mass extinction disguised within the finite pool of worry that individuals can endure. In no other time in history have humans had the ability to access factual information so easily and yet there exist those who are ignorant to facts and reality. This last point is perfectly illustrated by the reality that there were some people who remained skeptical of the existence of the COVID-19 virus even while tens of millions of people around the world contracted the disease and more than one million died from it. The lack of leadership from the White House led to the United States having, by far, the highest number of confirmed COVID-19 cases and deaths among all the nations of the world.

Let's take a closer look at human skepticism and the negative role it plays in trying to save the environment.

The Role of Politics and Big Business

Residing in an environment where all living species thrive should be a goal of all people, and such a goal should be devoid of political overtones, right? Presumably, all of us, regardless of political affiliation, want to breathe clean air and drink clean water. And yet, politics plays a role in how one perceives environmental issues. *Politics* refers to the activities associated with the governance of a country, state, or some other region, especially with regard to the debate or conflict among individuals or parties having or hoping to achieve power and control of governmental policy. Political leaders attempt to sway public opinion on matters of political interest and, conversely, the members of a political party hope to elect officials that set policy that is favorable to their concerns. With regard to politics and the environment, issues such as climate change and global warming, human-made climate change, the burning of fossil fuels, and whether policies to protect the environment or empower big business are among the most important topics of interest.

As a general rule, the Democratic Party not only strongly believes that climate change and global warming are a matter of fact, but they also believe that human activities, especially the burning of fossil fuels are the leading contributing cause of it. As a result, Democrats attempt to pass legislation that promotes the use and further development of renewable forms of energy and policies designed to protect the environment (e.g., protection for national parks and endangered species; anti-fracking; going green;

and the development of hybrid automobiles). Democrats want to expand the economic sector by promoting green initiatives and supporting green industries.

Republicans tend not to believe in climate change or global warming and they certainly do not believe that the burning of fossil fuels or any other human activity contributes to global warming. Republicans and their political leaders are generally far more concerned about economics and an expanding marketplace than they are about the environment. Republicans tend to be pro-fossil fuel industry and support such things fossil fuel extraction and quasi-green initiatives so long as they are good for business. They worry about government oversight on such industries as fossil fuel extraction and logging. Keeping people employed so that big business flourishes is a key concern with Republican politicians. "Big Business" is a term used to describe financial organizations considered as a group that are powerful, generate large profits, and exert a significant influence on social and political polity (Delaney 2020).

During the eight years of the Obama presidency a great deal of legislation was passed that favored protecting the environment. Obama made protecting the environment and combating climate change as one of the cornerstones of his presidency. Obama said many times that he "believes that no challenge poses a greater threat to our children, our planet and future generations that climate change—and that no other country on Earth is better equipped to lead the world towards a solution" (Nelson 2016). Protecting the environment for future generations is a goal that would seem hard to argue with, that is, if one cares about the future of America's (and the world's) children. Viewing the United States as a world leader in the political arena is another cornerstone of the Democratic political platform. As detailed by Angela Nelson (2016), the top fifteen top things Obama did for the environment are the establishment of the largest marine reserve in the world (he expanded the Pacific Remote Islands Marine National Monument to encompass more than 490,000 square miles); signed a bipartisan ban on microbeads; rejected Keystone XL oil pipeline; appointed a tough new EPA administrator; voted against the Clear Skies Act (Obama did this as an Illinois senator because Democrats believed the act would actually weaken clean-air regulations); expanded the California Coastal National Monument; established the largest ocean sanctuary on the planet; raised fuel-efficiency standards; unveiled the Clean Power Plan; established America's Great Outdoor Initiative; signed the Omnibus Public Land Management Act of 2009; used the Antiquities Act more than any president in history (allows U.S. presidents to make a presidential proclamation to create national monuments from public lands to protect significant cultural or scientific landmarks); invested in green energy during the Great Recession; started a plan to save bees and pollinators; and established America's first National Ocean Policy.

As illustrated previously in this text, President Trump has attempted to dismantle of all Obama's environmental achievements in favor of Big Business and especially the fossil fuel industry. Among Trump's anti-environmental policies: he pulled the United States out of the 2015 Paris Agreement on climate change (turning the U.S. from a global leader to a global predator); signed legislation that would weaken one of the nation's most aggressive efforts to combat climate change via the promotion of new fuel-efficiency standards for cars and trucks; gutted an Obama-era rule that compelled the country's coal plants to cut back emissions of mercury and other toxins produced by coal- and oil-fired plants; appointed a former lobbyist as head of the EPA; repeatedly mocked climate science via Tweets; altered the cost-benefits formula in order to justify continued dependency on fossil fuels; and pushed to eliminate or soften the Endangered Species Act. As reported in *The*

New York Times, the Trump Administration is reversing nearly 100 environmental rules (as of May 6, 2020, he had completed 64, 34 were in progress, and succeed in 98 rollbacks) (Popovich, Albeck-Ripka, and Pierre-Louis 2020). In sum, Trump has, or has attempted, to undercut eight years of science-driven, environmental-friendly policies from President Obama in favor the whims and desires of Big Oil and other Big Businesses. Is there any wonder why environmentalists universally show such disdain towards Trump?

The Role of Religion

The role of religion in environmental issues is diverse. There are spiritual people who believe that humans are a part of nature and therefore it is people's ethical duty to protect the earth. There are other religious people who believe that God made "Man" to be superior to all other living species and therefore it is the human right to reign supreme, to take the resources necessary, at any cost. If there are consequences to be had, it will be God's decision, and not us mortals. Some religious people believe that only God can influence the weather and therefore discount any notion that human activities influence the climate. Thus, whatever happens to the climate is a part of God's plan and not for humans to interfere with.

In his 1967 essay, "The Historical Roots of Our Ecological Crisis," Lynn White described how people's view of the environment is correlated with their religious beliefs: "What people do about their ecology depends on what they think about themselves in relation to things around them. Human ecology is deeply conditioned by beliefs about our nature and destiny—that is, by religion." White (1967) adds that before Christianity conquered paganism, it was a commonly held belief that all forms of life, including plants, had spirits, or souls. "In Antiquity every tree, every spring, every stream, every hill had its own genius loci, its guardian spirit." White argues, that while these spirits were accessible to men, they were unlike men and before humans cut a tree, mined a mountain, or dammed a brook, it was important to placate the spirit in charge of that particular situation, and to keep it placated. In other words, humans attempted to please the souls of nature before taking what they needed for their own purposes. This would change with the introduction of Christianity. "By destroying pagan animism, Christianity made it possible to exploit nature in a mood of indifference to the feelings of natural objects.... The spirits in natural objects, which formerly had protected nature from man, evaporated. Man's effective monopoly on spirit in this world was confirmed, and the old inhibitions to the exploitation of nature crumbled" (White 1967).

Many Christians believe that it is their responsibility to love all of "God's creatures." They cite numerous passages in the Bible that implore us to treat Mother Earth with care and not to demand more of her resources than our fair share. Consider the following Biblical passages (as cited from Christian Ecology):

> Ezekiel 34:17–18: As for you, my flock.... Is it not enough for you to feed on good pasture? Must you also trample the rest of your pasture with your feet? Is it not enough for you to drink clear water? Must you also muddy the rest with your feet?
>
> Isaiah 24:4–6. The earth dries up and withers, the world languished and withers, the exalted of the earth languish. The earth lies under its inhabitants; for they have transgressed the laws, violated the statutes, and broken the everlasting covenant. Therefore a curse consumes the earth; its people must bear their guilt.
>
> Jer. 2:7. I brought you into a fertile land to eat its fruit and rich produce. But you came and defiled my land and you made my inheritance detestable.

> Luke 16:2, 10, 13. And He called him and said to him, "What is this I hear about you? Give an account of your stewardship, for you can no longer be steward. He who is faithful in a very little thing is faithful also in much; and he who is unrighteous in a very little thing is unrighteous in much. You cannot serve both God and mammon."
>
> James 5:5. You have lived luxuriously on the earth and led a life of wanton pleasure; you have fattened your hearts in a day of slaughter [Christian Ecology 2012].

A couple of additional Bible references (as cited from Bible Gateway) indicate God loves not only people but every living thing on Earth from amoebas to whales:

> John 3:16. For God so loved the world that he gave his one and only Son, that whoever believes in him shall not perish but have eternal life.
>
> Colossians 1:16–2. For in Him all things were created: things in heaven and on earth, visible and invisible, whether thrones or powers or rulers or authorities; all things have been created through him and for him.

Despite the skepticism among many Christians about the role of humans on the environment, Pope Francis made his feelings clear. In 2015, Pope Francis warned about climate change and its grave implications for the environment and the socio-economic livelihood of humanity. In his Papal Encyclical "*Laudato Si*: On the Care of our Common Home" Pope Francis said of global warming: "A very solid scientific consensus indicates that we are presently witnessing a disturbing warming of the climatic system … accompanied by a constant rise in the sea level and, it would appear, by an increase in weather events.… Most global warming in recent decades is due to the great concentration of greenhouse gases … released mainly as a result of human activity.… The problem is aggravated by a model of development based on intensive use of fossil fuels [and] deforestation for agricultural purposes" (*The Citizen* 2015).

While it is the Christian belief that dominates the socio-political ideology of the United States, there are, of course, other belief systems with their own interpretations of the role of humans and the environment. Writing from the perspective of Judaism, Daniel B. Fink (2009) reminds us of the teachings of the Torah and its beliefs of how God created the earth in six days but how humans were the last thing he created after the land, oceans, plants and other animals. A possible implication is that all beings were created for their own sakes and not for the sake of humanity. The idea that "Man" shall take dominion over all the living things is open to interpretation as well, as the Hebrew word for "take dominion" (*v'yirdu*) comes from the same root as "to descend" (*yarad*). Thus, Rashi, a distinguished commentator on the Torah (nine hundred years ago), declares, "When humanity is worthy, we have dominion over the animal kingdom; when we are not, we descend below the level of animals and the animals rule over us" (Fink 2009). Fink (2009) suggests that abusing the environment is a sign of debasement rather than dominion and thus, if we destroy human life on earth through some sort of tragedy (e.g., nuclear accident or war), the cockroaches will, in all likelihood, succeed humans as the "masters" of the planet.

The Islam religion is similar to Christianity and Judaism in that human beings are seen as stewards of the environment but is more like Judaism in that humans are "only a manager of the earth and not a proprietor" (Oliver 2008). The Prophet Muhammad's stance on the environment is made clear in "Environmental Protection in Islam" (published by the Meteorology and Environmental Protection Administration of the Kingdom of Saudi Arabia), "Created beings are dependents of God, and the creature dearest unto God is he who does most good to God's dependents" (Oliver 2008). In addition, the

Islamic stance on the environment includes the notion that good deeds of humans "are not limited to the benefit of the human species but rather extend to the benefit of all created beings" (Oliver 2008).

Hinduism is a complex and diverse religion and shares certain beliefs with Buddhism but as it pertains the environment, Hindus are expected to live a simple life, learn to enjoy spiritual happiness, not to pursue material pleasures, and not to disturb nature's checks and balances. Hinduism would be a de facto supporter of renewable fuels as it is in line with sustaining the natural order of things. Furthermore, if Hindus use anything belonging to nature, such as fossil fuels or the forest, they are to return to nature at a greater rate so that nature can be replenished (Oliver 2008). One of the core beliefs of Hindus—karma—is certainly applicable to the need to protect the environment (the natural world) as the manner in which we live our lives leads to the establishment of our own destiny. Thus, harming the environment will lead to harm in the next life. If the concept of "karma" holds true, there are many politicians and people associated with Big Oil that are in deep spiritual trouble. The principle of non-violence embraced by Hindus would also lead to treating all living things in a sacred manner because they are part of God and the natural world.

Buddhism also has a number of subsets of adherents but in general, is considered the most environmentally friendly religion of them all because it believes in the fundamentally equality of all sentiment beings: "We are all born, we all age, then we all die" (Oliver 2008). Buddhists believe that humans are no more important than any other living creature and that all living creatures have the right to thrive. Two of the five precepts (guidelines about how to act properly) have a direct bearing with the environment. The first Precept of Buddhism states that we should abstain from taking life, any form of life, and this idea is connected to the belief of rebirth wherein it is possible for a person to be reborn as an animal. Buddhist are taught o believe in compassion to all creatures, as in the principle of "do no harm." The second precept of abstaining from taking what is not given (meaning, that which belongs to someone else, including the environment) leads to the belief that humans should not destroy the natural habitats and that we are to protect the environment from destruction and exploitation.

Anti-Science Rhetoric and the Conservative and Radical Right Media

A number of people, including politicians, claim that human-caused climate change is a hoax. They attempt to discredit the data presented by global climate scientists, biologists, and meteorologists, among other scientists, as political rhetoric. An anti-science bias, however, is fueled by such factors as a lack of education, lack of intellect, ignorance, the realization that facts disprove a belief, and/or justification for harming the environment. Science does not care about one's beliefs. Instead, science comes about as the result of the intellectual and systematic study of human behavior (social science) or of the natural world (natural science) that leads to new knowledge that is supported by data. Because Republicans tend to support Big Oil and profiting from the natural environment, they have put themselves in the unenviable position of being anti-science. No reasonable person wants to be put in the anti-science camp as it makes them look foolish, especially in today's technologically driven world. When ancient humans could not comprehend science, or something as simple as how to explain an eclipse, it was easy to forgive them for their ignorance; but contemporary people have no excuse for not embracing science, as science drives our understanding of the world and is directly connected to technological progression. President Trump, for example, does not believe in the science that clearly shows how humans have negatively impacted the climate and environment and yet he will have to rely on science in order to develop and maintain his "Space Force"—the "sixth branch of the military," as he puts it. Even a grade school child would have to question how one can believe in science to develop a space force and yet not believe in science that manages to simply explain how humans have greatly contributed to global warming and a deteriorating environment.

In Barack Obama's first inaugural address, he promised to "restore science to its rightful place," a not-so-subtle shot at the Republican Party that had come to be known as anti-science (Fisher 2013). There are certainly a number of Republicans with very unenlightened and irrational forms of thinking with regards to science. Running against Obama and his second term bid, Republican presidential candidate, Rick Santorum proclaimed in February 2012, "I refer to global warming as not climate science but political science" (Babington 2012). Santorum, who, not surprisingly, wanted to aggressively produce more oil and natural gas in the United States, said that Obama was "pushing a radical environmental agenda" (Babington 2012). Another Republican with very unenlightened and irrational forms of thinking with regards to science includes Representative Paul Broun of Georgia (an M.D.), who in 2012 declared that evolution, the Big Bang theory, and embryology are "lies straight from the pit of hell" (Fisher 2013). Among his other bizarre ideas, Broun claims "as a scientist" that he has found data that shows the Earth is no more than 9,000 years old and was created literally in six days (Williams 2012). Certainly Broun, who served as representative from 2007 to 2015, would not believe that there were five previous mass extinctions and would not be able to fathom that the earth is approximately 4.5 billion years old. Of course, this is mild stuff compared to Senator John McCain's 2008 vice presidential running mate Sarah Palin who claimed that men and dinosaurs coexisted on an Earth created 6,000 years ago. Wisconsin Representative Jesse Krember (R) also believes that the earth is only 6,000 years old. More importantly, Donald Trump as president has repeatedly ignored science in critical decision making, compromising our nation's ability to meet current and future public health and environmental challenges (The Union of Concerned Scientists 2019). Sadly, there are countless other examples of politicians who possess an anti-science bias, and sadder than this is the reality that there are millions of Americans who actually believe the same rhetoric. And yes, we can make matters even worse by pointing out the United States is not alone in its anti-science rhetoric. It's enough to makes us wonder if we are in the 2020s CE or 2020s BCE.

The anti-science rhetoric is fueled by the conservative nature of the mainstream media and the radical right media fuels the righteous right and their militant counterparts (e.g., paramilitary groups, Aryan Nation, American Nazis and Confederates). By saying mainstream conservative media, we are talking about the major networks (i.e., ABC, CBS, and NBC) that are too afraid to offend mainstream Americans by pointing out just how dire things are in society. Sure, they provide stories on racism, sexism, global warming, and so on, but never to the extent that liberals would like to see. Radical far-right mainstream media would be highlighted by FOX News, a network that is anything but "fair and balance" and essentially serves as a marketing arm of the Republican Party to spread their right-wing slant on issues at a level that can be understood by FOX viewers; and One American News Network (OANN) that promotes falsehoods and conspiracy theories that even FOX News will not air. CNN and MSNBC would be considered left-leaning mainstream cable networks. Of course, it is in the social media platforms where we will find the true far right and far left propaganda broadcasts.

The funny thing about perspective is the vast prism of viewpoints that exist on any particular topic, including the social overtones of the media. Former vice president Al Gore, who has made quite a name for himself as a crusader of the environmental movement to fight global warming, argues that the media is actually very conservative. Gore (2011) believes that the media is confused about whether it is in the news business or the entertainment business. Gore claims that the media doesn't appear to notice that the "Polluters" and "Ideologues" are trampling all over the "rules" of democratic discourse. According to Gore (2011), the media "are financing pseudo scientists whose job is to manufacture doubt about what is true and what is false ... spending hundreds of millions of dollars each year on misleading advertisements in the mass media; hiring four anticlimate lobbyists for every member of the U.S. Senate and House of Representatives."

Gore (2011) points out that the media have a conservative history when it comes to confronting scientific data and evidence. For example, a half-century ago when science and reason established the clear linkage between cigarettes and lung diseases, the tobacco industry hired actors, dressed them up as doctors and paid them to look into television cameras and tell people that the data presented by the Surgeon General's report was a lie. Gore (2011) claims that the scientific consensus is even stronger with regard to climate change than it originally was about the harmful effects of smoking tobacco. Today's conservative media treats climate change and global warming as if it's the contemporary version of tobacco and a cash cow product that needs to be protected from scientists and left-leaning social thinkers who dare to challenge the Big Business that runs the industry. And isn't that the reason the conservative and radical right media protects Big Oil, because it is a cash cow?

An Over-Reliance in the Ability of Science and Technology

While some skeptics question the legitimacy of environmental science as "real" science, and instead prefer to call it political science, there are other skeptics (with regard to the seriousness of global warming) who so strongly support science that they believe future technology will find a way to solve any and all of our environmental problems.

A faith in technology is reflected in the IPAT formula (described in Chapter 2). The IPAT formula reflects the skeptical, conservative viewpoint of the environment. In brief, it is believed that technology can solve whatever problems the environment presents us. Conservatives point out that since industrialization, humans have imposed their will on nature via such means as tunneling through mountains, damming rivers, irrigating deserts, and drilling for oil beneath the ocean floor (Macionis 2010). Furthermore, conservatives argue that the planet supports six times as many people as when Malthus lived and there is much greater affluence among most inhabitants than ever before (Simon 1995). "Because this optimistic position opposes the pessimistic view of Malthus, it is often described as *anti–Malthusian*" (Macionis 2010: 434).

American skeptics of human-caused environmental problems strongly support the use of fossil fuels as the prime source of energy. They rely on advancements in technology as the means to secure enough fossil fuels to keep the United States independent of foreign supplies of fossil fuels, especially crude oil. In the first edition of this text, we wrote of the optimism of skeptics in their belief that the United States would become the world's top oil producer of oil by 2015, a net exporter of the fuel by 2030, and nearly self-sufficient in energy by 2035 (Hsu 2012). The United States has already met the first goal as 2019 marked the sixth straight year it held the position of the top oil producing country in the world with an average of 17.87 million barrels per day, which accounts for 18 percent of the world's production (*Investopedia* 2020). Interestingly, past predictions by skeptics believed the U.S. would be producing 11.1 million b/d by 2015 but as we can see, production is much higher. The next leading oil-producing nations are Saudi Arabia, Russia, Canada, and China (*Investopedia* 2020).

Finite Pool of Worry

Who among us has not felt overwhelmed at some point in their lives? Many people endure a great deal of personal troubles and drama that cause stress, others are overwhelmed by world events, and some people feel besieged by both. We believe that there is a "finite pool of worry" that each of can endure; and that, a given person has the capacity to worry about a limited number of things before their heads figuratively explode or they suffer from physical pain (e.g., stress headaches or in extreme cases, have a mental breakdown). Even the so-called "worrywart" can only handle so much worry. Eventually, we just stop worrying about things because our pool of worry is already full.

Is it any wonder there are times when most of us would just like to take a break, go on a vacation and forget about it all, at least temporarily? Perhaps one of the reasons Alfred E. Neuman, the fictional mascot and cover boy of *Mad* magazine, remains so enduring to folks over multiple generations is his approach to life that is expressed in the famous quote "What, me worry?" Seemingly, life would seem so much easier if we could eliminate stress from our lives. As if personal, professional and social lives are not complicated enough, we have people telling us the "sky is falling" and the environment is beyond saving. Enough already, people may shout. But why are some people so passionate about trying to "save" the planet while others just don't want to hear about it?

Psychologists provide an explanation why so many people don't seem to care about such global matters as climate change and global warming. Dan Ariely, a professor of behavioral economics at Duke University, states, "We are collectively irrational" (Fahrenthold 2009: A-11). Psychologists say many Americans still don't feel close to the issue of climate change and rely on a "system justification" wherein humans resort to what is comfortable, the status quo and a willingness to defend it. We fret more about our own immediate lives than we do with matters beyond our immediate control in part, because of the finite "pool of worry" concept (Fahrenthold 2010).

Many people reason that the big issues, such as climate change, are out of their realm of control so why fret about it. Some people make modest efforts to help the environment thrive—they recycle, turn off electronic devices when they are not in use, fix leaky faucets, avoid using air conditioning except when it is really hot, and walk instead of driving whenever possible. Walking or biking is a good way to save on fossil fuel energy and it's a healthy way to stay in shape. Working out is also a good way to reduce stress. So, walking and running reduce stress and reduce one's carbon footprint. Carbon footprint, as you will recall, is the sum of emissions of greenhouse gases (carbon dioxide, methane and other carbon compounds) emitted due to the consumption of fossil fuels. Typically, as individuals, we think in terms of our carbon footprint with regard to the type of car one drives and how many miles one drives, the amount of heat or cooling used within the home, the CO_2 emitted as a result of the food we eat (because of production and transportation CO_2 costs), and so on. It is estimated that for each U.S. gallon of gasoline fuel consumed, 8.7 kg CO_2 is emitted (Time for Change 2012).

Dr. Seuss to the Rescue: Spare That Final Tree

How can one combat such skepticism about humanity's role in environmental destruction? None other than the famous children's author Dr. Seuss gives us a possible answer. One of the most beloved children's authors of all times, Theodor Seuss Geisel (1904–1991), was popularly known as "Dr. Seuss." A gifted cartoonist as well, his poetic skills were legendary—he had the ability to use simple rhymes that nonetheless stayed in permanently in one's head. Millions of people have had the experience of reading one of his classic works such as *The Cat in the Hat*, *Horton Hears a Who*, or *Green Eggs and Ham* when they were children and still remember them by heart decades later.

While he eventually wrote almost fifty books for children, most people do not know that that was not Geisel's original vocation. He began his career as a commercial illustrator, creating such memorable ads as "Quick Henry, the Flit!" for a brand of insect repellant, which became a national catchword, as well as a political cartoonist. During the Depression era he created many political cartoons for the radical publication *P.M.*, most of which attempted—in as humorous a way as possible given the subject matter—to alert Americans to the growing danger of Adolf Hitler and Nazism and encourage them to begin preparing for eventual war against the Fascist powers. This advocacy of preparedness went against the prevailing view, known as Isolationism, which predominated in America at the time. Having only just recently experienced the horrors of World War I, most Americans were in no mood to contemplate going to war yet again, and many—including such stalwart figures as the national hero Charles Lindbergh—urged his fellow citizens to stay out of European affairs. Lindbergh in fact downplayed the horror stories being told about the Nazis, and even accepted a medal from Hitler's right-hand man Hermann Göring.

Geisel took it upon himself to use his cartooning skills for an explicitly didactic purpose—to awaken Americans to the realities of the Nazi regime, and to appeal to their inherent sense of decency by showing them the outrages occurring in Germany and the lands being occupied by Hitler's regime. After the U.S. entered the war in December of 1941, Geisel joined the Army and became the head of the First Motion Picture Unit of the United States Army, creating cartoons that were used to boost the morale of the armed services and the general public.

After the war's end, Geisel devoted the rest of his long career to producing the children's books for which he is famous. But his commitment to public advocacy, while not as obvious as his work for *P.M.* or the First Motion Picture Unit, nonetheless remained an important part of his career. For instance, his most famous work, *The Cat in the Hat*, came about in part because of a growing national concern over childhood illiteracy. He took it as a personal challenge to see if he could write, using the simplest language possible, works that young people—as well as their parents, who would of course be reading these works aloud to their children—would genuinely enjoy. And, just as in his political cartoons for *P.M.*, there is a sly subversive wit in many of his children's books, such as when the Cat in the Hat continually goads the goody-goody goldfish for being too rule-bound and stuffy, and creates havoc in the house, only to bring everything to a nice resolution in the end. *Horton Hears a Who*, another beloved tale, has been interpreted by many as a call for toleration and a criticism of mob mentality—the very sort of issue that had concerned Geisel during the pre-war years when he studied Hitler's rise to power and the techniques he used to destroy individuality.

One can perhaps read too much into such interpretations, and no doubt most people delve into Dr. Seuss's works for the sheer fun of them, not to find hidden political messages. But the point is, one must always remember Geisel's underlying humanistic philosophy. And, at the height of his fame, he published a book that combined the righteous political advocacy of his early career with his whimsical cartooning skills and gift for light verse of his later career: *The Lorax*. Published in 1971, right at the beginning of the Environmental Movement, the book is a forthright denunciation of reckless consumer policies which wreak havoc on the environment, as well as an obvious paean to conservationism.

In the story, a mysterious figure known as Once-ler relates the sad tale of the missing Lorax, a wizened figure who had spoken on behalf of the trees and other living things. Most of the illustrations in the book evoke a spooky, darkened world where almost nothing lives, let alone thrives, and the Once-ler—a former factory owner who had mastered the art of creating all-purpose garments (Thneeds) from the abundant Truffula Trees—describes how, as his factory grew bigger and bigger, the number of Truffula Trees dwindled. He is warned by the Lorax that his greed and unconcern for the consequences of his actions is causing great harm. "I am the Lorax. I speak for the trees. I speak for the trees, for the trees have no tongues." He points out to the Once-ler that not only are the Truffula Trees being decimated, but the various creatures whose lives are interconnected with the trees, such as the Brown Bar-ba-loots, the Humming-Fish, and the Swomee-Swans, are also being adversely affected. But the Once-ler refuses to listen to these warnings, until at last there are no more trees at all, or any other living things. All that remains is a bad-smelling sky and smoke-smuggered stars. It is a horrific tale, and the illustrations of a devastated landscape are haunting. The Lorax, looking out upon at this horror scene, gives a sad backwards glance, and then takes his leave. The Once-ler has wised up but too late—or is it? He relates his tale to a small child who has found his hidden lair and tells him that the Lorax's final word to him was "UNLESS." What does that mean? After mulling it over for years, he finally understands, and relates the message to the child:

> "But *now*," says the Once-ler,
> Now that *you're* here,
> the word of the Lorax seems perfectly clear.
> UNLESS someone like you
> cares a whole awful lot,
> nothing is going to get better.
> It's not [Seuss 1991: 54].

This fable is a powerful rejoinder to environmental skeptics. It continues to sell well, was made into a popular 3-D computer animated motion picture in 2012 (with Danny DeVito voicing the Lorax), and has become one of the most effective metaphors encouraging conservation efforts. Random House, the publisher of the book, in conjunction with Dr. Seuss Enterprises, has initiated the Lorax Project to raise public awareness on environmental issues and inspire earth-friendly action worldwide.

Theodor Seuss Geisel himself remains an inspirational figure and shows how thrivability can be advocated through artistic representation. For more information on his life and influence, see Donald E. Pease's *Theodor Seuss Geisel* (2010)—or best of all, read or re-read *The Lorax*.

Popular Culture Section 5

Death by Volcanic Eruption

Volcanoes are among the most potentially deadly forces of nature on our planet. Volcanoes were the cause, or a contributing cause, of the last three mass extinctions. From a safe distance, the sight of a volcanic eruption with its awe-inspiring force and brightly colored lava is quite spectacular. Then again, hurricanes, tornadoes, wildfires, snowstorms, and other forces of nature are also quite magnificent when merely viewed from a safe distance. Volcanoes hold a particular fascination, however, because they can both destroy and create landscape. The fact that volcanoes are huge mountains, seemingly immovable, that can suddenly explode and spill deadly toxins into the air and cause ash thick enough to bury an entire civilization more than adds to our enthrallment of them. The threat of "super volcanoes," such as the one under Yellowstone National Park, adds to the allure and concern over volcanic eruptions.

The Random History website (2010) has chronicled 10 "great" volcanic eruptions that have occurred on earth since the time of humans. It begins with the colossal eruption of a volcano on the ancient Greek island of Thera that is said to have caused the collapse of the Minoan civilization in 1600 BCE. Archaeologists estimate that the explosion sent a 22-mile-high volcanic ash cloud into the atmosphere, as traces of this cloud can be found as far away as mainland Turkey, Egypt, and the Black Sea. It is also estimated that the fallen ash reached 23 feet and buried the Minoan settlement of Akrotiri, and that the amount of hot ash flows into the Aegean Sea displaced so much water that an enormous tsunami would have devastated the coastal region of other Greek islands. To put the lava flow in perspective, such an explosion would be up to 40 times the volume of the magma produced by the twentieth-century eruption of Mount St. Helens. Undoubtedly, the Thera volcano eruption altered the global climate. Haraldur Sigurdsson, a leading specialist on the study of volcanoes believes Plato's famous tale of the Lost City of Atlantis was inspired by volcanic eruption on Thera (Sigurdsson 1999). The island of Thera is known as Santorini today, and it is the only inhabited caldera (volcano cauldron) in the world (in2Greece 2010).

Among the more dangerous volcanoes in the world is Mount Vesuvius in Naples, Italy (Random History 2010). The distinction is attributed to the realization that in 79 CE a monumental eruption destroyed the Roman cities of Herculaneum and Pompeii, and that today approximately 3 million people reside in close proximity. The 79 CE volcanic eruption resulted in a column of ash and pumice that rose more than 15 miles and was then wind-carried to Herculaneum and Pompeii (Random History 2010).

Other deadly volcanic eruptions include the 1902 Mount Pelee eruption on the Caribbean island of Martinique that killed an estimated 30,000 people and the 1985

Nevada del Ruiz Volcano (Colombia) eruption that killed an estimated 25,000 people (Random History 2010). The Nevada del Ruiz volcano is ice-capped, and the 1985 eruption of hot ash meeting the ice cap led to a huge flood of water that streamed down the steep side of the volcano destroying trees and anything else in its path.

There are dozens of feature films that depict volcanic eruptions and their deadly consequences, ranging from *Krakatoa: East of Java* (1969) and *The Island at the Top of the World* (1974) through *Stonehenge Apocalypse* (2010) and including *St. Helens* (1981) and *Dante's Peak* (1997). *St. Helens* is a film about the Mount St. Helens eruption of 1980 wherein a 5.1 magnitude (Richter scale) earthquake at 8:32 a.m. triggered a series of explosions that only increased in frequency and intensity (Mount St. Helens Institute 2013). The eruption created an ash cloud that rose 15 miles into the atmosphere, killed 57 people, and caused more than $1 billion worth of damage, mostly to the lumber and agricultural industries (Random History 2010).

St. Helens is loosely based on fact, but it is an entertaining film that includes some footage of the real Mount St. Helens eruption in Washington State. The story begins on March 20, 1980, the date of the volcano's awakening, and ends on May 18, 1980, the date of the eruption. The story centers on activities of volcanologist David Jackson (played by David Huffman). The Jackson character is based on David Johnson, the actual volcanologist sent by U.S. Geological Survey to study the Mount St. Helens eruption. Another central character in the film is Harry Randall Truman (Art Carney), a lodge owner who refuses to abandon his business and home despite the warnings of deadly consequences when the volcano erupts. His decision not to leave is, of course, his last. Jackson too, would be among the dozens killed. The film ends with a nod to the circle of environmental life as the camera shows a small plant growing to life amid the death and destruction of the surrounding desolate landscape.

In *Dante's Peak*, we once again have a volcanologist, Harry Dalton (Pierce Brosnan), who is affiliated with the U.S. Geological Survey and who sent to investigate a volcano about to erupt. The film begins with Harry and his partner, Marianne, attempting to escape the ash and falling debris of the erupting volcano in Colombia. Four years later, he is sent to the small town of Dante's Peak, Washington, home of a dormant volcano. Dalton befriends the town's mayor, Rachel Wando (Linda Hamilton), a single mom. Before long, Dalton discovers that the local lake's acidity is higher than it should be. He also notices dead squirrels. Just as one of Wando's children is about to jump into a natural hot springs, Dalton stops him. He has noticed two bodies have been boiled to death by the water.

Naturally, Dalton tries to warn the townspeople of Dante's Peak that he suspects the volcano is no longer dormant. Dalton's boss, Paul Dreyfus (Charles Hallahan) appears in town and tells him he is overreacting and that he needs to calm down. As Dreyfus and his team set up to monitor the volcano, Harry and Rachel become closer to one another. After a week, Paul decides that the volcano is not a threat to erupt and informs Harry that it is time to leave. At the last minute, Harry discovers there is something wrong with the local drinking water and they determine that volcanic activity is the cause. With Rachel's mayoral help, they order an immediate evacuation of the town. The top of Dante's Peak explodes, causing panic. Earthquakes and aftershocks rock the surrounding area. Eventually, the main characters except for Dreyfus make it out alive. The film ends with a helicopter view of the destroyed town and a crater where the top of the Dante's Peak once existed as a volcano.

Dante's Peak is one of the best volcano movies made, it has a great cast, and the cinematography is fantastic. That people die, towns and vegetation are destroyed, however, provides the darker side of the reality that is the deadly consequences of volcanic eruptions.

Summary

In this chapter, we looked at the role of nature as the potential cause of the sixth mass extinction and examine the skepticism held by some toward the role of humans in the sixth mass extinction.

Nature, both worldly and outer-worldly, has been responsible for compromising the earth's environment in the past. In fact, the first five mass extinctions were all the result of forces of nature originating on the planet or bombarding our planet from outer space. Presently, there is ample evidence for nature's adverse role in the environment, including volcano eruptions; lightning strikes; wildfires; storms, including superstorms, hurricanes, droughts/water shortages, floods/tsunamis, snowstorms, and tornados; the influx of invasive species from one ecosystem to another; vapors emitted into the air from sulfur springs and the decay of dead vegetation and animal species; and outer-worldly forces.

There are a number of people who skeptical that we are currently in the sixth mass extinction era, and that mass extinction is largely caused by human action. There are skeptics because the role of politics and Big Business; conservative religious beliefs; the possession of an unenlightened and anti-science bias; the influence of conservative mainstream and far right radical media; an over-reliance in the ability of science and technology to conquer any threat to human existence; and a finite pool of worry.

Chapter 6

Environmental Ethics and Thrivability

Thales (circa 624–546 BCE) is usually credited as the "First Philosopher" (a term meaning lover of wisdom) of the Western World. A pre-Socratic thinker, he is most famous for his cryptic assertion that "everything is made of water." Ever since, generations of scholars have attempted to make sense of this strange claim. As the noted 20th-century philosopher Bertrand Russell put it in his *History of Western Philosophy*: "This is discouraging to the beginner, who is struggling—perhaps not very hard—to feel that respect for philosophy which the curriculum seems to expect" (Russell 1948: 43). But, as Russell adds, there may be more to this assertion than meets the eye, for in many ways Thales was also one of the world's earliest scientists, someone who wished to know how the world actually works, and who thought it possible for human beings to come to an understanding of nature through the use of their reasoning powers.

Thus, while no one is quite sure just what Thales really meant by saying "everything is made of water," he is usually taken to be the founder of the field of philosophy known as metaphysics, the study of ultimate reality. Thales seemed to be saying that, while there appears to be many different things in the universe, in actuality all of these apparently distinct things are really one and the same, namely water. This is usually called "The One and the Many Problem" and philosophers ever since have been debating whether or not the universe consists of only one thing (monism) or many things (pluralism). Even most monists, though, do not accept Thales' position—almost immediately after he made his assertion, there were those who claimed other singular realities, such as fire, air, atoms, or numbers. Nonetheless, Thales is credited with having gotten the ball rolling, as it were, by getting us all to think about the question of what *is* reality, and by emphasizing the importance of water to existence. Perhaps he can be considered not only the First Philosopher but the First Environmentalist, since, as has been shown throughout this text, humans who take water for granted do so at their own peril.

In addition, Thales was also considered to be what today we would call a meteorologist, someone who studies natural patterns in order to predict what weather conditions will be like and how these will impact the harvest. It was said of him that he was able to corner the olive market when he realized that a bumper crop was likely that year due to the probability of rain coming after a long drought. He purchased all the olive presses, and thereby had a monopoly when his fellow citizens of Miletus wanted to make olive oil. It is also said of him that he predicted an eclipse and tried to show that rather than being a sign of the gods' displeasure it was a purely natural phenomenon.

There is also a cautionary point to the story of Thales, however, that still applies to the modern world. Legend has it that he scoffed at a lowly young woman for her ignorance and encouraged her to watch the skies to learn about the movement of the stars. Shortly after, while walking forward and looking upward, he fell into an open cesspool. The young woman looked down at him, flaying away in the muck, and said that while she might not know about the stars, she knew enough to watch where she was going! So, among his other attributes, Thales is the original embodiment of the "absent-minded professor."

As philosophy began to grow as a discipline, another of its central areas became the study of moral behavior, known as ethics. *Ethics* is defined as "the moral principles that govern a person's behavior" (*Oxford Dictionary* 2020). In particular, ethics is concerned with value judgments: it looks at not only why we act in certain ways but asks whether or not such actions are right. At least in the Western World, it can be argued that until fairly recently the primary focus of ethics has been on human beings. Traditional believers in the supernatural have often argued that there is nothing wrong with humans asserting a privileged position over all other life on Earth, since we alone have been created in God's image, and all other forms of life are here to serve us. This attitude, by the way, is now being challenged by many theists who maintain that God wishes humans to be stewards, or protectors, of life on Earth—thereby showing the malleability of scriptural interpretation (for further discussion on this issue, see the book *Environmental Stewardship in the Judeo-Christian Tradition: Jewish, Catholic, and Protestant Wisdom on the Environment,* published by the Acton Institute in 2007).

In the history of philosophy, such ethical thinkers as Aristotle (384–322 BCE), Thomas Aquinas (1225–1274) and Immanuel Kant (1724–1804), in their different ways, basically argued that only humans are moral creatures since only humans have the ability to think. It is thinking which makes us accountable for our actions, in that we do not act merely from instinct, and can thus be held morally responsible and accountable for what we do. For most ethical thinkers, animals and other forms of life have no moral standing, but rather can be used as means to an end for human happiness. This approach is often called "anthropocentrism" and it has for the most part been unchallenged until contemporary times. In this chapter, we will look at some of the forerunners to today's environmental ethical movement, and then show how the concept of thrivability is intimately connected with an evolutionary approach to understanding life. If one does not hold that the complete destruction of humanity is at least a possibility, then arguments in favor of environmental thrivability will lack urgency and persuasiveness. Furthermore, if one does not at least contemplate the interconnectedness of all existence and the fragility of life on Earth, then one's moral framework will likely lack a commitment to a sustainable worldview.

Ancient Cynics

While philosophy as a discipline has usually been intricately connected with an otherworldly metaphysics, which places emphasis on the human soul as something unique and eternal, one particular movement, identified with Diogenes of Sinope (404–323 BCE) is often held to be the precursor to modern environmentalism. Diogenes and his followers, known as the Cynics, were the original naturalists, seeking to live a virtuous life according to nature rather than pining for a perfect world in the next life. It is said that when the noted philosophers of Plato's Academy, in attempting to come up with a definition of the essence of being human, proudly asserted that a human being is "a featherless biped," Diogenes rushed into their midst waiving a plucked chicken. "Here is Plato's man," he declared. Diogenes was also famous for walking through the streets of Athens in broad daylight with a lantern and announcing that he was looking for "a real human being." The point of this silliness was to show that reality does not exist theoretically, but in day-to-day experiences. There is no abstract "man"—there are only concrete human beings. To quote Bertrand Russell again, Diogenes "proclaimed his brotherhood, not only with the whole human race, but also with animals. He was a man about whom stories gathered, even in his lifetime" (Russell 1948: 255).

In perhaps the most famous of such stories told about him, Diogenes was once visited by Alexander the Great, the conqueror of the known world, who had heard that the greatest living philosopher lived in Athens. Himself a former student of Aristotle's, Alexander came across Diogenes while he was sunning himself. Towering over the recumbent figure, Alexander said in his most powerful voice: "Tell me what you'd like and I will give it to you." Diogenes looked up and asked him if he could ask for *anything* he'd like. "Anything at all," replied Alexander. "Then," he replied, "if that's the case, please get out of my light. You're blocking the sun." In other words, Diogenes was perfectly happy before Alexander came his way, and desired nothing from him.

Cynicism proudly asserted itself as a way of life, not merely an academic pursuit (indeed, the Cynical movement was set up explicitly to oppose Plato's Academy, an elitist organization specifically set up to train "the best and the brightest" to rule over others). The basic message of the Cynics was that one should live according to nature. Civilization is artificial, and the more one gets caught up in its clutches, the less one is true to oneself. The Cynics held that one should follow the model of the animals, which eat when they are hungry and sleep when they are tired, rather than living their lives according to time schedules and rules of etiquette. Their opponents therefore began to refer to them as "dogs" (*kunikos*), from which the word "Cynic" is derived. While the term "Cynic" was meant as an insult, the first Cynics, beginning with Diogenes, "embraced their title: they barked at those who displeased them, spurned Athenian etiquette, and lived from nature. In other words, what may have originated as a disparaging label became the designation of a philosophical vocation" (*Internet Encyclopedia of Philosophy* 2020).

The Cynics felt that happiness is achievable during one's lifetime. Idealists like Plato scorned this notion and looked for true happiness in the next world. The Cynics also felt that humans should be comfortable with their own bodies, and not ashamed of their animal nature. In the words of the philosopher Friedrich Nietzsche (1844–1900), "For the happiness of the animal, that thorough cynic, is the living proof of the truth of cynicism." The bodily functions are natural and should not be taboo. The contemporary philosopher Peter Sloterdijk, in his 1987 book *The Critique of Cynical Reason*, goes so far as to call the Cynics the precursors of the modern-day ecological movement. By living close to nature, and by spurning the trappings of a consumer society, the Cynics did their best to maintain ecological balance. "Cynical philosophers," says Sloterdijk, "are those who do not get nauseated" (1987: 151).

However, one of the strongest critics of the Cynics was Aristotle, who felt that, while their emphasis on understanding and living according to nature was sensible, their disavowal of civilization was unrealistic. Indeed, Aristotle went so far as to say that living in society was basic to humans, since we are social creatures by nature, and civilized society—with its emphasis on technological means of better life—is the best way for humans to both endure and to thrive. It was Aristotle's development of the concept of "natural law" which had a much more profound effect upon Western Philosophy, especially during the so-called Middle Ages (roughly from the 4th to the 16th century) when much of his thought was adapted by Jewish, Christian and Muslim thinkers, especially Thomas Aquinas (who was often said to have "Christianized" Aristotle).

Modern Philosophy and the Darwinian Revolution

So-called "Modern Philosophy" is said to have begun with René Descartes (1596–1650), a French rationalist who broke with tradition—particularly the tradition of reading Ancient thinkers like Aristotle and Aquinas as if they somehow possessed all the knowledge necessary for life—and urged everyone to think for themselves. In particular, he stressed the importance of experimentation as the best means of understanding reality. Descartes lived during the time now known as the Scientific Age, and he himself made major contributions to knowledge about human anatomy, especially how the heart and brain work. As a scientist, Descartes argues that our senses do not really tell us everything about the world and if we wish to attain genuine knowledge, we must utilize different methods to acquire it. Descartes made note of the new physics developing during the sixteenth and seventeenth centuries and felt that such scientific discoveries would reveal "another world" wholly unlike our sensory world (Crumley II 2016). We know today that science has in fact discovered "another world" that cannot be observed by our senses and examples of this abound (e.g., viruses and parasites that cause illness and yet are not visible by the naked eye nor directly observable by our other senses) (Delaney 2019). The global COVID-19 pandemic was often referred to an "invisible enemy" because we could not go to war against it in any sort of traditional warring way.

While Descartes was a mathematician and scientist, he is best known for his thought experiments regarding human reason, and his argument (known in Latin as *cogito, ergo sum*, meaning in English "I think, therefore I am") that each and every one of us can know for certain that we exist, and from this knowledge can build a solid foundation of certainty. The importance for this in regard to Environmental Philosophy, though, is the way in which he used this argument to continue to assert the unique position that humans have in the world, as—in his view—the only thinking animals on earth. Our ability to think, he held, makes us fundamentally different from all other life forms, and thus our chief concern, from both an ethical as well as a metaphysical perspective, should be with human beings, not with other animals.

A later reaction against this Rationalist view, known as Romanticism, was led by philosophers such as Jean-Jacques Rousseau (1712–1778) and poets like Percy Shelley (1792–1822), who both argued that humans had a moral obligation not to make the world anew but rather to appreciate the great handiwork that is nature. But such a nature-oriented approach to philosophy is especially linked to an evolutionary viewpoint, which only began to be understood in the late eighteenth and early nineteenth centuries. It is thus the naturalist Charles Darwin (1809–1882) whose theories revolutionized the field of environmental philosophy by showing that the human species, like all other living things, has adapted over time, and is not—as disparate thinkers like Aristotle,

Aquinas, Descartes and Kant had all held in different ways—essentially different in kind from other life forms.

One reason ethicists have perhaps not taken this debate to heart as much as possible is that applying evolutionary principles to ethical thinking is still controversial, even more than one hundred and fifty years after Darwin's *On the Origin of Species* first appeared. To quote from philosopher Suzanne Cunningham's book *Philosophy and the Darwinian Legacy*: "Darwin has ... offered us a new account of *ourselves*. He has argued that human beings, along with the rest of nature, need to be understood as the product of completely natural forces. And his theory asserts not only the natural origin of our bodies, but also the natural origin and development of our mental powers and our moral sense. To this extent, I suggest, his views have a singular significance for philosophy" (Cunningham 1996: 7). In many ways, Environmental Ethics is still grappling with this fundamental change in emphasis, as the following sections will show. One of the best-known modern thinkers who has tackled this issue head-on and who argues that Darwinism has changed the very nature of ethical reasoning is the Australian philosopher Peter Singer (1946–).

Speciesism

Singer, one of the prime movers in the growing animal rights movement, is in the forefront of those arguing that the very nature of ethics has been evolving, and that it should no longer primarily be concerned with human beings but should rather take a more holistic approach. He has coined the term "speciesism" to denote a prejudice toward the members of one's own species as against the members of other species. While admitting that it is an unattractive word, he nonetheless feels that it gets across the main point he is trying to make: speciesism is analogous to racism and sexism and is as morally unacceptable. Whatever else great philosophical thinkers like Aristotle, Aquinas, Descartes, and Kant might have written, they were in basic agreement that not all human beings were truly equal, nor should they be treated alike. It was only in the 19th century, Singer argues, that arguments in favor of abolishing human slavery (which had been based upon the view that some humans were born to be servants and could not by nature be free) started to be taken seriously, and only in the twentieth that a general acceptance that racism is immoral began to predominate in ethical theorizing. Likewise, arguments in favor of women's equality, while possible to trace back to the time of Plato, were not truly accepted as valid until roughly the same time, and much still remains to be done to actually put such views into practice. The point is, racism and sexism are today in general considered immoral, and the burden is now upon those who would argue otherwise. In the early 21st century, arguments against speciesism are only now beginning to be taken seriously, but future generations, Singer holds, will likely look back upon this as another example of unenlightened thinking.

In his book *One World: The Ethics of Globalization* (2nd edition) Singer phrases his view on the changing nature of ethics in the following way: "Our value system evolved in circumstances in which the atmosphere, like the oceans, seemed an unlimited resource, and responsibilities and harms were generally clear and well defined" (Singer 2002: 19). But, with our increasing awareness of the fragility and interconnectedness of all living things, he further asks, "How can we adjust our ethics to take account of this new situation?" (Singer 2002: 20).

The point that Singer raises is pertinent to environmental thrivability, which by and large takes an evolutionary stance. We now recognize that Homo sapiens exist thanks to the process of evolution, and differ from other life-forms by degree, not by kind. Does this imply, therefore, that we must incorporate these other life-forms within our ethical system? And if we do not, are we guilty of speciesism?

Singer argues that just as it was once inconceivable for whites to consider other races as equal in virtue, and for men to consider women as non-subservient, so we have reached a new conception in our ethics. While it has traditionally been the case that ethics—the study of right and wrong—can only be applied to humans, since only humans are deliberative, rational beings, this view is being overridden by a concern for alleviating suffering for all beings capable of feeling pain. Furthermore, taking an evolutionary approach, Singer maintains that humans must move from an anthropocentric (human-centered) attitude to one of biocentrism (life-centered), and extend the concept of ethical obligation beyond our own species in order to truly be moral agents.

The anthropocentric approach then, ranks human life above any other species and views mankind as separate and superior to the natural world (Sivinski and Ulatowski 2019). When humans embrace this perspective, it leads to many potential ethical and moral concerns, as humans feel entitled to use the earth's resources (e.g., fossil fuels and treat animals as food) for their own personal gain alone. Those with an anthropocentric approach tend not to be concerned with the effects that their actions have on the environment (Sivinski and Ulatowski 2019). Cocks and Simpson (2015) point out that while it may seem anthropocentric proponents do not see value in nature and other species, that is not true, but any such value is determined by whether or not they are beneficial to humans. Thus, as Kopnina and associates (2018) indicate, nature serves as a means to ends for humans. "Anthropocentrism, in its original connotation in environmental ethics, is the belief that value is human-centered and that all other beings are means to human ends" (p. 109). This leads many environmentalists to view anthropocentrism as ethically wrong and at the root of the ecological crisis (Kopnina et al. 2018). Washington (2013), for example, views anthropocentrism as a significant driver of ecocide and the environmental crisis and a philosophy that downplays humanity's dependency on nature.

Two Types of Biocentrism

In contrast to anthropocentrism is biocentrism. Broadly speaking, there are two types of biocentric approaches. The so-called Gaia principle is a term coined by the biologist James Lovelock (1919–), who at the time of this writing has reached the grand old age of 101. He named it after the Greek Earth goddess and says that we must take into consideration the interests of the biosphere itself in all our actions. Our actions must be based on what benefits the whole of life. This is the ultimate version of utilitarianism, a philosophy concerned with the greatest good for the greatest number. As Peter Singer has urged, a true utilitarian would apply this ethic not just to what benefits the greatest number of human beings, but what brings about the most pleasure and least pain for all sentient beings. The other approach to biocentrism is called "environmental individualism." It holds that we must take into consideration the good of not only species in general but each individual member of a species.

The prime example of "environmental individualism" is Albert Schweitzer's "reverence for life" philosophy: Schweitzer (1875–1965), a philosopher, theologian and social advocate, was awarded the Nobel Peace Prize in 1952 for his humanitarian work. He considered his greatest contribution to humanity to be his concept of "reverence for life" (in German *Ehrfurcht vor dem Leben*). This held that all things that are alive are deserving of esteem and must be treated with respect. This view can be connected with the deontological, or duty-based, ethics developed by Immanuel Kant (who was, by Singer's standards, an ardent speciesist as well as a racist and sexist, but who nevertheless came up with an approach to ethics which theoretically treats all beings as equal in moral worth).

Each of these biocentric approaches raises several issues. Taken to its ultimate extreme, the Gaia principle could be used to justify the extinction of those species that cause untold suffering for other species. The whole of life—Gaia itself—might benefit by the removal of certain of its constituent parts. This outlook is best expressed in a chilling poem by the occultist Aleister Crowley (1875–1947), which he ironically titled "The Optimist," and advocated for the killing off of mankind.

> Kill off mankind
> And give the earth a chance.
> Nature may find
> In her inheritance
> Some seedlings of a race
> Less infinitely base.[AU: Date? May be protected by copyright.]

We will return to this extreme anti-humanist position later in the chapter. But the Gaia principle can also be connected with the more radical members of groups such as Earth First! which often seem to take, at best, a dim view of humankind. Martin W. Lewis, in his book *Green Delusions: An Environmentalist Critique of Radical Environmentalism* refers to this as "antihumanist anarchism" (1992: 22).

In addition, the comparison of speciesism to racism and sexism has been strongly criticized by philosophers such as Bonnie Steinbock, who argue that "racism" and "sexism" are terms that relate specifically to human beings, and that the analogous connection with "speciesism" is at best a category mistake. While there are certainly great similarities between humans and other animals, especially the primates, how close is the comparison with beings such as snails and sunflowers? Does the analogy break down at any point and, if so, where?

As for the environmental individualistic approach, much depends upon just how one defines "reverence" or "respect for life." Even Schweitzer, while he may have felt saddened by his actions, advocated killing mosquitoes that spread the typhus disease. As George Orwell might have put it, all life is equal, but some life is more equal than others.

Edward O. Wilson and Anthropocentrism

When discussing environmental thrivability, one can ask if it is even possible for human beings to truly transcend an anthropocentric perspective. A frequent criticism of the anthropocentric attitude is that humans have no right to determine the course of events for other species. In reply, one might argue that it is those who take a biocentric approach who are guilty of arrogance or hubris. Unless we can become like Doctor Doolittle and learn the languages of all other species, who are we to take up their proxy in moral matters? Is not this a secular version of "playing God?" Furthermore, at the heart of many biocentric philosophical arguments is the presumption that there is a bona fide "balance of nature," a harmonious interconnectedness of all living things, which the human species is causing to go off course. But this is similar to economic arguments that there is an invisible hand that should guide all transactions, which governmental regulation interferes with. It seems closer to the mystical than to the scientific worldview.

At any rate, those concerned about environmental sustainability need to be sensitive to the anthropocentric versus biocentric ethical debates. We recognize that protecting the planet and life on it is a human concern. Furthermore, evolution shows us that all of life is indeed interconnected, and the destruction of one species could well cause harm to countless others. Environmental humility is an important part of sustainability. This is the argument used by people such as Harvard University's Edward O. Wilson (1929–), who has pointed out the utility of saving the world's rainforests, since they may well contain plant life which, while presently undocumented, could hold the cure for many human ailments. But Wilson is quick to add that his is not merely a crude anthropocentric utilitarianism. He also argues for what he calls *biophilia,* a natural desire to feel as one with the world, coupled with an aesthetic appreciation of the beauties of nature.

Wilson is the Pellegrino University professor emeritus at Harvard University, where he was also curator of entomology at its Museum of Comparative Zoology. He is best known for his theory of "sociobiology," which holds that behavior—including human behavior—is primarily caused by genetic and evolutionary processes rather than being culturally determined. This view, first explicated in his 1975 book *Sociobiology: The New Synthesis*, continues to generate much discussion and controversy, particularly in the related field of evolutionary psychology. Wilson first came to prominence for his work in entomology, the study of insects, particularly on the understanding of ant societies. He has received the Pulitzer Prize twice, once for his writings on sociobiology and once for his writings on ants.

Another major topic that Wilson has popularized is *biodiversity*, the importance of appreciating the diverse forms that life has taken and the interconnectedness of all living things. He has been an important voice in the environmentalist movement, especially defending the preservation of the world's rainforests. Science, he argues, has still not grasped the underlying mechanisms that maintain the ecosystem, and until such a time, if ever, that this is understood it is imperative that all living things be maintained. In addition, he points out that the destruction of the rainforests might be leading to the destruction of not only human life but potentially all life on earth.

Wilson has also advocated the concept of *consilience,* the idea that all knowledge can be unified under a single framework of understanding. This concept goes against the prevailing trend, in which different disciplines no longer interact or feel they are even dealing with the same topics. Wilson harkens back to the Enlightenment ideals of such thinkers as the Marquis de Condorcet (1743–1794), and he is proud to be a modern exponent of the Enlightenment. Wilson is not himself a professional philosopher, and indeed has been criticized by professionals in the field who question the depth of his understanding of philosophical concepts. But he is an important public intellectual and remains one of the most prominent exponents of evolutionary theory. He has been called by many people "Darwin's Natural Heir," and his writings offer a practical guide for preserving the environment both as an end in itself and as a beneficial goal for humanity as a whole. Furthermore, his hopeful approach that we both can and will confront the challenges currently facing us is very much in line with an environmental thrivability approach—it is not enough to merely sustain things as they are, he holds. We must also do all we can to improve the possibilities for all living things to thrive as best they can.

Evolutionary Ethics

Suzanne Cunningham makes an excellent point regarding the relevance of Darwin's work to philosophy. But the title of her book—*Philosophy and the Darwinian Legacy*—is slightly misleading, in that she shows how for the vast majority of philosophers working in the 20th century, Darwin was mostly evaded. The two seminal traditions which dominated the field, namely analytic philosophy and phenomenology, for all their differences, continued to pursue the Cartesian dream of certain knowledge grounded in unchanging metaphysical truths.

The main exception to this was John Dewey (1859–1952). America's public philosopher for most of his long career, he had a profound influence on such fields as education, politics, ethical theory, and aesthetics. Interestingly enough, Dewey was born the same year that *On the Origin of Species* was published. He lived through momentous changes, from the U.S. Civil War to the U.S. Civil Rights Movement. But most of all, Dewey was himself a philosopher of change, who consistently sought to apply Darwin's evolutionary theories to all the areas of philosophy.

Dewey argued that all knowledge is derived from experience. Ideas must be referred to their consequences—it is important to break down distinctions between theories and their applications. The movement which he was identified with, *Pragmatism*, comes from the Greek word meaning "action." For Dewey, philosophy's role is to assimilate the impact of science on human life.

Dewey was therefore one of the first philosophers to take Darwin seriously. Unlike the many philosophers—including G.E. Moore, Bertrand Russell and Edmund Husserl—discussed in Cunningham's book, Dewey did not evade the implications of evolution. Fifty years after *On the Origin of Species* was published, Dewey wrote an essay titled "The Influence of Darwin on Philosophy." In it, he points out that the combination of the words "origin" and "species" embodied an intellectual revolt as well as a biological advancement. Previous thinkers had held that species were unchanging and eternal—fixed and final. Ever since Plato, they had been in search of either eternal forms or some world beyond that of physical appearance. What corresponds to this? As Dewey points out, it leads to a "search for certainty": immortal souls, unchanging knowledge, all-powerful creators. The pursuit of wisdom was identified with eternal life and fixed ends. The end result was dualism—the belief that the mind and body are distinct, coupled with a fundamental belief that humans differ from animals in kind, not degree.

How to heal this split? Darwin gave us the method. But Darwin was himself following in the footsteps of other scientists—empiricists like Francis Bacon and John Locke, who insisted that any theory must be supported by hard evidence and must have more than just explanatory power. The empirical school, from which Darwin descended, holds that the scientific method, not speculation, is our best road to knowledge. "Without the methods of Copernicus, Kepler, Galileo, or their successors in astronomy, physics, and chemistry," Dewey stated, "Darwin would have been helpless in the organic sciences" (Hickman and Alexander 1998: 41).

In his essay, Dewey also astutely points out that there is a need for a Darwin in philosophy—one who can break down the hold of the past in intellectual matters much as Darwin did for biology. What is the bearing of Darwinism on philosophy? "Philosophy," he writes, "forswears inquiry after absolute origins and absolute finalities in order to explore specific values and the specific conditions that generate them" (Hickman and Alexander 1998: 43). Intelligence is not some absolute power but rather our species' survival tool. It has evolved and adapted itself over time. It is not a god-like substance or supernatural gift—other animals have forms of consciousness, too, and examining the similarities as well as the differences between us can have fruitful results. Human behavior is to be explained in terms of and related to behaviors similar to but more complex than the behavior of other animals. In this essay as well as throughout his many other writings, Dewey called for an empirical study of human's place in nature.

Why is Darwin so controversial to philosophers? One reason is that ideas give way slowly. Darwin had cleared away old ideas in biology—now the time has come to clear away old ideas in philosophy. Pragmatism spelled out the view that modern science was changing our relationship with nature, by giving us a dynamic rather than a static model. We can alter our environment, not merely accommodate ourselves to it. In conservative times change is feared. But we live in dynamic times—change should be welcomed. Dewey was influenced by positivistic philosophers like G.W.F. Hegel (1770–1831) and Auguste Comte (1798–1857) but made it clear that he did not share their views that progress is inevitable. Darwin had showed that regression and catastrophe are also factors in the lives of species. Humans can give up on critical intelligence. Seeking certainty is one way of doing so, in that it is a fruitless hope for solid ground in a world where everything is in a state of flux. There is no utopia—the goals we achieve provide us with new problems to solve. This view was vitally connected with Dewey's lifelong advocacy of universal education and democracy, the form of government which allows for the most personal freedom and opportunity. Education should give us the tools to sharpen our intellectual facilities and better prepare us for dealing with an environment that is constantly moving. Like Darwin, Dewey studied the ways in which young animals struggle for survival. He drew connections from this with the need for experimental schools. Education sharpens our instrument for survival—intelligence—but the curriculum to be taught and the means to do so are not set in stone.

Also like Darwin (albeit more explicitly), Dewey was critical of organized religions, especially the ways in which they sought a timeless and perfect reality. He was interested in the here and now, not the hereafter. If one can give up the quest for certainty one can better relate to the world as it is, not as we would have it to be. Dewey calls for a re-examination of Darwin's own nuanced views on the evolution of the mind and ethics. Darwin was himself ambivalent about Herbert Spencer's ethical views, and there was much difference between his metaphysics and economics and those of Bergson or Marx, regardless of how much the Bergsonians and Marxists might seek to identify their philosophies with Darwin. Like the philosophers that Cunningham writes about in her book, Dewey understood the dangers of crudely connecting Darwin's biological findings with speculative philosophies, particularly when the latter were unscientifically grounded. Even today, such fields as evolutionary psychology can still use a good dose of analytic rigor to avoid appearing as new versions of Rudyard Kipling's "Just-So Stories" in their explanations of why we act the way we do.

Anti-Human Ethics

Interestingly enough, taking an evolutionary view does not necessarily lead to extolling the continuation of the human species. As mentioned above, the Gaia principle could well be used to argue that humans have served their purpose and are now more of a hindrance than a help in the grand scheme of things. The German philosopher Eduard von Hartmann (1842–1906) came up with an evolutionary view not only antithetical to anthropocentrism, but to biocentrism as well. *Anti-human ethics*, then, is the realization that there is no real need for the continuation of humanity. Following in the footsteps of the philosopher Arthur Schopenhauer (1788–1860), he argued that the basis of ethics should be grounded in a realization of the utter futility of life. Combining Schopenhauer's pessimistic metaphysics with the ethical writings of Immanuel Kant and the process philosophy of Hegel, von Hartmann published a series of books which received wide recognition and made him one of the best-known thinkers of his day (*Internet Encyclopedia of Philosophy* 2020).

Von Hartmann's first and most popular work, *The Philosophy of the Unconscious*, was published in 1869, and became an immediate bestseller. In it, he advocated what he called "Spiritual Monism," the view that matter is both idea and will. The opposition of these two forces accidentally created the existence of consciousness.

The existence of consciousness creates a problem, though, as conscious beings become increasingly aware of the misery of the world around them. The evolution of human beings, with their highly developed awareness of the world around them, led to three stages in the attempt to combat pessimism. The first stage claimed that happiness is possible in the here-and-now (and was identified with the rise of the Roman Empire and its claims to have pacified the world). When this was found untenable, the second stage arose, which claimed that happiness would be found in the next world (as found in the doctrines of Plato, Christianity and Islam). The third stage, identified with the rise of Positivism and Socialism, claimed that true happiness would be found in a future state of earthly utopia.

Yet, von Hartmann argued, increasingly consciousness shows that all three stages are mere wishful thinking. This is indeed the best of all possible worlds, but it would be much better if no world existed at all. The growing awareness of this fact will eventually lead to a cooperative effort among all conscious beings to bring about the end of existence. Suicide would not be sufficient, as this would merely eliminate individuals, not the entire human species. And even the complete destruction of human life would not suffice, as other life forms would eventually become conscious and experience the same futility. Therefore, the proper goal would be for the full development of all particular wills, to the point where consciousness would be developed to its fullest potential and where the final solution—the termination of the universe itself—could be arrived upon through cooperative efforts.

Von Hartmann thought this was the epitome of ethical thinking, as it gave a common end (teleology) that could truly unite all beings. Previous attempts to ground ethics upon formal principles, pursuit of pleasure, or transcendental wish-fulfillment were doomed to failure, because they are false and the evolving consciousness would see through such falsehoods eventually. Surprisingly enough, von Hartmann's advocacy of (eventual) mass destruction met with an enthusiastic audience, and his unique approach to pessimism made him a popular writer until his death in 1906. Amongst other things, von Hartmann was a strong critic of liberal Protestant Christianity, which he felt to be the last gasp of the Christian transcendental ethics. The Social Gospel movement, which tried to shift the emphasis of Christian teachings from the next world to bettering the present earthly condition, was merely a move from the second to the third stage of optimistic illusion.

A few reasons for von Hartmann's popularity include the clarity of his writing style, his usage of up-to-date scientific findings, and his rejection of the shallow religious and secular utopian movements of his time. In addition, the paradoxical view that human beings—and indeed all life forms—could be united under one common project was appealing to those of a universalist point of view, even if that project was the end of all life itself. Still, after his own cessation of being, von Hartmann's writings fell into obscurity. His advocacy of the destruction of all life certainly puts even Crowley to shame and is a *reductio ad absurdum* of the view that intelligence is futile. Environmentalists can nonetheless benefit from his criticisms of transcendental escapism and reflect upon his objections to secular utopian visions. The realization that not just human but all life on earth could come to an end leads one to shift one's moral focus to truly trying to understand the importance of sustainability.

There is an ongoing body of work which speculates upon what the world would be like if human beings should disappear entirely. There is a long-standing joke, alluded to in the 2008 animated feature film *Wall-E*, that after a nuclear holocaust the only form of life that will survive will be the cockroach, which is indestructible. In 2007, Alan Weisman wrote a book—which was also made into a *NOVA* television special in 2009—titled *The World Without Us*. In it, he imagines in great detail how various life forms would evolve and thrive, before the earth itself finally ceases to be. He writes: "Look around you, at today's world. Your house, your city. The surrounding land, the pavement underneath, and the soil hidden below that. Leave it all in one place, but extract the human beings. Wipe us out, and see what's left. How would the rest of nature respond if it were suddenly relieved of the relentless pressures we heap on it and our fellow organisms? How soon would, or could, the climate return to where it was before we fixed up all our engines?" (Weisman 2007: 5). Surprisingly enough, given its rather bleak thesis, Weisman's book was a *New York Times* bestseller.

Following along the line of the quote above, the popular Marilyn vos Savant, in her *Parade* magazine column of May 1, 2011, addressed the question of just how long it would take the earth to return to its "pristine state" if all human life were to vanish. She writes: "Buildings, roads, dams, and bridges would become ruins in just a few centuries, but they'd take thousands of years to disappear entirely. Meanwhile, nuclear waste in long-term storage would gradually become harmless. Without human attention, our hundreds of active reactors would catch fire or melt down and release radiation, but even that wouldn't stop nature's rapid return…⊠. Excess carbon dioxide would be cleansed by the oceans over tens of thousands of years. By then, added methane would be long gone. The toxic impact of pollutants such as DDT wouldn't last even a century" (vos Savant 2011: 12). She goes on to say, "So the planet would forget all about us in maybe 50,000 years—far less time than humankind has existed. And if an unaltered atmosphere isn't an essential part of what you call a pristine state, our influence would be gone in less than half that time—and maybe much less" (vos Savant 2011: 12). Chilling words indeed for those who don't wish to contemplate the end of humanity or a world without us in it.

But notice, it is still *human beings* who are doing the speculation. And while some people such as Crowley or von Hartmann might take a cold comfort in thinking about, and actually hoping for, the end of humanity, others will feel a motivation to do what is necessary to avoid such a calamity. As vos Savant herself notes, the very meaning of the earth's "pristine existence" is quite debatable. Those who secretly—or perhaps not even so secretly—wish to see Crowley's vision come true are engaging in what Friedrich Nietzsche perceptively called a "cheerful pessimism." It is a strange thing to actually desire the destruction of one's own species. David Wallace-Wells, in his chilling 2019 book *The Uninhabitable Earth: Life After Warming*, goes into detail about all the many reasons why human survivability is highly unlikely in the next 500 years or so, due to climate change, economic collapse, unbreathable air, and continuous natural disasters. But he adds: "Since I first began writing about warming, I've often been asked whether I see any reason for optimism. The thing is, I *am* optimistic … we may conjure new solutions, which could bring the planet closer to a state we would today regard as merely grim, rather than apocalyptic" (Wallace-Wells: 31). Cold comfort, perhaps, but at least not nihilistic.

Those who are concerned about environmental sustainability need not postulate a perfect balance of nature nor a pristine existence. They must, however, be motivated by the hope that human actions can truly make a difference when it comes to maximizing the possibilities for life on earth for—as the old hymn puts it—all creatures great and small.

A Pragmatic Appreciation of Environmental Ethics

In conclusion, the pessimism of von Hartmann and the optimism of utopianism both seem to lack what John Dewey would call a pragmatic basis. If human beings are necessarily going to cause their own extinction, or if no matter what we do we will somehow be rescued by divine intervention, then sustainability is really not our primary concern. Such views, for pragmatists like Dewey, do not address the very real environmental concerns that face us. For Dewey, human intelligence must always be applied to problems at hand. Even in his own time he was able to see that our growing awareness of environmental issues is coupled with the knowledge that species evolve over time, and most species ultimately become extinct. This provides us with real challenges to contemporary ethical thought. In the words of Larry Hickman, the former director of the John Dewey Center and one of that philosopher's most noted commentators, "As a committed evolutionary naturalist, Dewey argued for the view that human beings are in and a part of nature, and not over against it. It was his contention that human life constitutes the cutting edge of evolutionary development (but not its telos), because, as he put it, humans make self-reflection a part of evolutionary history when they come to consciousness by means of social intercourse" (2007: 132). Most of all, Dewey attempted to break down dualisms which led merely to antagonisms and misunderstandings. Perhaps "biocentrism" and "anthropocentrism" are such dualisms which need a Deweyan analysis. Dewey's call for an evolutionary ethical approach is still a worthy cause, and one which can give philosophical support to the continuing efforts to understand the meaning of "thrivability" and the practical effects that follow from it.

Evolutionary ethics is a controversial field. It draws a very close connection between humans and other members of the animal kingdom—for many religious believers, a connection too close for comfort. As Dewey points out, it is hard to imagine human morality without such tendencies as sympathy-related traits; internalization of rules and regulations; reciprocity; and conflict-resolution abilities. Other species besides our own demonstrate these characteristics. For humans, the transmission of such tendencies occurs through culture.

Charles Darwin brought about a new way to understand the human species, grounded in science. His hope was for the advancement of science coupled with an increasing cooperation among the people of the world. Such a hope does not accord with the views of those who feel morality means adhering to commandments from a mysterious being from on high who damns all who refuse to kowtow to him. But it is in harmony with those who feel that the Golden Rule—regardless of where it comes from—is worth following, for both its intrinsic value and for the benefits which result. One need not be a theist to appreciate the wisdom of the Good Samaritan story. John Dewey, for one, would have concurred.

What role can philosophy play in helping us to better address the environmental realities of the present-day? Contemporary philosophers Andrew Light and Eric Katz, in their book *Environmental Pragmatism*, following in Dewey's footsteps, argue that environmental pragmatism can take at least four forms:

- Examinations into the connection between classical American philosophical pragmatism and environmental issues;
- The articulation of practical strategies for bridging gaps between environmental theorists, policy analysts, activists, and the public;
- Theoretical investigations into the overlapping normative bases of specific environmental organizations and movements, for the purposes of providing grounds for the convergence of activists on policy choices and among these theoretical debates,
- General arguments for theoretical and meta-theoretical moral pluralism in environmental normative theory [Light and Katz 1996: 5].

In the following chapters we will explore in more detail, from a pragmatic perspective, how human beings can effectively confront the realities of environmental crises and do all they can to maximize their abilities so that, rather than merely "sustaining" present situations, they can hopefully thrive within the changing ecosystem.

Popular Culture Section 6

The Darwin Awards: Weeding Out the Stupid

One of the main themes of this chapter has been the importance of taking Darwin seriously. But, in popular culture, there is also a very well-known website and series of books which takes Darwin—or more properly natural selection—in a jocular fashion. This is the so-called "Darwin Awards," an internet phenomenon which has documented various bizarre ways in which human beings have lost their lives by not using their common sense or rational skills. Many of these have been compiled by Wendy Northcutt, a graduate of UC Berkeley with a degree in molecular biology, who began collecting such stories in 1993. She started a popular website, www.DarwinAwards.com, and in 2000 put together a compilation of such stories in her book *The Darwin Awards: Commemorating Those Individuals Who Ensure the Long-Term Survival of Our Species by Removing Themselves from the Gene Pool in a Sublimely Idiotic Fashion.*

The website and book are chock full of anecdotes about humans who accidentally blow themselves up, get killed by dangerous (and sometimes not so dangerous) animals, blithely avoid warnings about natural disasters, and overall engage in acts of breath-taking stupidity. She writes: "Darwin Award Winners plan and carry out disastrous schemes that an average child can tell are a really bad idea. They contrive to eliminate themselves from the gene pool in such an extraordinary idiotic manner, that their action ensures the long-term survival of our species, which now contains one less idiot" (Northcutt 2000: 2).

A stickler for details and a strong advocate of avoiding urban legends and tall tales, Northcutt demands that all stories on the Darwin Awards site must be verifiable, documented cases of stupidity. She has five basic rules:

- Death: The candidate must remove himself/herself from the gene pool. Merely being grievously harmed is not sufficient for inclusion.
- Excellence: The candidate must exhibit an astounding misapplication of judgment. For instance, people—usually in an inebriated state—who somehow think they can emulate William Tell and shoot an apple off of someone's head (with an arrow, gun or other instrument of destruction) would not be eligible for the award, but a person willing to put said apple on his or her head most certainly would.
- Self-selection: The candidate must be the cause of his/her own demise. Someone unfortunately mauled to death by a lion who escapes from the zoo would not apply; someone who chooses to "run with the bulls" in Pamplona, Spain, and gets gored to death would fit this category nicely.
- Maturity: The candidate must be capable of sound judgment. "This means no children, Alzheimer's disease sufferers, or Downs Syndrome patients" (Northcutt 2000: 5). But this still leaves a huge number of possible candidates, all of whom should know better but who freely choose to rush in where even fools would fear to tread.
- Veracity: As mentioned above, the event must be verified—actual names, dates, and descriptions have to be provided as evidence, to avoid the "Friend of a Friend" phenomenon that haunts the urban legend world (see the popular website www.Snopes.com whenever you come across a weird story that seems too good to be true. Nine times out of ten, it usually is a fabrication or exaggeration).

And, of course, it's important to add that unlike most such honors like the Oscar, the Emmy, or the Tony, the Darwin Awards are always bestowed posthumously!

Here is a rather typical example of a Darwin Award, dated August 15, 1999, from Germany. The name has been withheld to protect the relatives of the "winner," but Northcutt assures us she can provide proper documentation for those who might doubt it: "A hunter from Bad Urach was shot dead by his own dog after he left it with a loaded gun. The fifty-one-year-old man was found sprawled next to his car in the Black Forest. A gun barrel was pointing out the window, and his bereaved dog was howling inside the car. The animal is presumed to have pressed the trigger with its paw, and police have ruled out foul play. Since it happened in a hunting preserve, the dog may elect to have the trophy head mounted on a wall in its doghouse" (Northcutt 2002: 36).

Northcutt calls herself a "thanatologist," meaning one who studies death. And, while demonstrating a rather barbed if not downright wicked sense of humor, she is also able to laugh at herself. As interviewer Vicki Haddock relates, Northcutt came close at one time to earning one of her own awards: "It almost happened when a recent heat wave gave her the idea to 'air-condition' her sweltering home. She pried up an oubliette floor grate in the hallway, intending to install a fan to suck up the basement's cooler air. But she left to answer the phone, and hours later strode back down the hall and obliviously stepped into the gaping hole. As her body swooshed down, she thought, 'Oh no! I'm gonna win my own award!'" (Haddock 2013: 1). Luckily, she lived to tell the tale—but it is interesting that, like Thales at the beginning of this chapter, she fell into a hole. She too can be an absent-minded thanatologist.

The Darwin Awards are funny, but also disturbing. They are a type of cruel humor, laughing at others' expense, what the Germans call *schadenfreude*. No doubt the family and friends of the hunter above, as well as the friends and relatives of the thousands of other "award winners" would not find these anecdotes very amusing, to put it mildly. There seems to be a lack of empathy to the awards—surely, as Northcutt herself demonstrated, in many of these documented cases we can all think, "that could have been me."

Perhaps more importantly, the name of the award is a misnomer. It is not really Darwinian at all, if by that one means that stupidity (however that is defined) is an inherited trait. This smacks more of a Lamarckian view than a Darwinian one. And as Northcutt herself stresses, it is essential that a person, in order to win the award, be capable of acting otherwise. That is to say, true stupidity rules one out of the running. And perhaps most important of all, even if one grants the central premise that such acts "eliminate one from the gene pool" in many, if not most such cases (since children are explicitly ruled out of the running), this is a moot point, since it's likely that the person may well have already bred.

The Darwin Awards are fun, in an admittedly dark sort of way. But for environmental ethics and thrivability, there is a lesson to be learned. By willfully ignoring warning signs, and blithely acting as if there is no real environmental crisis at hand, human beings might be exemplifying on a grand scale the kind of stupidity that the Darwin Awards highlights.

Summary

From its very beginnings over twenty-five hundred years ago, philosophy has been concerned with understanding the role of humans in nature, and how our ability to reason can guide us to be wise. Traditionally, this has led to the view that humans are fundamentally different from all other living things, and because of this ethics—the concern over how to morally judge our actions—is only pertinent to humans, and not to other animals.

However, at least since the mid–19th century, thanks primarily to the writings of Charles Darwin and other biologists who have given us a better understanding of the interconnectedness of all life, we now have a view of how relatively recent the human species is as a life form on the planet Earth. In addition, we now can see that it is more likely than not that most species, including our own, will cease to exist if they do not adapt themselves properly to environmental changes.

While some continue to ignore these facts and hold onto the hope that humans are apart from the natural world, others take an almost gleeful view that it might in fact be best for the planet if humans—due either to their own faults or due to natural catastrophes beyond their making—cease to exist.

These two extremes, though, can be challenged by a pragmatic ethical philosophy grounded not only in Darwinian theory but also the moral writings of philosophers like John Dewey and the scientific writings of biologists like E.O. Wilson, who urge us to face facts and apply our reasoning powers to coming up with strategies that will allow us to not only endure but find new ways to thrive in changing circumstances.

Since philosophy, it is often said, begins in wonder, and since the first credited philosopher was so concerned about grounding humanity within the natural world, it might well be that it is time for us to go back to Thales.

Chapter 7

Helping the Environment Thrive

There is a great deal of evidence that the earth's environment has been compromised. Some may question the idea of human-caused climate change while others warn of an impending environmental meltdown due primarily to human dependency on burning fossil fuels, but we can all agree on certain areas of concern. For example, the rainforest is being depleted and everyone recognizes that trees producing oxygen are critical for all life forms. Some people argue that we have an ethical obligation to serve as stewards of the earth while others question, who are humans to think that we should "play god" and interfere with nature? Despite the skepticism of some, there have been pro-environmental social movements designed to help sustain the Earth since 1970. The authors of this text argue that striving for sustainability is not an adequate enough response to the existing compromised environment; instead, humans need to help the environment thrive.

While a segment of the general population remains unconvinced that humans need to exercise any, let alone drastic, preventive steps to assure that the environment flourishes, a majority of folks are conscientiously attempting to help the environment thrive. There are many things that conscientious people can do to help secure our natural environment. Conscientious people who want to do their part to save the ecosystem can heed the warning from scientists, including climatologists, about the many threats to our planet—the only one (that we are aware of) that sustains human life. As individuals, we can do a lot of little things, such as recycling, driving fuel-efficient automobiles, reducing waste, supporting organizations that are "green-friendly," and so on, but our biggest contribution toward saving the ecosystem comes in the form of demanding that our political leaders and big business take drastic steps to protect the environment. After all, the little things that people can do to save the environment pale in comparison to what we need businesses and governments to do.

When it comes to taking steps toward helping the environment thrive, the best place to start is the "going green" social movement. We must also protect and save natural resources, develop new technology, assure environmental rights, and pass and enforce pro-environmental legislation.

The Meaning of Going Green

What is the meaning of green? As a color, it can refer to issues of jealousy as in the "green-eyed monster" and "green with envy." But it is also a restful color, capable of calming as it has both a warming and a cooling effect. From an environmental standpoint, the color green is associated with the color of grass and lush forests. There are people in many parts of the world that are green with envy for the green ecosystems that others enjoy or take for granted. The color green has become so associated with the environment that it has become a cliché to say, "We are going green." Going green is a general expression that incorporates the wide variety of behaviors and activities that people can engage in to become eco-friendly. Going green is so popular it qualifies as a social movement. As described in Chapter 2, a social movement is a persistent and organized effort on the part of a relatively large number of people who share a common ideology and try to bring about change or resist change. Nearly everyone (e.g., businesses, schools, individuals) claims to be going green in one form or another. Going green means different things to different people but the core idea involves taking steps to help protect the environment and save our natural resources so that ecosystems everywhere can thrive.

For many environmentalists, the meaning of going green still refers to addressing basic concerns and needs such as developing and promoting clean transportation like public transit, hybrid and electric vehicles, car sharing and carpooling programs; encouraging the use of energy efficient light bulbs (both in the home and commercially); drying laundry on clotheslines instead of using a dryer; using reusable shopping bags instead of the paper or plastic bags offered at some retail stores (be sure to keep your reusable shopping bags clean by wiping them down with disinfected or by washing them by hand and not in a washing machine); recycling; utilizing new technology designed to reduce our collective carbon footprint; and creating and enforcing pro-environmental rights and legislation.

In the following pages we will look at some of the green efforts utilized in an effort to help the environment thrive. Among these efforts are recycling; green roofs, rooftop gardens, green ice, green homes, green warehouses; LEED certification; waste reduction; using renewable energy sources; activist gardening; porous surfaces; fallen fruit programs; and going green for good health.

Recycling

Younger people today may believe that the idea of recycling is new, but in reality, people used to recycle regularly in the past. For example, decades ago everyone returned milk, soda and beer bottles to the store and the store sent them back to the plant to be washed and sterilized and refilled, so we could use the same bottles repeatedly. Notice the key word—bottles. The implication being, bottles are made of glass, not plastic. That people used washable and reusable products (e.g., cloth diapers and glass containers) before plastics and throw-away products is an important reminder that consumers were actually environmentally friendly prior to going green social movements. It was Big Business and the lure of easy convenience that necessitated the eventual going green ideology. Consider, for example, the contrast in drinking coffee today compared to the past. In the past, people drank coffee at home or brought coffee in a reusable thermos to work, and if they stopped for coffee on the way to work, they sat down and drank from an actual cup. As people became increasingly mobile their coffee drinking habits changed. People on the go needed coffee "to go" and necessitated the need for "portable" cups and lids. "In 1964 on Long Island, N.Y., convenience chain 7-Eleven became the first chain to offer fresh coffee in to-go cups. The company quickly expanded to-go coffee to the rest of its Northeast chains, and then nationwide" (Park 2014). The disposable, throw-away paper, usually lined with a membrane of polyethylene (plastic) or Styrofoam-type containers were convenient for consumers but a nightmare for the environment as land and sea became littered with these unsightly cups. "It's estimated that 500 billion coffee cups are produced globally each year and if they were placed end to end they'd circumnavigate the globe 1360 times. Planet Ark says that about 60,000 kilograms of plastic waste from coffee cups is directed to landfill each year in Australia, where it can take about 50 years to break down" (Potter 2017). Today, most to-go coffee cups are made with compostable (a mixture of various decaying organic substances) and biodegradable paper cups.

Green Roofs and Buildings

Logic dictates that the most essential natural resource is oxygen, or more generically speaking, air, preferably, clean, breathable air. An increasing number of people are embracing the "going green" mentality when it comes to saving and preserving water, while also addressing the concerns of ground water contamination, stormwater drainage, wastewater, and protecting watersheds and wetlands. For example, a number of buildings now include "green roofs," also known as "living roofs" or "eco roofs" that are partly or completely covered with vegetation. A roof with solar thermal collectors or photovoltaic modules (jointly connected solar cells that can convert light into electricity) may be considered green roofs. Some people consider rooftops that maintain plants in pots as a variant of a green roof, but there is a debate about this classification.

The National Park Service (NPS) (2020e) describes a green roof as "a layer of vegetation planted over a waterproofing system that is installed on top of a flat or slightly-sloped roof. Green roofs are also known as vegetative or eco-roofs." Similar to green roofs and rooftop gardens are "rain gardens." Rain gardens are another way to help prevent rainwater runoff. Rain gardens divert rainwater runoff from streets and sewers. According to the Rain Garden Network (2011), a rain garden is a shallow depression that is planted with deep-rooted native plants and grasses positioned near a runoff source like a downspout, driveway or sump pump to capture rainwater runoff and stop the water from reaching the sewer system. A green roof needs a structural sound foundation to handle the extra weight of the vegetation and the people who tend to them. Green roofs provide shade and remove heat from the air through evapotranspiration, which reduces temperatures of rooftop surfaces and the surrounding area. In addition to reducing the heat on rooftops, green roofs reduce energy use, reduce air pollution and greenhouse gas emissions, improve human health and comfort, enhance stormwater management and water quality, improve quality of life for many species, and generally last longer than traditional roofs (EPA 2012n; Kulkus 2011). Green roofs also provide aesthetic beauty, and for those who work in tall buildings that overlook rooftops, the sight of green vegetation is a welcome alternative to the drab, conventional rooftops. Green roofs are increasingly popular in the United States and around the world with more than a hundred million square feet of green roofs in all. Chances are, there is a green roof nearby where you live, go to work or go to school.

Consider, for example, the green roof atop the Oncenter Convention Center in Syracuse, one of the largest in New York State. The conversion process took six weeks and involved ripping off a layer of the old Oncenter roof, lifting raw materials 4 stories in the air and shooting 670 tons of soil up and across an expanse of about the size of a football field and half (Kulkus 2011). The 2,400 pounds of plants on the green roof will absorb 1 million gallons of rainwater a year, that's one million gallons of water that won't enter the sanitary sewer system, which is often overwhelmed during heavy storms (Kulkus 2011).

The same Oncenter complex plays host to the AHL Syracuse Crunch hockey team, where hockey players skate on "green ice." The ice itself is not green but rather, the ice is made from rainwater. A $1 million system collects, treats, and freezes ice for the hockey rink at the War Memorial Arena at the Oncenter (Oncenter 2011; Weaver 2011). The system is expected to capture 250,000 gallons of rainwater per year. Green roofs, green ice and green homes (e.g., through such initiatives as refurbishing run-down homes instead of razing them and sending the debris to a landfill) and green warehouses (e.g., refurbishing old buildings and meeting standards to become LEED-certified) are all found in the greater Syracuse area, and that helps to explain why the EPA ranked Onondaga County and Syracuse in the Top Ten Greenest Communities (Weaver 2011).

LEED Certification

What criteria are used to determine what is a green building you might ask? Becoming green-certified, or more accurately speaking, LEED-certified, is the answer. LEED certification is the "gold standard" in the construction industry and refers to buildings that have been designed, built, and maintained using green building and energy efficiency best practices; and involves a rigorous third-party commissioning process to assure pro-environmental standards have been achieved. This third party is the U.S. Green Building Council (USGBC), creators of the LEED green building rating system, as a way to spotlight the buildings that are most friendly to the environment and save energy dollars. "The LEED rating system offers four certification levels for new construction—Certified, Silver, Gold and Platinum—that correspond to the number of credits accrued in five green design categories: sustainable sites, water efficiency, energy and atmosphere, materials and resources and indoor environmental quality" (Natural Resources Defense Council 2012a). LEED standards are applicable to new commercial construction and major renovation projects, interiors projects and existing building operations (NRDC 2012). There are more than 160 cities and communities around the world that participate in the LEED certification program. Promoting one's self as LEED-certified is valued achievement among businesses, as they are able to advertise such a designation in hopes of attracting customers and business partners. City school districts can also achieve designation as LEED-certified schools. The idea of green roofs, green homes, green buildings and warehouses are fine examples of going green.

Waste Reduction

Another method of going green involves reducing the amount of waste we create. The EPA (2011g) encourages a number of practices that reduce the amount of waste needing to be disposed of, such as waste prevention, recycling, and composting:

- Waste prevention—Designing products to reduce the amount of waste that will later need to be thrown away and also to make the resulting waste less toxic.
- Recycling—The recovery of useful materials, such as paper, glass, plastic, and metals, from the trash to use to make new products, reducing the amount of new raw materials needed.
- Composting—Involves collecting organic waste, such as food scraps and yard trimmings, and storing it under conditions designed to help it break down naturally. This resulting compost can then be used as a natural fertilizer.

There was once an adage that people abided by that seems appropriate here, "Waste not, want not." In other words, if you don't really need it, or want it, but you attain it anyway and just end up throwing it away, you have created waste—and you have harmed the environment.

Using Renewable Energy Sources

Going green also involves the use of renewable energy sources (see Chapter 2) as a way of reducing dependency on fossil fuels. Businesses and residential homes can go green by getting off the power grid. Solar parking lot canopies, for example, have become a lucrative business opportunity for entrepreneurs and represent a financial savings for customers. A decade ago, the Cincinnati Zoo activated a nearly four-acre solar canopy over its parking lot, and it generates enough electricity to power 200 homes (Clubb and May 2011). A number of large stadiums utilize renewable sources and that is a good thing as they require a huge amount of power, especially on game days. Even when not in use, stadiums and other sporting arenas need large amounts of power for maintenance (Unwin 2019). Among the stadiums that are becoming increasingly sustainable are the solar-powered pavilion at Chase Field (Arizona Diamondbacks) which generates clean energy for the grid as well as the stadium itself, contains LED concourse lights, low-flow sinks, EV charging stations, and concession stands that save grease (from their churro dogs and other food items) in order to recycle it into a biodegradable diesel fuel (Hunt 2019). Other energy-smart stadiums include Lusail Stadium, Qatar, an 80,000-seat capacity stadium will use solar panels to reduce temperature to 27 degrees Celsius even when it is 40-plus outside; Antalya Stadium, Turkey, the country's first solar-powered stadium, which will generate power via solar panels and divert it back to the grid on non-game days; Tokyo Olympic Stadium, Japan, scheduled site of the 2021 Games (because of the COVID-19 pandemic the Games were postponed for a year); Mercedes-Benz Stadium, Atlanta, in which more than 4,000 solar panels were installed at the stadium and surrounding areas (primarily above car parks); Stadio della Roma, Italy, expected to be completed by 2023 will be built based on LEED standards; and T-Mobile Arena, Las Vegas, which uses high-efficiency LED lighting throughout to reduce energy usage were among the criteria used to help this desert stadium attain LEED gold certification (Unwin 2019).

There exists a growing and thriving solar industry in the United States, one that supports an estimated 250,000 jobs and has helped reduce planet-warming emissions. The Desert Sunlight facility, a few hours east of Los Angeles, is the world's largest, and one of five, solar projects that have received federal loan guarantees, totaling $4.6 billion. The funding is the result of the American Recovery and Reinvestment Act; an economic stimulus bill passed by Congress and signed by President Obama during the depths of the Great Recession (Roth 2020a). Further investment in solar energy would benefit the environment and the economy (via jobs).

A variation of solar energy is wind. Utilizing wind power is one of the oldest human energy technologies. The process of creating wind energy begins with wind turbines that are used to generate electricity. The wind turns the propeller-like blades of a turbine around a rotor, which spins a generator, which creates electricity (Energy Efficiency & Renewable Energy 2020). Certain weather patterns, such as wind flow patterns and speeds, create more favorable conditions to produce wind power in some areas of the world than compared to others. The terms "wind energy" and "wind power" both describe the process by which the wind is used to generate mechanical power or electricity. Wind turbines can be built on land or offshore in large bodies of water like oceans and lakes (EERE 2020). Generally, wind turbines are clumped together in a desirable area and form what are known as "wind farms" (WF).

Five of the ten largest wind farms in the world are operated in the United States, two of the top ten wind farms are offshore and the remainder onshore. According to Power-technology.com (2019) the largest wind farm is Jiuquan Wind Power Base, Inner Mongolia, China, it features 7,000 wind turbines. The third largest wind farm overall and the largest American WF is the Alta Wind Energy Centre in Techachapi, Kern County, California, with a capacity of 1,538 megawatts. The largest offshore WF is the Walney Extension Offshore WF located in the Irish Sea with a total capacity of 659 MW.

Activist and Community Gardening

A guerrilla approach to going green known as activist gardening has taken hold in many cities, including Portland, Detroit, Baltimore and Washington. Activist gardening involves planting "seed bombs," golf ball-sized lumps of mud packed with wildflower seeds, clay and a little bit of compost and water (Wax 2012). Conducted typically by young urbanites who are trying to redefine the seed-bombing pastime as a tool of social change, activist gardening is civil disobedience with a twist: "Vegetable patches and sunflower gardens planted on decrepit medians and in derelict lots in an effort to beautify inner-city eyesores or grow healthful food in neighborhoods with limited access to fresh food" (Wax 2012). The controversial aspect of activist gardening is connected to use of graffiti by the activists, which they view as an important way to get the word out about why they are conducting seed-bombings. The idea of using graffiti in connection with beautifying the environment seems odd, as graffiti itself is generally viewed as blight on the urban environment and is, of course, against the law. Authorities are not quite sure about the legal ramifications of people growing urban gardens on plots of land not owned by the activists. The activist gardeners believe that whenever one sees a space in the city that is "yucky" it can be made more beautiful with sunflowers and vegetables (Wax 2012).

Garden activists employ a less guerrilla approach to urban gardening and promote community gardening programs as a way to improve people's quality of life. As found on their website, Garden Activist (2020) states, "Community gardening improves people's quality of life by providing a catalyst for neighborhood and community development, stimulating social interaction, encouraging self-reliance, beautifying neighborhoods, producing nutritious food, reducing family food budgets, conserving resources and creating opportunities for recreation, exercise, therapy, and education."

One example of a community food garden is Seattle's "Beacon Food Forest," located in a working-class neighborhood of Beacon Hill (15th Avenue South & South Dakota Street). The Beacon Food Forest is a "community-driven garden project that uses a gardening technique that mimics a woodland ecosystem using edible trees, shrubs, perennials, and annuals make up the lower levels.... Fruit and nut trees are [found in] the upper level" (Seattle.gov 2020). The Beacon Food Forest has three main priorities: (1) create a community around growing and sharing food; (2) improve local food security by empowering the community to grow and harvest food on public land; and (3) rehabilitate the local ecosystem and biodiversity (Seattle.gov 2020). In addition to providing fresh fruits, vegetables and nuts, this community garden features accessible raised beds, demonstration gardens, giving gardens, honeybee, outdoor meeting space (with seating and cover), an orchard, and public art (Seattle.gov 2020).

As the COVID-19 pandemic raged on throughout 2020 the authors were left to wonder, wouldn't this be a great time for people across the nation and world to start community gardens? It would give new meaning to the famous closing lines of the philosopher Voltaire's 1759 book *Candide*: "Tend your garden." Imagine if every person with a yard grew vegetables and fruit trees. Such a concept is known as "foodscaping"—a fancy way of saying lawns with gardens. Neighbors could consult with one another so that they could trade the food that is grown. Imagine if every person that lived in an environment that could support such gardens did so. (Desert areas with limited water supplies might find this to be a challenge. But colder areas could grow gardens even in the winter with local greenhouses.) Such a notion as foodscaping could help us to prepare for the next global pandemic.

Fallen Fruit Programs

In cities across the United States, there exist variations of a "fallen fruit" program (also known as "urban foraging") that encourages citizens to take, for free, fruit grown on public property. All too often these fruits are wasted because they are not picked, turn sour and die. In Los Angeles, for example, there are numerous trees that bear limes, lemons, avocados, bananas, and figs. In Buffalo, New York, there exists an "endless orchard" wherein anyone anywhere can plant, map, and share fruit. "The Endless Orchard is a real living fruit orchard planted by the public, for the public—a movement of citizens transforming their own neighborhoods" (Fallen Fruit of Buffalo 2017).

The Fallen Fruit organization aspires to be the most comprehensive and open geographic dataset of urban edibles; it provides site location via maps it maintains, and constantly searches for new free fruit programs that pop up across the U.S. "'Community maps' are built by foragers and freegans as they pursue their communities for things to eat" (Falling Fruit 2020).

In 2013, Long Beach, California, created a novel approach—"The Great Wall of Mulch"—to combat freeway noise and air pollution. The city built a 12-foot barrier of shredded tree clippings held together by two chain-link fences resulting in a low-cost structure designed to dampen the sound and block the sight of diesel trucks from the heavily traveled Terminal Island Freeway. The city came up with the idea of using mulch from its tree-trimming operations because it's more visually pleasing, graffiti-proof and practically free. The city also plans to plant trees and shrubs along the wall to absorb some of air pollutants, such as fine particulates in diesel exhaust (Barboza 2013).

Porous Surfaces

The use of porous surfaces is another technique that can be utilized to help the environment thrive. The term "porous" refers to a surface that has minute spaces or holes through which liquid or air may pass. A porous surface allows the passing of water, liquid, or vapor. Porous pavement—also known as permeable pavement or pervious paving—has become increasingly popular as a construction alternative. "Porous pavement is a permeable pavement surface with a stone reservoir underneath. The reservoir temporarily stores surface runoff before infiltrating it into the subsoil. Runoff is thereby infiltrated directly into the soil and receives some water quality treatment" (Perkiomen Watershed Conservancy 2011).

Porous pavement will resemble traditional asphalt, but it lacks some of the "finer" materials and instead, incorporates void spaces that allow for infiltration. According to the National Ready Mixed Concrete Association (NRMCA) (2011), "By capturing stormwater and allowing it to seep into the ground, porous concrete is instrumental in recharging groundwater, reducing stormwater runoff, and meeting U.S. Environmental Protection Agency (EPA) stormwater regulations."

There are a variety of porous surfaces, including sidewalks, walkways, driveways, and bricks which are all connected to water and water runoffs. Porous pavement is an efficient way of protecting waterflow and preventing erosion. "By stopping stormwater from pooling and flowing away, porous paving can help recharge underlying aquifers and [it] reduces peak flows and flooding. That means that streams flow more consistently and at cooler temperatures, contributing to healthy ecosystems. Stormwater pollutants are broken down in the soil instead of being carried to surface waters" (Environmental Building News 2009). It can also help to contribute to healthy trees, which in turn aids in reducing urban heat.

Going Green Is Good for Your Health

In Ireland, a handful of junior doctors and medical students in the Irish Healthcare System got together and founded Irish Doctors for the Environment (IDE) in 2018. "The IDE is driven by three key realizations: First, the biggest global health issue we're facing is climate change. Second, we have a window of opportunity to change things. And third: tackling climate change could be the greatest global health opportunity of the 21st century" (*Irish Examiner* 2019:4). The IDE offers a number of great ideas about how we can improve our personal health while also helping the health of the environment. For example, IDE promotes what they call "active travel"—walking, cycling, public transportation. Active travel is great for a number of health reasons, such as reducing the chance of obesity, stroke, ischemic heart disease, chronic obstructive lung disease, and diabetes. The IDE promotes a number of "small actions" (actions that we have promoted throughout this text) that can lead to big changes in personal and environmental health: Choose not to use a fossil-burning vehicle for journeys (opt for active travel instead); eating less meat, particularly red meat (meat is not a sustainable food source and we can help reduce greenhouse emissions); lobby for better access to drinking water from fountains (reduces the number of plastic water bottles); spend more time in outdoor spaces, such as parks and beaches (this helps to promote conservation); eat a plant-based whole food diet and whole grains, vegetables and fruit (reduce processed food consumption); and vote for green policies (that benefit personal and environmental health) (*Irish Examiner* 2019:5).

Protect and Save Natural Resources

The discussion on "going green" works in conjunction with the idea of protecting and saving Earth's natural resources. Natural resources are finite, and they always have been despite humans' past proclamation to the contrary. And as the world's population continues to increase the drain on natural resources becomes more pronounced. A number of these natural resources are critical for human survival, let alone all the other land and marine species. Even skeptics of human-caused environmental problems should be concerned about the reality that fossil fuels are finite and eventually they will disappear, a fact that should stimulate people into action to find alternative sources of energy. If we run out of fossil fuels, humans and other living species can survive, but it will take the development of new technology to help humanity enjoy their current lifestyles. But even new technology is no substitute for clean air, the rainforests, fresh drinking water, and sustainable land to grow food and allow for the grazing of livestock. Clearly, it is in our best interest, skeptics included, to protect and save our natural resources.

Sustaining Clean Air

Logic dictates that the most essential natural resource is oxygen, or, more generically speaking, air, preferably, clean, breathable air. The EPA is trying to secure clean air by such means as keeping an eye on the emissions we burn and controlling the use of toxic chemicals. But we need to do more than that; we need to make sure there is enough oxygen in the air. And to assure that necessity, we need to save the trees. Trees, and other forms of vegetation, produce oxygen through the process of photosynthesis. Plants take in carbon dioxide and water and produce glucose and oxygen (O_2) as a by-product. Oddly, oxygen can be viewed as a "waste product" that is essential to human, animal, and plant life.

Oxygen (designated by the letter "O" in the periodic table), is a Group 16 element. A periodic table description of oxygen reveals that two-thirds of the human body, and nine-tenths of water, is oxygen. The gas is colorless, odorless, and tasteless. Liquid and solid oxygen are pale blue and strongly paramagnetic (contains unpaired electrons) (WebElements.com 2011). On the planet Earth, oxygen constitutes about one-fifth of the atmosphere by volume (Answers.com 2011). Conversely, the atmosphere of Mars contains only about 0.15 percent oxygen, making life, as we know it, impossible to exist. Certainly, humans could not live on Mars (without building a livable and sustainable protective home station that could protect us from Martian microbes, radiation, and the lack of oxygen). There is more oxygen on the sun, but we cannot live there either, as it is too hot. It would appear that humans are best suited to survive on Earth—a good reason to protect the planet, true?

Another allotrope (a variant of a substance consisting of only one type of atom) of oxygen (O_2) is ozone (O_3). Ozone is formed from electrical discharges or ultraviolet light acting on O_2. In fact, as the National Aeronautics and Space Administration (NASA) (2001) explains, "Ozone is easily produced by any high-voltage electrical arc (spark plugs, Van de Graaff generators, Tesla coils, arc welders)." NASA (2001) states, "Each molecule of ozone has three oxygen atoms and is produced when oxygen molecules (O_2) are broken up by energetic electrons or high radiation." As explained by WebElements.com (2011), the ozone is an important component of the atmosphere (in total amounting to the equivalent of a layer about 3 mm thick at ordinary pressures and temperatures) which is vital in preventing harmful ultraviolet rays of the sun from reaching the Earth's surface. The ozone layer absorbs about 97 to 99 percent of the sun's high frequency ultraviolet light, light which is potentially damaging to life on Earth. Every 1 percent decrease in the

Earth's ozone shield is projected to increase the amount of UV light exposure to the lower atmosphere by 2 percent (NASA 2011). The burning of fossil fuels harms the ozone layer because it breaks down the protective shield. Even aerosols in the atmosphere have a detrimental effect on the ozone layer. This helps to explain why aerosols in hair spray are so dangerous. Large holes in the ozone layer are appearing at the polar regions of earth and they are increasing in size annually.

A contaminated ozone layer will compromise our oxygen supply on earth. As a result, it is critical to save all sources of oxygen production—vegetation, trees and especially rainforests.

Sustaining the Rainforest

Life and natural resources go hand-in-hand. Unfortunately, humans have taken nature's bounty for granted and are guilty of many eco-unfriendly practices such as cutting down the rainforest (deforestation), draining lakes and marshlands, and covering productive agricultural land with pavement and cities. When humans do this, we remove the capacity of nature to provide for us. Globally, one of the top priorities to saving the environment centers on protecting the rainforest. The tropical rainforests help to balance the earth's eco-system by recycling carbon dioxide into oxygen. In effect, the rainforests cleanse the atmosphere to help us breathe. The Amazon rainforest is such a critical producer of oxygen that it is described as the "Lungs of our Planet." As such, the Amazonian rainforest produces about 20 percent of the Earth's oxygen. To underscore the obvious, we need oxygen, or more simply, air, to breathe in order to live and all the other species share this same need.

We discussed the importance of the rainforest in far greater detail in Chapter 4 and we identified four main causes of deforestation: agriculture; logging; nature (e.g., wildfires); and overpopulation. SavetheRainforest.org (2011) acknowledges these four primary causes of deforestation and add a few additional sources: fuel wood (trees provide a source of heating fuel); large dams (in India and South America, hundreds of thousands of hectares of forests have been destroyed by the building of hydro-electric dams); mining and industry (rainforests are destroyed and indigenous people moved so that industrialists can mine and establish other businesses); colonization (partly related to overpopulation and in an attempt to cut down on overcrowding in urban areas, governments have established a number of schemes involving moving indigenous people from their homes and setting up colonies in the rainforest—however, the land in the new colonies is generally not fertile enough to sustain agriculture for the "colonists" to live on); and tourism (the creation of national parks in rainforests has led to the destruction of huge amounts of acres of rainforests).

We will underscore the obvious, the destruction of the rainforest must be stopped or all living species risk extinction. Humans need to realize that we are interconnected to all life forms found on this planet. This connection is not literal, of course, such as the portrayal of the Na'vi on the planet Pandora depicted in the film *Avatar*, but it exists nonetheless because human actions affect all other living species on Earth. In turn, humans are dependent upon all that Earth has to offer and it is therefore in our best interests to protect the environment and allow it to flourish. To learn more about the literal (and fictional) connection between the Na'vi and the other living species on Pandora, see the popular culture section at the end of this chapter.

Sustaining Clean Water Supplies

If oxygen is the most important natural resource provided by the earth's environment, and how could it not be, water must be second. With the importance attached to clean water, why would humans risk contaminating it? Most likely, the answer is the same as the answer to the question why are humans destroying rainforest—because of economic greed, ignorance, and skepticism.

Despite the importance of water, humans dispose of a great deal of garbage in groundwater, rivers, lakes and oceans. Water pollution is a serious global problem. It is one of the leading causes of death and disease. Water is considered "polluted" when it is impaired by contaminants (e.g., chemicals) and can longer support human use. When humans drink polluted water it often has serious effects on their health. Water can become polluted by nature via volcano eruptions, algae blooms, storms, and earthquakes. Humans also pollute water by dumping chemicals, deadly toxins, and garbage into it. Undoubtedly, everyone has seen polluted water. The sight of polluted water alone should make someone ill; after all, how can people have such a careless disregard for such a valuable commodity?

We have already discussed water pollution and garbage islands (i.e., the Great Pacific Garbage Patch) in Chapter 4, so our concern here rests primarily with sustaining clean drinking water. The EPA has established the Office of Ground Water and Drinking Water (OGWDW) to help safeguard the drinking water of Americans (EPA 2011c). The OGWDW, together with states, tribes, and its many partners, work in conjunction with the EPA's ten regional drinking water programs. The OGWDW and its affiliates oversee:

1. Ground Water—Among the programs in existence to guarantee quality ground water are the Underground Injection Control Program, the Ground Water Rule and source water protection.
2. Lakes—Lakes and reservoirs cover nearly 40 million acres of the United States. Among the programs designed to protect our lakes are the Clean Lakes Program, the Nonpoint Source Control (restoration of lakes), and the creation of "Lakes Awareness Month."
3. Oceans, Coasts, Estuaries and Beaches—The Oceans and Coastal Protection Division envisions and strives for clean and safe oceans and coasts that sustain human health, the environment, and the economy.
4. Rivers & Streams—There are over 3.5 million miles of rivers and streams in the U.S., covering an enormous and diverse landscape. Because of the large number of jurisdictions that rivers and streams may run through, the OGWDW relies on a great deal of cooperation and data from affiliated organizations. Among these reporting organizations and data sources are National Aquatic Resource Surveys, Wadeable Streams Assessments, National Rivers and Streams Assessment, Water Quality Conditions Reported by the States, and the National Water Quality Inventory Reports.
5. Stormwater—Stormwater runoff is generated when rain and snowmelt do not soak into the ground but flow over land or impervious surfaces like paved streets, parking lots, and building rooftops, along the way accumulating debris, chemicals, sediment or other pollutants that could adversely affect the water quality if left untreated. The OGWDW works to make sure this does not happen.
6. Wastewater—A watershed is an area that drains to a common waterway such as a stream, lake, estuary, wetland, aquifer, or even the ocean. Because we all live in a

watershed, our actions directly affect it. The primary role of the OFWDW is to educate people on their role in controlling wastewater.

7. Wetlands—Wetlands are areas where water covers the soil, or is present either at or near the surface of the soil all year or for varying periods of time during the year, including the growing season. The EPA's Wetlands Program is dedicated to protecting the wetlands via surface and groundwater protection programs; resource management programs, such as flood control and control of stormwater; and a variety of education programs.

It remains in the best interests for humans and all other species to have access to clean water and this is why the work of OGWDW is so important.

Safeguarding and protecting the health of Americans is (or should be) the top priority of the EPA (under the Trump administration, the EPA often demonstrated its stronger commitment to politics and economics than the environment) and addressing the quality of water during the COVID-19 pandemic was among their duties. In April 2020 the EPA (2020d) assured us that the virus was not detected in drinking-water supplies. The public was encouraged to continue using water from their tap as usual. The EPA (2020c) also "encourages the public to help keep household plumbing and our nation's water infrastructure operating properly by only flushing toilet paper. Disinfecting wipes and other items should be disposed of in the trash, not the toilet." These are words to heed even when we are not in a pandemic.

Sustainable Land

Although many species can live on a variety of land surfaces, such as sandy beaches, arid deserts, and snowy mountains, sustainable land serves its greatest role for humans via growable topsoil. Without farmland, most of the foods humans enjoy would disappear. Consequently, sustainable land management is very important. "The United Nations defines sustainable land management (SLM) as 'the use of land resources, including soils, water, animals and plants, for the production of goods to meet changing human needs, while simultaneously ensuring the long term productive potential of these resources and the maintenance of their environmental functions'" (FAO 2020c). There are a number of variables that can interfere with proper sustainable land management including human activities (e.g., land neglect and harmful agricultural practices) along with nature itself. (Recall that topsoil refers to the uppermost layer of soil which has the highest concentration of organic and microorganisms which plants have most of their roots and which the farmer turns over in plowing.) Nature can compromise the quality of topsoil through occurrences like droughts and wildfires. Some land does not have suitable topsoil that allows for food crops to grow; and in many cases, vast numbers of people live in such areas (e.g., deserts).

The late Sam Kinison, a one-time Pentecostal preacher turned radical comedian who was known for his intense, high octane and politically incorrect observational humor, punctuated by his trademark scream, delivered a very thought-provoking routine on world hunger and how humanity chooses to address hunger. In a mid–1980s appearance on *Letterman*, his first time on the show, Kinison asked the audience, "Do you want to do something about world hunger?" Stop sending starving people food, he said. Send them moving vans and boxes to have them move to where the food is. Kinison then shifts into his hyper-mode of delivery and yells, "You live in a desert! Nothing grows in a desert. Nothing will ever grow here."

Although his approach is harsh, there is substance in his tirade. Ask yourself, why do people live in deserts, or any place where food does not grow, and multitudes of people starve to death because the land cannot sustain adequate food supplies for living species? Decades after Kinison made this routine famous, or infamous (and one of the authors caught his performances while living in Las Vegas, where he would dramatically reach to the floor and pretend to pick up desert sand and yell, "It's sand, it will always be sand, you can't grow food in sand"), it seems as though governmental agencies are finally figuring this out. For example, in rural Malawi (Southern Africa), an area characterized by farm fields of cemeteries of cornstalks, the United States Agency for International Development (USAID) has been working with local farmers to grow crops using new methods instead of simply sending them food—a temporary solution at best. It finally dawned on food agencies such as USAID that for a half of century, agriculture has been one of Africa's biggest failures. African agriculture yields only one-third the global average and has risen more slowly than in the rest of the world (Kristof 2012). The land in Africa is often unsustainable because only 3.5 percent of its cropland is irrigated, compared with 39 percent in South Asia; although it should be noted that Asia applies almost 20 times as much fertilizer as Africa (Kristof 2012). The FAO (2015) claims that agricultural growth is more important for Africa than for any other continent as about 70 percent of Africans and roughly 80 percent of the continent's poor live in rural areas. Africans are increasingly unable to meet their basic food needs as the growing population has increasingly taken over land to build homes and other structures while water resources become scarce or degrade and agricultural productivity stagnates (FAO 2015).

While quality land suitable to grow food is a big concern in Africa, here in the United States land represents another natural commodity that is essential to our very survival and yet it often falls victim to human abuse. For example, we are witnessing the disappearance of farmland and ranchland due to urban sprawl. According to the American Farmland Trust (AFT) (2011), every minute of every day, we lose two acres of farm and ranch land to sprawling development. The AFT (2020) indicates that America's irreplaceable farmland supports a trillion dollar/year agriculture economy and that it's the foundation of our rural communities, provides jobs, recreational opportunities, and a deep connection to the land. The AFT (2020) states, "Well-managed farmland supports wildlife and biodiversity, cleans our water, increases resilience to natural disasters like floods and fires, and helps to combat climate change. It's now clear that we can't realize global climate goals only by reducing emissions, that we also need to retain farmland and actively manage it to draw down carbon from the air." It is refreshing to realize that the AFT recognizes the reality of climate change and the need to help combat global warming via SLM techniques.

In addition to the American Farmland Trust, there exist a wide variety of organizations attempting to save land.

- The Mountain Area Land Trust has a mission to "save scenic vistas, natural areas, wildlife habitat, working ranches and historic lands, for the benefit of the community and as a legacy for future generations" (Mountain Area Land Trust 2011).
- The Ecology Fund raises money to save 86 square feet of land every day. This

includes the South American rainforest and the United States Wilderness (Ecology Fund 2011).

- Save Our Land, Save Our Towns is a grassroots organization founded by Tom Hylton dedicated to reversing the car-dominated culture and urban sprawl into rural areas. This organization also promotes the development of urban life via safe, verdant, walkable neighborhoods (Save Our Land, Save Our Towns 2011).
- World Wildlife Fund, committed to reversing the degradation of our planet's natural environment.
- Forest Stewardship Council, a non-governmental nonprofit established to protect forests.
- The Nature Conservancy, dedicated to saving lands and water.
- Friends of the Earth, focused on political and economic decisions as they affect public health and land conservation.
- Wildlife Conservation Society, operates a system of urban parks and works to save wildlife and wild lands through careful science, international conservation, education, and the management of the world's largest system of urban wildlife parks.
- A large number of local organizations have attempted to preserve open lands; perhaps there is such a movement in your hometown.

Of concern to environmentalists is the need to save marshlands. As the name might indicate, a marshland is neither dry land—a solid part of the earth's surface—nor water. A marshland is a type of wetland that is subject to regular or continuous flooding. Generally, the water is shallow and is inhabited by a number of animal and plant species. The ecological value of marshlands, such as shrubs, tall grass, and forbs, is that of a giant filter and sponge that maintains water quality and cleanses pollutants which pass through them (Slesinger 2009). The filter and sponge function is created by the marshland's thicket of plant life with deep roots. This mesh also helps to stop flooding on adjacent dry lands. In addition, marshlands attract an array of beneficial insects, fish, birds and animals that are a critical aspect of the ecological cycle.

Develop New Technology

A good way to help protect and save our natural resources so that the environment can thrive involves developing new technologies that halt the drain of these precious commodities. If we hope to help the environment thrive, we need to lessen our reliance on fossil fuels as our chief source of energy and stop seeking new ways to speed the process of corrupting local ecosystems. We need to develop new and greener technology. Among the technology growth areas designed to help save natural resources are advancements in gas-saving automobiles, the development of a low-cost biogas filter, LED light bulbs, innovative recycling efforts, and replacing Styrofoam and plastic packaging materials with natural products such as mushrooms.

Advancements in Gas-Saving Automobiles

Developing new technology to lessen our dependency on fossil fuels is just one, albeit one very important, step in sustaining the environment. This can include pursuing alternative energy sources such as solar, wind, and nuclear power to replace our reliance on crude oil (and oil-related products such as gasoline), coal and natural gas. We can reduce our addiction to gasoline by driving fuel-efficient automobiles, car-pooling, taking mass transit, and working more from home whenever possible. Many businesses discovered during the COVID-19 pandemic that a great deal of work can be done remotely, from people's homes; thus, reducing automobile emissions and raising the amount of work that can be completed when people are not stuck in traffic.

Some people are installing various gas-saving devices to their existing cars, especially those that make use of hydrogen. For example, the "Hydrogen Hurricane," a gas-saving gizmo available via the Internet, can increase gas mileage by nearly 40 percent. The main component of this device is a steel cylinder filled with distilled water. With the electricity supplied from the car's battery, the unit makes bubbles of hydrogen and oxygen through a process called electrolysis. A hose carries the hydrogen and oxygen to the engine's air intake (Knauss 2008). A Chevrolet Prizm that gets 35 miles per gallon (on the highway) can get 50 miles per gallon after the installation of the Hydrogen Hurricane (Knauss 2008). Jay Leno, former host of NBC's *Tonight Show*, boasts about driving hydrogen-powered automobiles that he owns which do not use gasoline and emit zero pollution into the air. Although we don't all own such cars, perhaps the time will come when the auto industry will mass produce such gas-saving vehicles. The fact that, as of now, filling stations do not sell hydrogen-boosted fuel has hampered the potential mass production of vehicles that could benefit from this hydrogen to the fuel/air mixture in a conventional gasoline engine.

Another fuel alternative for automobiles is ethanol. Ethyl alcohol, ethanol for short, and is the same type of alcohol in alcoholic beverages. It is primarily a biofuel additive for gasoline. Most standard cars can run on an up to 10 percent mixture of ethanol in regular gasoline (E10) without modifications of the engine. E85 is a mixture of up to 85 percent ethanol and at least 15 percent regular gasoline. This mixture is becoming increasingly common in the United States, Brazil and Europe. On the plus side, burning ethanol is better than burning gasoline, therefore helping people to reduce their collective carbon footprint, which helps to save the environment. On the negative side, besides difficulty starting engines in cold weather with a high ethanol ratio, the demand for corn to produce ethanol is contributing to world hunger. That's right: stop burning fossil fuels but take grain away from starving people. The increasing demand for biofuels has placed a burden on those trying to produce enough corn and grains to feed people around the world while supporting the "green" effort to lessen dependency on fossil fuels. "Each year, the world demands more grain, and this year the world's farms will not produce it. World food prices have surged above food crisis levels of 2008. Millions more people will be malnourished, and hundreds of millions who are already hungry will eat less or give up other necessities. Food riots have started again" (Searchinger 2011: E-1). Once again, the question bears asking, "What is an environmentally-conscientious person to do?"

The Ford Motor Company is combining recycling efforts with the development of new technology in some of its automobiles. For example, the 2012 Ford Focus will include shredded pairs of old blue jeans under the floor. That's because Ford is using recycled cotton clothing such as denim in the sound-deadening material and carpet backing in the next generation of the compact car. Ford said that the amount of recycled cotton in each Focus is equivalent to two pairs of blue jeans. As reported in the *Associated Press*, Ford is looking at other discarded materials that can also be used in cars as part of its strategy toward recycling to help divert waste from landfills. In the past, Ford has used recycled resins for underbody parts, soy-based foam for seat cushions and recycled yarn for seat covers (*The Post-Standard* 2011a).

Gasoline-powered cars and trucks are often the leading source of climate-changing emissions and that is case in California. The emissions are bad for people's health and the primary cause of Los Angeles' infamous, lung-damaging smog. Emerging research also suggests that breathing dirty air makes people more likely to die of COVID-19 (Roth 2020b). Clean vehicles not only help reduce dangerous greenhouse emissions they are a proven job creator with companies like "BYD, Proterra and Tesla operating manufacturing plants in California and most major automakers planning to invest billions of dollars in cars and trucks that don't use gasoline" (Roth 2020b: A8).

Electric cars are another way to decrease the amount of carbon emissions pumped into the atmosphere. Tesla (a company named to honor Nikola Tesla, the eccentric inventor and engineer who came up with the alternating current electronic system) was founded in 20003 by Elon Musk, after he learned that General Motors had recalled all of its electric cars and "crushed them in a junkyard" (Musk 2017). Since all electric vehicles rely on rechargeable batteries, there are virtually no pollutants released into the atmosphere where the vehicle is active. The motors on such vehicles are quiet and require little to no maintenance as compared to gas-guzzling internal combustion engines. And they are obviously more energy efficient. Proterra, Inc. is an American automotive and energy storage company based in Burlingame, California, which designs and manufactures electric transit buses and electric charging systems. And BYD is a Chinese manufacturer which has become the world's largest electric vehicle company. One sign of its success is that "more than 500,000 electric buses ply Chinese roads, compared with fewer than 1,000 in the U.S." (Campbell and Tian 2019).

The Development of a Low-Cost Biogas Filter

Among the chief concerns of livestock farms is the methane output. Recent technological advancements have led to the creation of a low-cost biogas filter that removes harmful chemicals in methane from farm waste. Here's how it works: manure and other organic farm waste goes through an anaerobic digestion process in an on-site digester; the resulting biogas is filtered through naturally occurring bacteria housed in a specially designed plastic silo that separates sulfuric acid and sulfur from the combustible methane; the sulfur can then be added to other nitrogen-rich products of the digestion process to be reused as fertilizer to grow crops (Baker 2012). The larger the farm, the greater the economic and environmental savings, as the biogas filter removes corrosive hydrogen sulfide from the biogas at one-tenth the cost of conventional removal methods (Baker 2012). Conscientious persons who enjoy dairy products or the occasional steak or hamburger can enjoy their eco-friendly products.

Filters of all sorts can help benefit the environment by removing harmful emissions into the air; although it is still important to properly dispose of the used filters or you are just creating a different type of environmental harm. But back to the benefits of filters. Clean furnace and air conditioning filters, for example, can help reduce energy costs as a dirty filter restricts air flow into your system causing it to work harder; improve airflow efficiency and filter allergens and other particulates our of your air; and help to maintain the life of your system, thus reducing the need to throw away old pieces of equipment and replacing them with new ones.

LED Light Bulbs

Perhaps the favorite light bulb of many folks for home use were the 100-watt incandescent bulbs. But these were banned from stores on January 1, 2012, because they were energy hogs. Well, there is good news and bad news for 100-watt bulb fans. The good news, available now in stores everywhere are light-emitting diodes that shine as brightly as incandescent 100-watt bulbs, according to the maker, Osram Sylvania (Svensson 2012). LED 100-watters not only shine as bright as the old bulbs, but they do also so at a fraction of the cost; thus, they represent an energy saver. So what's the bad news? The cost. The 100-watt LED bulbs cost around $50 each (as of late 2012). There are LED bulbs equivalent to 60-watt regular bulbs available for around $25 each (Svensson 2012).

The switch to LED light bulbs represents a technological advancement over traditional bulbs because they save on energy and last longer. The development of LED bulbs coincides with the federal government's ban on energy inefficient bulbs. "Much the way it forces car manufacturers to improve fuel efficiency, the government is forcing the lighting industry to move away from incandescent bulbs because they convert relatively little of the electrical input into light. Most of the energy is dissipated as heat" (Svensson 2012). The old-styled 100-watt bulb was the first to be banned, the 75-watt bulbs followed at the start of 2013, and the 60-watt and 40-watt bulbs at the start of 2014 (Svensson 2012). In the long run, sacrificing traditional light bulbs for the new LED bulbs will be worth it, especially for the environment.

Innovative Recycling of Wastepaper

In Chapter 3, we stated that paper and paperboard products made up the largest percentage (25 percent) of municipal solid waste (MSW) generated by Americans. While we are in the electronic age, a high number of tons of paper are produced annually. At the World Counts website is a running counter of the tonnage of paper produced yearly, monthly, weekly, and daily. On May 13, 2020, there were already over 205 million tons of paper produced to-date for 2020. Put another way, well over 1.5 million tons of paper are produced daily across the globe. It takes a lot of trees to make that much paper. It takes twice as much energy to produce paper as it does to produce a plastic bag. Then again, we have learned about the environmental problems associated with plastic use. Deforestation, as we also know, harms the environment. Most paper is not recycled resulting in MSW at landfills or litter on land and sea.

What can we do reduce paper waste and pollution? There are a number of suggestions from different organizations. The Denmark-based World Counts (2020) has ten recommendations:

- Recycle all paper waste.
- Be a conscious consumer and buy "100% post-consumer waste recycled" paper products.
- In the office, reuse paper (e.g., if you've only used one side) instead of throwing it away.

- If you have scanned a copy of a file, don't print it too (unless really needed).
- Use email instead of paper when communicating with clients and customers.
- Reduce the use of paper cups and disposable paper plates by keeping reusable items in the office pantry.
- Encourage your officemates and friends to recycle their paper by putting them in recycling bins.
- Insist on "Process Chlorine Free" paper materials.
- Buy products with the least paper packaging.
- Take advantage of the latest technologies like tablets, computers and smart phone to keep your files and notes.

In India, a country that produces over 100,000 tons of MSW every day, which sums up to about 36.5 million tons annually, has encouraged its citizens to recycle wastepaper. Sangeetha Murali (2016) has a number of creative and crafty ideas on how to reuse wastepaper: Papier-mâché decoration items; Papier-mâché paper bowls; newspaper baskets; recycled paper photo frames; recycled paper bags; recycled ice cream sticks bookmarks; pen and pencil stand; recycled paper wall hangings; recycled paper bracelets; and upcycled paper earrings.

One of the more unique and practical approaches to recycling wastepaper involves turning the wastepaper into bedding for dairy cattle. One such company involved in this innovative recycling effort is the Syracuse Fiber Recycling (SFR). SFR takes 1,400 tons of wastepaper a week and turns it into bedding for cows, shipping seventy-five, 72-yard loads a week to local farms, and farms throughout New York State and Pennsylvania (Groom 2012b). For those unfamiliar, bedding is important because if cows are comfortable, they will produce more milk, produce longer and be healthy longer, according to Shawn Bossard, farm manager at Morrisville State College (Groom 2012b). Typically, cow bedding comes from straw. The recycling process works like this: paper waste is delivered to the SFR plant, where it is chopped and cooked into something like a compost pile. Every seven minutes, some of the compost-like mixture is fed onto a conveyor belt that takes it to an area to be mixed with a lime additive (Groom 2012b). The lime additive raises the pH of the bedding. The treated compost is sent to a hammer mill where it is chopped into small pieces and dried again. SFR adds secret ingredients to make the bedding material soft to the touch and devoid of smell. Because of the eco-safe additives that serve as a fertilizer, after the cows use the bedding it can be spread on soil to help crops grow in an eco-friendly manner (Groom 2012b).

Replacing Styrofoam and Plastic Packing Materials with Mushrooms

The past few decades have played witness to an increasing number of products wrapped in packages that contain more packing materials than actual product. Purchase almost any electronic item and receive any item via mail or shipping delivery service and undoubtedly it will be packaged in a hard plastic bubble or protected by Styrofoam peanuts. In almost every situation, the packing material must be thrown away as waste, non-biodegradable waste to be specific. These packing materials are a biohazard for our landfills.

For years now, innovative persons have been developing new ways to pack items in an eco-friendly manner. For example, Ecovative Design, the brainchild green business of two former mechanical engineering and design students at Rensselaer Polytechnic Institute, Eben Bayer and Gavin McIntyre, have developed a technology that allows them to use the hidden "roots" of mushrooms as packing material. Mushrooms are the key ingredient in the pale, soft blocks produced by the thousands by Ecovative that are used to cushion products ranging from Dell, Inc., servers to furniture for Crate and Barrel (*The Citizen* 2012b). Bayer and McIntyre discovered that the sticky substance on the bottom of mushrooms called mycelium could be turned into a glue and when that glue is combined with corn husks and other food byproducts it takes on a form similar to Styrofoam (Weir 2012). The mix of mushroom roots and other natural products can be shaped into any form to be utilized as packing materials.

The company has grown quickly, and Bayer and McIntyre are ambitiously attempting to take over the packing materials market. At present, the costs of Ecovative's products are slightly more than expanded polystyrene, but their costs are expected to decrease as production continues to increase. As companies such as Ecovative Design grow, so too does the number of green jobs.

The obvious eco-benefits of materials such as mushroom and other natural food byproducts-based products rest with the realization that the material is biodegradable and therefore contributes to up-cycling (meaning that the resulting product is of a higher value than the original item).

Hemp Products

Making products out of hemp is not a new idea; in fact, hemp, a member of the cannabis family (but nearly completely devoid of tetrahydrocannabinol—THC), is one of the world's oldest cultivated crops. Although there has been documentation of hemp products (e.g., as rope and paper) in the United States predating the Revolution by a century and half (Able 1980), it has only recently begun to reach its full potential. Developments in technology have greatly expanded hemp's value as a green energy source. Hemp serves many green functions, and it is time to tap into the full potential of this plant. Listed below is a sampling of hemp products that can help the environment thrive, if only we used them instead of current products.

1. Hemp as Paper Materials—Hemp has long been used a paper making material. Utilizing hemp for paper products would help save the rainforests, as we chop billions of trees each year to make paper worldwide. For nearly 100 years we have been aware of the fact that one acre of hemp can produce 4.1 times the amount of timber material that an acre of trees yields (Barry 1995). Among the first documents revealing this information was the United States Department of Agriculture's *Bulletin No. 404*, "Hemp Hurds as Paper-Making Material," wherein it was reported that hemp production would be an economic benefit to the United States, especially because of its yield factor.

Hemp is a fibrous cash crop with a large variety of purposeful uses that can yield around 20 tons of dry plant material per hectare (Gonzalez-Garcia et al. 2010). Large-scale hemp production requires fertilizing the plot before sowing and then harvesting the plant after about 5 months of growth. According to a lifecycle assessment of hemp, the "crop is rarely threatened by dangerous pests so no pesticides or herbicides are required. Supply of irrigated water is unnecessary, due to sufficient annual rainfall" (Gonzalez-Garcia et al. 2010: 925). Originally grown in Asia, hemp has been naturalized and cultivated for its fiber throughout the world and continues to be grown as a renewable resource, especially in Europe. Why not in the United States, you might wonder? We wonder the same thing.

We should value hemp as a source of paper material because it would help save the rainforests and because of this; it is an environmentally friendly life form. We know that saving the rainforests is important because it produces oxygen while absorbing CO_2. Hemp, too, absorbs CO_2 and converts carbon dioxide into oxygen and improves the soil on which it is grown. Sometimes used as a "mop crop" and planted on some farms to restore the soil's nutrients during agricultural crop rotation, hemp has the phytoremediation potential to remove toxins in the soil. Following the nuclear reactor meltdown in Chernobyl, experiments found that hemp not only reduced radioactive contaminants in the soil, but parts of the contaminated hemp plant could still be used without health concerns (Vandenhove and Van Hees 2005). Hemp also helps to prevent soil erosion, an environmental problem that is manifested by deforestation. Saving the trees, and the rainforests specifically, producing oxygen, reducing CO_2, assisting the soil as a natural fertilizer and preventing soil erosion is just the beginning of what hemp production can do.

2. Hemp Construction—In 2007, the Centre for Alternative Technology (CAT) hosted the Hemp and Lime conference in England to investigate whether hemp-and-lime-insulated homes have the potential to be carbon neutral. Evidence presented at the conference revealed that materials used with hemp hurds (small fibers removed from the core of the hemp plant, similar to that used for animal bedding) mixed with lime—creating a product known as Hempcrete—is a material that potentially causes no carbon emissions at all (Latona 2007). CAT spokesperson Jessa Latona said in a conference press release, "Houses built from hempcrete have been found to create less waste and need less fuel to heat than conventionally constructed homes, both saving carbon dioxide emissions associated with a building. The hemp crop already has a multitude of uses, although using hemp hurds for buildings is a potential new and large market for builders and farmers in the U.K." Latona (2007) also stated, "Fast growing hemp captures and stores carbon from the atmosphere during growth and overall the CO_2 balance of the hemp crip means that CO_2 may actually be removed from the atmosphere and locked away into the fabric of hempcrete homes." Furthermore, although hempcrete (a substitute for concrete) contains a small amount of cement (approximately 15 percent), it is still a porous material—meaning that moisture can travel through the walls, regulating humidity and avoiding moisture problems to an extent (Centre for Alternative Technology 2012).

3. Foam Insulation—Traditional polyurethane and polyisocyanurate foams for insulation include toxic materials; the containers these products are packed in must also be disposed as toxic waste; and, during the application process, the vapors are subject to combustion and the person spraying the insulation foam must be covered from head to toe in a HazMat suit. Hemp can be substituted for polyurethane and polyisocyanurate foams without risk of harmful toxic exposures to the person applying the insulation and those who have to breathe the air in the insulated rooms.

4. Animal Bedding—Ecofibre Industries Operations, an Australian agricultural industry company that produces sustainable hemp fiber for national and export industry, has a created an animal bedding product—Agrisorb—from hemp. The animal bedding comes from the soft core of the industrial hemp stem and is promoted as the most absorbent bedding material on the market. The bedding material stays dry longer than straw or wood shavings and will absorb four times its own weight. Other benefits of the

Agrisorb bedding material are low dust and low palatability (which limits respiratory and other health risks); chemical free (industrial hemp requires no pesticides or herbicides when grown, consequently preventing chemical residues being transferred to the animals or humans applying the bedding); and being economically and environmentally friendly (the longer life of hemp bedding equates to dollar savings and reductions on labor, transport and storage costs) (Ecofibre Industries Operations 2010).

5. Clothing and Carpets—Bast fibers from the hemp plant can be used to make carpets, clothes and fabric that are stronger than cotton. As an organically grown crop, hemp does not require the billions of dollars in annual pesticide use that the cotton crop requires (Simpson-Holley and Law 2007). Unless made from 100 percent safe biodegradable materials, conventional carpets cannot be returned to the soil, they must be dumped at a waste landfill. Hemp carpets are biodegradable. Bast fibers can also be used to create plastics and composite materials that have been used to construct car panels. Because hemp produces more useful material per acre than cotton and many other crops, a reduction in agricultural land area need is feasible if only people would seriously consider the hemp industry as a viable economic and environmental alternative to current methods of making products.

6. Hemp as a Biofuel—Hemp is cultivated slightly differently when being processed as a biofuel than when it is used for its fiber. Generally speaking, the differences are a matter of when harvest occurs and how it is stored; it goes into a silage tube immediately after harvest until it is converted into a biofuel. One study found that 10.2 tons per hectare of dry matter (DM) could be harvested as biogas in about 4 months and 5.8 tons per hectare of DM could be harvested as biofuel. "Hemp has high biomass DM and good net energy yields per hectare. Furthermore, hemp has good energy output-to-input ratios and is therefore an above-average energy crop" (Prade et al. 2012). As corn ethanol is becoming more of an agricultural problem than it is worth, "hemp plants absorb 1.7kg of atmospheric CO_2 for every kilo of dry hemp material," making it an environmentally friendly biofuel and biogas resources (Simpson-Holley and Law 2007: 19). Unlike some ethanol crops, hemp crops do not require pesticides or herbicides, is frost-tolerant, and can be used in crop rotation to condition the soil.

7. Nutritional Value of Hemp Seeds and Hemp Oil—Hemp seeds and oil have nutritional value that may be used as a food or dietary supplement. "Hemp seeds contain an edible oil, rich in essential polyunsaturated fatty acids (PUFAs) and Vitamin E, and low in saturated fats" (Simpson-Holley and Law 2007: 21). The two PUFAs are omega–3 linolenic acid and omega–6 linolenic acid, allowing the oil to be used as either a nutritional supplement for the treatment of some chronic diseases or as a culinary oil. The a-linolenic acid (ALA) in hemp oil can be converted into eicosapentaenoic acid (EPA) and then docosahexaenoic acid (DHA), a product found in fish oil. Fish oil is a supplement in many people's diets, but the hemp product would be more natural. Hemp oil may help to reduce cardiac disease and stroke as well as relieve arthritis and reduce the chances of likelihood of Alzheimer's disease, depression and bipolar disorder (Simpson-Holley and Law 2007). Although the DHA and EPA extracted from hemp oil is significantly less than what is readily obtained from fish oil without conversion, it could be substituted as a supplement for people who do not like to eat fish. G-linolenic acid (GLA) has been shown to have "beneficial effects in the treatment of premenstrual syndrome, atopic eczema, osteoporosis, cardiovascular disease and asthma" as well as preventing cancer progression (Simpson-Holley and Law 2007: 23). Second only to soy as a protein source, hemp seeds can be eaten raw, ground or baked into a meal, and used to make hemp milk and hemp

tea. As a food source, hemp seed has been used as animal feed, which further reduces our dependency on corn-based products.

A listing of other hemp products is provided by the Ministry of Hemp (2017): Foods & Drinks (hemp seeds, hemp seed oil, protein powder, hemp tea, energy bars, coffee, hempseed burgers, healthy flavored water—Hemp2O, hemp seed butter, hemp milk, hemp vodka, hemp beer, hemp hot dogs, hemp flour, and hemp granola); Clothing & Accessories (shirts, jeans, shoes, jackets, backpacks, yoga pants, sunglasses, hats, beanie, wallets, socks, totes, sandals/flip flops, belts, scarf, ties/bow ties, handkerchief/pocket squares, bracelets, robes, and overalls); Beauty & Skin (body lotion, balms/lip balms, shampoo and conditioner, body wash, facial cream and cleanser, hemp sunscreen, and hemp serum); Health & Medicine (hemp extract, hemp essential oil, aromatherapy candles, message oil, hemp heat muscle rub, and vape juice); Pets (dog toys, dog collar and leash, animal bedding, and CBD oil for dogs); Automobiles (an entrepreneur in Florida built a sports car out of hemp, biofuel, and thermoset compression molding); Home & Office (pens, hemp sheets, hemp towel, paper, hemp curtains, laundry detergent, hemp crafts, hemp chair, tablecloths, and hemp blankets); and Farming & Gardening (hemp growing mats and soil cleanup); Industrial & Others (ropes, hemp plastic, hempcrete—instead of concrete, oil spill cleanup, hemp flags and hemp batteries).

As this review of hemp products reveals, there is a wide range of economic and environmental benefits to maximizing this near miracle plant. What we need is a social movement to embrace hemp, and then this product can fully assist the environment's thirst to thrive.

So concludes our sampling of developments in technology designed to help the environment thrive. A number of people value the creativity of humans and their potential to create technology that will allow us an opportunity to protect and save the environment. It takes innovative and creative people to come up with ideas to stimulate technology. Who knows what source may serve as inspiration for future developments?

Along with millions of others, the authors are glad that there are private citizens like James Cameron willing to spend their own money in their search for knowledge. Cameron, a well-known Academy Award winning filmmaker (including *Avatar,* which is discussed in detail at the end of this chapter), is also an ocean enthusiast who made history in March 2012 as the first solo explorer to reach the deepest point in the ocean—almost seven miles down (35,756 feet)—when his custom-built one-man submarine touched down in the western Pacific Ocean's Mariana Trench (Vastag 2012). In addition to filming his fantastic voyage, Cameron spent six hours at the bottom of the trench collecting scientific samples (e.g., rocks, soil and any deep-sea creatures he encountered) using hydraulic arms attached to the sub. The samples are now being analyzed and only time will tell what, if any, secrets they reveal about our planet. Ideally, this information will someday help us to help the environment to thrive.

Environmental Rights and Legislation

In every society, citizens have rights, some more than others. And these rights are protected by the governments that establish them. Perhaps, if the earth is to thrive, its environment needs rights as well. After all, every time we hear of oil spills destroying marine life and vegetation, the ozone layer coming closer to deterioration, and the oceans playing unwilling hosts to garbage islands, it makes us cringe to think that humanity can get away with such extreme violations of misconduct against the environment. If you think the idea of the environment enjoying "rights" is a little far-fetched, how about the idea that humans should have the right to reside in a livable environment? That would seem reasonable. The nation of Slovenia feels so strongly about water as a fundamental right that in 2016 the Parliament adopted an amendment that declares the country's abundant clean water supply as a fundamental right for all (Agence France-Presse in Ljubljana 2016).

Placing a Dollar Value on Natural Resources

Perhaps in this "economics-first" world that we live in, the best solution in our attempt to save the environment involves putting a dollar value on nature itself. Attempts to put a dollar value on the natural world is referred to as "natural capital" or "ecosystem services" (Sutton 2015). In this regard we are less likely to "waste" our precious natural resources and are more likely to try and save them. With this idea in mind, the ecosystems around the world would become our most valued capital and nature's assets would become more highly valued. Some economists, such as Robert Costanza and associates (1997), have already attempted to place a value on the ecosystem. "In the most comprehensive estimate by a group led by Robert Costanza in the late 1990s, the value of the major ecosystem services was estimated to be in the range of U.S. $16–54 trillion per year, with an average of U.S. $33 trillion per year. They noted that this may still underestimate the value, but when compared with the global gross national product [at the time] of around U.S. $18 trillion, even this minimum estimate is staggering. This means that the natural ecosystem provides, on average, twice the amount of value as all the activities of humans around the world combined" (SuperGreenMe 2008).

In 2014, Costanza and some new associates recalculated their famous study from the 1990s and found that a better annual global estimate of the value of the environment would have been much higher—US$145 trillion in 1997. Unfortunately, Costanza and associates (2014) valued today's natural world at a figure lower than the 1997 figure, US$125 trillion a year. The reason for the decline is attributed to the damage to Earth's ecosystems in the interim (Sutton 2015). Even though Nature is losing value due to human neglect, it is still the most valuable commodity we have, both in terms of economics and ecology.

As a specific example of valuing nature, City Plants, a nonprofit that works with city departments to plant and care for public trees has appraised the value of Los Angeles' trees (including privately owned ones) at $12 billion. The City Plants reports states that Los Angeles should be spending three times as much on its urban forest (*Los Angeles Times* 2019).

Certainly, natural resources are valuable, and it seems reasonable that the environment should have rights; if for no other reason, than to secure the future of humanity and other living things. What do you think?

The Human Right to Live in a Healthful Environment

In the past, the United States Congress has listened to proposals for an amendment to the Constitution recognizing the rights of citizens to live in a healthful environment, including 1968 and 1970 (Cusack 1993). Both attempts at adding such an amendment failed. Imagine that, Congress deemed it unimportant for citizens to have the right to live in a healthful environment! Of course, it wasn't the case that congresspersons were against the idea of a healthy environment for people to reside in; instead, it was more of a matter that they did not want to go against politically powerful entities (e.g., the fossil fuel industry) that feared having their profits compromised if safeguards for the environment had to be built into their business practices.

In the wake of these two unsuccessful attempts to provide American citizens with the federal right to a clean environment, a number of states in the 1970s (the "environmental Seventies" as the decade was known) passed legislation designed to protect the environment on behalf of humanity (Meltz 1999). Meltz (1999) provides a review of state provisions creating a personal right to a healthful environment:

- *Hawaii Const. art. XI, 9:* Each person has the right to a clean and healthful environment, as defined by laws relating to environmental quality, including control of pollution and conservation, and protection and enhancement of natural resources. Any person may enforce his right against any party, public or private, through appropriate legal proceedings, subject to reasonable limitations and regulation as provided by law (effective January 1, 1979).
- *Illinois Const. art. XI 2:* Each person has the right to a healthful environment. Each person may enforce this right against any party, governmental or private, through appropriate legal proceedings subject to reasonable limitation and regulation as the General Assembly may provide by law (effective July 1, 1971).
- *Massachusetts Const. article of amendment XLIX:* The people shall have the right to clean air and water, freedom from excessive and unnecessary noise, and the natural, scenic, historic, and esthetic qualities of their environment (effective November 7, 1972).
- *Montana Const. art II, 3:* All persons are born free and have certain inalienable rights. They include the right to a healthful environment and the rights of pursuing life's basic necessities, enjoying and defending their lives and liberties, acquiring, possessing and protecting property, and seeking their safety, health, and happiness in all lawful ways. In enjoying these rights, all persons recognize corresponding responsibilities (effective July 1, 1973).
- *Pennsylvania Const. art. I, 27:* The people have a right to clean air, pure water, and to the preservation of natural, scenic, historic and esthetic values of the environment [effective May 18, 1971].

The Montana law has as much to do with guaranteeing citizens their right to bear arms as it does protecting the environment, but the other state legislations were clearly attempts to provide a certain quality of life for citizens. While the states have granted people certain environmental rights, the U.S. Constitution still does not. Such a reality should be shocking that a professed leader of the world does not have in its constitution a mandate assuring basic environmental rights for its citizens.

Globally, there are more than 100 nations that provide its citizens the right to a healthy environment through constitutional recognition and protection. About two-thirds of the constitutional rights refer to a healthy environment; alternative formulations include rights to a clean, safe, favorable, wholesome or ecologically balanced environment (UNEP 2020b). Two nations that are holdouts in supporting recognized international law that provides constitutional rights of a healthy environment to its citizens are Canada and United States. The right to a healthy environment is recognized in five provinces and territories in Canada. "Quebec put the right into its Environmental Quality Act in 1978 and added it to its provincial Charter of Human Rights and Freedoms in 2006. Ontario enacted a comprehensive Environmental Bill of Rights in 1993. The Yukon, NWT, and Nunavut have modest environmental rights legislation" (Boyd 2013).

Earth Day

The pro-environment social movement began in earnest on April 22, 1970, the date of the first Earth Day. Sociologically-speaking, 1970 was quite a fascinating year, as it followed the height of hippie and flower-child counterculture of the 1960s—a decade characterized by a wide variety of social movements and collective action amid an unpopular and continuing war in Vietnam and Americans driving massive V-8 sedans powered by leaded gas, all under the shadow of industry belching out smoke and sludge with little fear of legal or moral consequences or bad press. Air pollution was common and generally associated with the smell of prosperity; after all, air pollution caused by industry meant that Americans were at work and American industry was strong.

The idea of Earth Day came to founder Gaylord Nelson, at the time a U.S. senator from Wisconsin, who was inspired to do something about protecting the environment following the ravages of the 1969 massive oil spill in Santa Barbara, California (Earth Day Network 2012). Another variable that led to the first Earth Day was the fact that a river, the Cuyahoga River in Cleveland, actually caught on fire in 1968 as a result of an oil slick and decades of industrial waste being dumped into the river. This was not even the first time the river caught on fire as dating back to the beginning of the 20th century, the river had caught fire on several other occasions (Rotman 2020). The river fire actually drew little national attention initially because "water pollution was viewed as a necessary consequence of the industry that had brought prosperity to the city" (Rotman 2020). However, this attitude would change as environmentalism took shape. Senator Nelson, meanwhile, was galvanized by the student-led anti-war movement and felt that if he could infuse that type of energy with an emerging public consciousness about air and water pollution, he could stimulate a pro-environment social movement (Earth Day Network 2012). Nelson persuaded Pete McCloskey, a conservation-minded Republican congressman, to serve as his co-chair for a national teach-in on the environment. With both Democrats and Republicans on board, Nelson sold his pro-environment social movement concept to the media and as a result of these actions, on April 22, 1970, the first Earth Day took place.

Approximately 20 million Americans celebrated the first Earth Day by taking to the streets, parks, college campuses and auditoriums to demonstrate for a healthy, sustainable environment in massive coast-to-coast rallies (Dunaway 2008; Earth Day Network 2012). At the State University of New York at Oswego, students buried a Chrysler near Culkin Hall, the administration building, as a protest against air pollution caused by cars. Sadly, more than 40 years later, many students drive SUVs on the Oswego campus and campuses across the country. The pro-environment movement remained relatively strong in the 1970s and led to the creation of the Environmental Protection Agency and the passage of many pro-environmental laws (i.e., the Clean Air Act, the Clean Water Act, and the Endangered Species Act). However, the momentum of the social movement stalled during the 1980s in the United States under the Reagan-era mantra of conspicuous consumption. "Environmentalists ('tree huggers') became subjects of ridicule" (Pederson 2011).

While "tree huggers" in the United States were being negatively portrayed by some during the 1980s, the movement had extended worldwide. In 1990, Earth Day mobilized 200 million people in 141 countries, lifting environmental issues onto a world stage (Earth Day Network 2012). Earth Day 1990 gave a huge boost to recycling efforts globally and helped pave the way for the 1992 United Nations Earth Summit in Rio de Janeiro, and it prompted President Bill Clinton to award Senator Nelson with the Presidential Medal of Freedom (1995)—the highest honor given to civilians in the United States—for his role as Earth Day founder (Earth Day Network 2012). By 2000, Earth Day events were highlighting the need for clean energy and warned of global warming. Earth Day 2010 was encumbered by a growing anti-environmental movement of climate change deniers, well-funded oil lobbyists, reticent politicians, a schism among environmentalists as to their focus, and an often-disinterested public (Earth Day Network 2012).

During the first decade of the 21st century, the environmental movement was both strongly embraced by millions of people who actively participate in methods designed to help the environment thrive and countered by legions of people who either don't want to be bothered by conversation efforts or skeptics to human-caused environmental problems such as global warming. As Kermit the Frog is fond of saying, "It's not easy being green." This phrase uttered by Kermit, a Muppet on the show *Sesame Street* (which itself came on the air in 1969, just before the Earth Day social movement began), has been uttered since the first season of this favorite children's show. In a 2006 Super Bowl ad, Kermit is shown traversing a mountainside when he comes across a silver Ford Escape Hybrid and as he turns to face the camera, Kermit exclaims, "I guess it *is* easy being green!" (House 2007).

In 2020, Earth Day reached its 50th anniversary, having accomplished many things but still facing backlash from conservative anti-environmental forces. This 50th anniversary of Earth Day was also supposed to play witness to huge gatherings of pro-environmentalists but such rallies were mostly non-existent due to COVID-19 and mandates to maintain physical distancing of 6 feet in between people and avoidance of gatherings of 8 or more people (the number of people allowed to congregate varied by state), (Note: For some odd reason, the term "social distancing" was used instead of "physical distancing" even though people were free to engage in social networking as a means to maintain closeness with others while mandates to maintain a physical distance from others without protective measures were employed and enforced.) While Earth Day 2018 and 2019 had a focused theme on the dangers of plastics, the 2020 Earth Day theme was to center on the need to take action on climate change and global cleanup. As previously stated, however, collective action did not occur because of the pandemic. Nonetheless, Earth Day 2020 will always be memorable because we learned first-hand how human activities are directly (and indirectly) responsible for harming the environment. No longer could skeptics deny the fact that humans are a primary cause of pollution. Consider for example, how the Earth turned wilder and cleaner as most humans burrowed inside their homes waiting and hoping for the pandemic to pass. During the short time leading up Earth Day and just after, the air cleaned up globally. The ability to view not-so-distant sights in New Delhi, one of the most polluted cities in the world, became visible for the first time in decades. Nitrogen dioxide pollution in the northeastern United States was down 30 percent compared to normal in April. Air pollution in Rome from mid–March to mid–April was down 49 percent from the previous year (Borenstein 2020a). More stars were visible in the night skies across the world. Animals started to roam freely in areas they dared not venture (because of humans) just a couple of months earlier. For example, coyotes were found meandering along Michigan Avenue in Chicago and near the San Francisco's Golden Gate Bridge. A puma roamed the streets of Santiago, Chile. Goats took over a town in Wales. Hungry monkeys entered people's homes and opened refrigerators looking for food in India (Borenstein 2020a). In just under two months, the Earth was cleansing itself and animals moved about unencumbered. We will discuss this phenomenon in greater detail in Chapter 9, but suffice it to say, Earth Day 2020 was very memorable and provided clear glimpses of the negative impact humanity has on the planet.

Clean Air Act of 1970

The U.S. Congress passed the Clean Air Act of 1970 in light of concerns for clean air but also in response to Earth Day 1970, a key turning point in the American public's consciousness about environmental problems (Rogers 1990). "The juxtaposition of Earth Day and the 1970 amendments was no accident. As a representative body, Congress was responding to the heightened public concern about the environmental pollution that was symbolized by the Earth Day demonstrations" (Rogers 1990). According to its website, the Clean Air Act is the law that defines the EPA's responsibilities for protecting and improving the nation's air quality and the stratospheric ozone layer. The Clean Air Act Amendments of 1990 were the last changes in the law. Legislation passed since then has made several minor changes. The Clean Air Act was incorporated into the United States Code as Title 42, Chapter 85 (EPA 2011a).

The EPA published *The Plain English Guide to the Clean Air Act*; a segment of this guide is titled "Why Should You Be Concerned About Air Pollution." The EPA states that, on average, each of us breathes over 3,000 gallons of air each day. However, what should be obvious to all is the EPA's position that if the air is polluted, it can make us sick. Air pollution can also damage trees, crops, other plants, lakes, and animals. In addition to damaging the natural environment, air pollution also damages buildings, monuments, and statues. It reduces how far we see in national parks and cities and it interferes with aviation (EPA 2008). The *Plain English Guide* explains that poor air quality compromises economic productivity as well. For example, because air pollution causes illnesses, people lose days at work. Air pollution also reduces agricultural crop and commercial yields by billions of dollars each year.

The passage of the Clean Air Act of 1970 resulted in a significant change in the federal government's role in air pollution control. This legislation authorized the development of comprehensive federal and state regulations to limit emissions from both stationary (industrial) sources and mobile sources. Four major regulatory programs affecting stationary sources were initiated:

- The National Ambient Air Quality Standards (NAAQS)
- State Implementation Plans (SIP)
- New Source Performance Standards (NSPS)
- National Emission Standards for Hazardous Air Pollutants (NESHAP)

The adoption of the Clean Air Act coincided with the passage of the National Environmental Policy Act that established the U.S. Environmental Protection Agency (EPA). In essence, the EPA was established to implement the various requirements in the Clean Air Act of 1970 (EPA 2010).

Three amendments to the Clean Air Act were passed in 1977. However, a number of significant amendments were added years later via the "1990 Amendments to the Clean Air Act of 1970." The 1990 legislation

- authorized programs for Acid Deposition Control;
- authorized a program to control 189 toxic pollutants, including those previously regulated by the National Emission Standards for Hazardous Air Pollutants;
- established permit program requirements;
- expanded and modified provisions concerning the attainment of National Ambient Air Quality Standards;
- expanded and modified enforcement authority; and
- established a program to phase out the use of chemicals that deplete the ozone layer [EPA 2010].

Air, our most precious commodity, has humanity looking out for it. And that should make us feel a little better about our future. Then again, it is humanity that is compromising the air we breathe, especially via the burning of fossil fuels. In fact, the running theme behind all the initiatives discussed in this chapter centers on the premise that we are trying to save ourselves from ourselves.

The struggle to protect ourselves from ourselves has become increasingly complicated over the past decade or so because of political partisanship. The Clean Air Act was passed while Richard M. Nixon, a Republican, was in office. Environmental laws that passed Congress so easily in the 1970s are now at the center of partisan dispute with dozens of bills having been introduced to limit environmental protections that anti-environment critics say will lead to job loss and economic harm (Cappiello 2012). Anti-environment lobbyists have attempted to discredit scientists that warn us about the compromised environment and discredit the EPA in the eyes of the public. Their efforts have been paying off as a 2010 Pew Research Center survey showed that 57 percent of those questioned held a favorable view of the EPA, compared to a 1997 poll that showed 69 percent with a positive view of the agency (Cappiello 2012). In a 2019 Pew Research Center survey, it was still 57 percent of Americans that held a favorable view of the EPA, 34 percent held and unfavorable view, 9 percent did not know (Pew Research Center 2019).

The 1972 Clean Water Act and the 1974 Safe Drinking Water Act

The 1972 Clean Water Act was an updated version of the 1948 Federal Water Pollution Control Act and was designed to protect our access to clean (non-polluted water) by establishing regulations against polluters who compromise the quality of our waters, including marshes and swamps, along with rivers, lakes and oceans. Although it is big business that causes the greatest harm to our waters, smaller companies and individuals are held accountable for their actions as well, and if found guilty, fines and incarceration are applicable.

The 1972 Clean Water Act was passed to ensure that our drinking water is safe, to restore and maintain oceans, watersheds, and their aquatic ecosystems, to protect human health, support economic and recreational activities, and provide a healthy habitat for marine life, plants and wildlife. The Clean Water Act gave the EPA the authority to set effluent limits on an industry-wide basis and on a water-quality basis that ensure protection of the receiving water. The act requires that anyone who wants to discharge pollutants to first obtain permit, or else "that discharge will be considered illegal" and subject to fines and prosecution (EPA 2011g).

In a further attempt to assure quality drinking water for consumption, Congress passed the Safe Drinking Water Act (SDWA) in 1974. The SDWA represents another 1970s attempt to provide citizens with an environmental "right." The SDWA is the main federal law that ensures the quality of Americans' drinking water. The EPA sets standards for drinking water quality and oversees the states, localities, and water suppliers who implement those standards (EPA 2011b). The law was amended in 1986 and 1996 and established a number of mandates to protect drinking water and its sources: rivers, lakes, reservoirs, springs, and ground water wells (EPA 2011b). Because of the SDWA, nearly all Americans receive high quality drinking water every day in their homes from public water systems.

Although everyone enjoys clean drinking water, the Clean Water Act is under attack by Republicans who, once again, argue that environmental safeguards are bad for business. In 2005, for example, at the behest of then Vice President Dick Cheney, Congress approved what came to be known as the Halliburton Loophole, a rider to the National Energy Policy Act which exempted hydraulic fracking from regulations prescribed by the Clean Water Act, the Safe Drinking Water Act, and the Clean Air Act. The Halliburton Loophole has allowed fracking companies to essentially do as they please at extraction sites. This includes the controversial drilling process that involves the use of toxic chemicals, followed by a waste-product dumping process that is mostly unregulated. As of 2018 hydraulic fracturing was still exempt from the Safe Drinking Water Act, Clean Water Act, Clean Air Act, Resource Conservation and Recovery Act, and the National Environmental Policy Act. The industry has no limits on how much water they are allowed to use and are still inexplicitly not required to disclose toxic and radioactive chemicals in the drilling fluids. That contaminating water during the fracking process is deemed more important that clean water is yet another indication of the power of the fossil fuel industry.

The Clean Water and Safe Drinking Water Acts were designed to assure our access to quality drinking water. Why would anyone be against this? Alarmingly, Republicans have introduced a number of bills to undermine the authority and spirit of the Clean Water and Safe Drinking Water Acts. In 2011, Representative John Mica (R-FL) introduced a bill titled the Clean Water Cooperative Federalism Act that would erode the federal government's ability to step in when state water quality standards are not strong enough to protect public health. The bill, according to the Natural Resources Defense Council, would make it easier for coal companies to dump their mine waste into rivers and streams. While the House passed the bill, it was doomed for failure as President Obama vowed to veto any such legislation. Under President Trump, attempts to undermine the Clean Water and Safe Drinking Water Acts were still under attack. In 2019 alone the Trump administration's attack on clean water

- wiped out protections for the streams that feed the drinking water for more than 1 in 3 Americans and the wetlands that filter pollution and protect communities from flooding;
- proposed a plan to make it harder for state and tribes to protect local water quality from big energy projects like pipelines and fossil-fuel terminals;

- finalized a rule to do absolutely nothing to safeguard communities from discharges of hazardous chemicals from industrial facilities;
- proposed a scheme to make it easier for companies to pave over streams or fill-in wetlands;
- released a proposal to allow coal-fired power plants to dump more toxic waste into rivers and other sources of drinking water; and
- announced a new polluter loophole that would put our surface waters at risk [Clean Water Action 2010].

Political arguments regarding how to create a federal budget and where money should be allocated is one thing, but how is that clean water is a matter of political divide?

Arbor Day and World Environment Day

While Arbor Day may not be as well celebrated as Earth Day, it is still an important day with great environmental sentiment. The Latin origin of the word "arbor" is tree and from this comes the idea of planting trees, as well as the upkeep and preservation of trees. In the United States, as with most nations, Arbor Day is celebrated on April 24, two days after Earth Day. (Note: Arbor Day is often celebrated on the last Friday in April, though some states have moved the date to coincide with the best tree planting weather.) It is a day to pay homage to nature in general and trees specially. The origins of Arbor Day date back to the early 1870s in Nebraska City when a journalist by the name of Julius Sterling Morton and his wife moved to Nebraska in 1854 (13 years before Nebraska gained statehood) and purchased 160 acres of land in the city. They planted a wide variety of trees and shrubs in what was a primarily flat stretch of desolate plain (History.com 2020). When Morton became the editor of the state's first newspaper, Nebraska City News, he used the print media as a platform to share his knowledge of trees and to stress their ecological importance to the state. His message resonated with his readers as many of them had recognized the lack of forestation in their community. On January 7, 1872, Morton proposed a day that would encourage all Nebraskans to plant trees in their community. Morton came up with the idea of Arbor Day to recognize all trees and shrubs. The very first Arbor Day was held on April 10, 1872, and Morton led the charge in planting approximately 1 million trees (History.com 2020). The tradition quickly spread to other areas of the country. Arbor Day became an official state holiday in Nebraska in 1885 and April 22 was initially chosen because of its ideal weather for planting trees and in recognition of Morton's birthday. On April 15, 1907, President Theodore Roosevelt, a supporter of the conservation movement, issued a proclamation—"Arbor Day Proclamation to the School Children of the United States"—essentially stating that on Arbor Day school children should plant trees as part of school projects (History.com 2020). As part of the "environmental Seventies," Arbor Day officially became recognized nationwide thanks to the efforts of President Richard Nixon (History.com 2020). Arbor Day is celebrated in many countries worldwide, again on different dates to coincide with ideal growing seasons.

In New Zealand, for example, Arbor Day is celebrated on June 5, which is also World Environment Day (WED). The government encourages New Zealanders to plant native trees to restore habitats damaged or destroyed by humans. World Environment Day is designed to draw attention to the foods we eat, the water we drink, the air we breathe, and the climate we live in, all in a manner designed to assist nature before its too late. The theme for World Environment Day 2020 was biodiversity—a call to action to combat the accelerating species loss and degradation of the natural world. According to the World Environment Day (2020) website, "One million plant and animal species risk extinction, largely due to human activities." As a result, WED recommends that people should urge governments to rethink their economic systems that have a negative impact on the environment. The five main drivers of biodiversity loss are land-use change; overexploitation of plants and animals; climate emergency; pollution; and invasive alien species (WED 2020). Among the recommendations of the WED organization are plan a cleanup activity with friends and neighbors; get outside and enjoy nature; write to your elected officials; use less water; walk more; carpool; use LED light bulbs; buy energy-efficient appliances; recycle e-waste; shut off your devices when you are finished using them; and switch to sustainable products. Since the authors wrote the first edition of this book after being inspired by the beauty of New Zealand and the desire to sustain its natural habitat, they particularly appreciate these recommendations.

This concludes our discussion on ways in which we can help the environment thrive; we will revisit the notion of "going green" in Chapter 9.

Popular Culture Section 7

Avatar and Becoming in Tune with the Environment

Avatar was released in December 2009. It earned over $1 billion in just 19 days and was the first film to make more than $2 billion in U.S. box office receipts and more than $3 billion worldwide. To say that *Avatar* constitutes an example of a popular film would be a gross understatement.

For those who did not see *Avatar*, a little background information is necessary. *Avatar*, a science fiction film written and directed by James Cameron, is set in the future (2154) on Pandora, a moon in the Alpha Centauri star system, where humans are mining a precious mineral called unobtanium (meaning, unobtainable precious metal). The mineral is so rare and valuable on earth that the RDA Corporation has established an imperialist mining industry on Pandora, threatening the indigenous Na'vi people (a humanoid species that grow to about 9 or 10 feet tall and are blue in skin color) and the thriving environment. One clan, the Omaticaya ("The People"), is especially vulnerable. The Omaticaya live in a gigantic and massive tree known as "Hometree." The significance of this comes with the realization that the largest deposit of unobtanium is located underneath Hometree. The RDA Corporation wants to destroy the tree and fails to heed the significance, or recognize the interconnection, between the Omaticaya and their home dwelling. In fact, except for the human scientists, who are sympathetic to the Na'vi, RDA does not even care about the local ecosystem or how it is interconnected to all living creatures.

Pandora is a spectacular world with a wide array of animal species (dinosaur-like) such as blue lemurs, a hammerhead type of primitive elephants, flying dragons, illuminating butterflies (the seeds of the "Sacred Tree") and so on (Delaney and Reed 2013). Like the Na'vi people, every animal species is oversized, no doubt because the flourishing environment in which they live allows them to grow big and tall. The physical environment of Pandora also includes floating mountains (the Hallelujah Mountains) with waterfalls, floating rocks, and a Sacred Tree that is interconnected with past and present Na'vi people. Eywa, the great spirit of Na'vi, serves as a deity, or goddess, of all life. The Na'vi believe that all life forms on Pandora are connected through a network of energy that flows through all living things and that all energy is only borrowed, then one day you have to give it back. Near the end of the film, one of the RDA scientists, Dr. Augustine, explains that the forest has an electro-chemical communication capacity via the interconnectedness of all the trees' roots. Augustine claims that the Pandora forest is like the syntax between neurons inside human brains (Delaney and Reed 2013).

The Pandoran ecology is essentially a vast neural network that expands throughout the entire lunar surface in which the Na'vi and other species can connect. The Na'vi have bones that are reinforced with naturally occurring carbon fiber and possess a distinctive tendril feature protruding from the back of their heads surrounded by hair that feeds directly into their brains. It is this organ that allows the Na'vi to literally connect with other organisms around them via the transfer of electrochemical signals such as thoughts and memories to the trees, plants, and other animal species. The Na'vi literally plug their braid into other creatures' nervous systems (Rottenberg 2009). To become a warrior, a Na'vi must tame one of the flying dragons (known as Ikran) by plugging their braid into the creature's nerve system. Furthermore, the Na'vi, who mate for life (through a process known as "tsaheylu") use their braids as a neural bonding system.

While it is true that each species on Pandora must meet its own survival needs, and that includes hunting and killing prey to survive (the food chain of survival), the Na'vi have adopted a philosophy that involves honoring the death of life forms that had to give their lives for the sake of the survival of the people. This philosophy is similar to that of Native Americans and their honoring of the buffalo. The buffalo was respected, but it was necessary to kill buffalo for food, clothing and shelter. They did not kill buffalo for sport. The Na'vi have the same type of relationship with many creatures on Pandora; that is, they do not kill for sport, but instead, for survival.

It is the Na'vi's healthy respect for nature and all that it offers that is the lesson from *Avatar* that humans should embrace here on Earth. Take only what is needed to survive, honor that which was taken, and allow other species to live and flourish. Working in harmony with all the elements of every ecosystem will stimulate a flourishing environment. While it is true that humans do not possess a literal connection with the ecosystem like the Na'vi and their tendril feature, we are indeed connected to the environment. Ideally, humans will come to recognize our figurative connection to the ecosystem before it is too late.

James Cameron has been hard at work on the sequels to *Avatar,* planning no less than five of them. *Avatar 2* is tentatively scheduled to be released in December of 2022, with the others following every other year until December of 2028. The original release date for *Avatar 2* was to be December of 2021, but production in New Zealand had to be shut down due to the coronavirus crisis.

Summary

In this chapter, we examined some of the many efforts designed to help the environment thrive. The primary topic areas include the meaning of going green; protecting and saving natural resources; developing new technology; and environmental rights and legislation.

The "going green" social movement began in earnest during the "environmental Seventies" when the idea of helping the environment thrive really resonated with the public and political policy makers. We discussed a number of specific examples of going green including recycling; developing green roofs; having buildings meet LEED certification requirements; waste reduction; promoting the use of renewable energy sources , such as wind farms and solar energy; activist and community gardening along with fallen fruit programs wherein fresh fruits and vegetables grown on private and public property could be shared freely with community members; the use of porous surfaces; and a reminder of how going green is good for your health.

The discussion on "going green" works in conjunction with the idea of protecting and saving Earth's natural resources. To that end, the topics of sustaining clean air, sustaining the rainforest, sustaining clean water supplies, and sustaining the land were reviewed. Developing new technology was recommended as a good way to help protect and save natural resources. Specific examples include advancements in the development of gas-saving automobiles; the development of low-cost biogas filters; LED light bulbs; innovative ways of recycling wastepaper; replacing Styrofoam and plastic materials with mushrooms; and utilizing hemp products.

While individuals can do many things to help the environment thrive, it is critical to pass legislation that requires Big Business to use greener business strategies. We also suggest that the environment should be considered as a type of entity deserving of rights. To those who think it may be a little far-fetched to think of the environment as deserving of rights, we suggest that, at the very least, humans should have the right to reside in a livable environment. Far too often, the mantra of economics as more important than ecology comes to the forefront. So perhaps, for the economics-first proponents the idea of placing a dollar value on natural resources would help to sway greener decision-making. It was the "environmental Seventies" when we witnessed the greatest legislative movements to protect the environment. This was the decade when both Democrats and Republicans agreed that environment was important enough to protect and to that end a number of significant events occurred including Earth Day; the Clean Air Act of 1970; the 1972 Clean Water Act; the 1974 Safe Drinking Water Act; Arbor Day; and World Environment Day.

Years ago, German sociologist Max Weber warned us that bureaucracy had created a society akin to an "Iron Cage." If we do not change our environmental habits, and soon, we face a world that could be characterized as a "Toxic Cage" of doom, gloom, and despair.

CHAPTER 8

Happiness Is a Thriving Environment

In this chapter, we will examine the meaning of happiness and describe the many different ways of achieving happiness. We will conclude with a discussion on the concept of Gross National Happiness (GNH). Peanuts comic strip creator Charles Schulz, who gave us the lovable character Snoopy, once famously said that "happiness is a warm puppy." While the authors agree that puppies (and other pets) can certainly bring happiness, what we are most concerned about is the happiness that comes about because of a thriving environment.

The Pursuit of Happiness

Bearing in mind that the concept of happiness has different meanings for different people, let's begin our discussion with a couple of definitions of the term. Ott (2010) defines happiness as the degree to which an individual judges the overall quality of his or her life, on the whole, favorably. From this perspective, the more favorably one views their own life, the happier they are. Sonja Lyubomirsky (2008) suggests that happiness can be measured in terms of experiences of joy, contentment, positive well-being "combined with a sense that one's life is good, meaningful, and worthwhile" (p. 32). Similar to the authors of this text, Brockmann and Delhey (2010) use a multi-disciplinary approach when examining happiness. They state, "Happiness has a long philosophical tradition, a biological core, a close match with economics, psychological standing, sociological significance, and political implications" (p. 1). In other words, there are many ways in which happiness needs to be explored.

Most people like to be happy, although it is likely that you have encountered some who seem to be the "happiest" when they are miserable, having given up any hope and desire of attaining happiness, or who seem to find happiness in ruining other people's pursuit of happiness. So, before we examine those who like to pursue happiness, let's take a quick look at those who believe happiness is overrated.

In early April 2012 news came out of Yale University that the pursuit of happiness can sometimes make you less happy, which, paradoxically, actually made quite a few unhappy people, well, happy. Craig Wilson, a writer for the *USA Today*, said as much in a column—"Off with their smiling heads!" (Wilson 2012). Although he denies being a curmudgeon, Wilson has ranted before about his distrust of overly happy people and especially the use of smiley yellow happy face icon that people use in emails (and presumably he hates the many variations of emojis that convey happiness as well) Wilson (2012) describes how he prefers using an ATM instead of going inside the bank; not because of the convenience, mind you, but rather because he doesn't want to go inside a bank and be greeted with smiling, happy bank employees wishing him a great day. Instead, Wilson describes finding happiness from the mechanical, non-human, non-emotional exchange of customer and machine. Wilson also describes being upset with a "perky" woman who worked for Gallup and called to ask him about the service he receives at his bank. One question in particular upset Wilson, "Did I feel 'special' when I left the bank?" Wilson pondered, how could I possibly feel special simply by going to the bank? The authors of this book, although persons who prefer being happy but who fully understand the meaning of unhappiness, have to agree with Wilson on, at least, that last point—why would someone feel special simply because he or she made a banking transaction?

The 2012 Yale study, headed by psychologist June Gruber, focused on four specific questions:

- Is there a wrong degree of happiness?
- Is there a wrong time for happiness?
- Are there wrong ways to pursue happiness?
- Are there wrong types of happiness?

Perhaps not surprisingly, the study answers "yes" to all four. It compares the search for happiness to the desire for food. Although necessary and beneficial, too much food or happiness can cause problems and may lead to bad outcomes. For example, the study concluded that cheerfulness can make people gullible, selfish and less successful, and that's just the beginning of the problems (Gruber 2012). The research also indicated that very high levels of positive feelings predict risk-taking behavior, excess alcohol and drug consumption and binge eating, and may lead us to neglect threats (Gruber 2012).

In a 2012 *Washington Post* article, Marta Zaraska reviewed a number of studies that question the full-out pursuit of happiness. For example, Zaraska described renowned happiness researcher and psychologist Edward Diener and associates' research on more than 16,000 people from around the world that indicated people who reported the highest life satisfaction (judging life satisfaction at 5 on a 5-point scale) early in life reported lower income later in life than those who felt less happy earlier in life. The happiest young people were more likely to drop out of school earlier than the unhappier ones. Diener's research also found that the happiest college students (in 1976) earned, on average, almost $3,500 a year less in their 30s than their not-so happy college peers (Zaraska 2012).

A 1994 study in the *Journal of Personality and Social Psychology* by psychologist Galen Bodenhausen and his colleagues at Northwestern University on 94 undergraduates who were asked to participate in a simulated "students court" found that "happy" mood participants were more prone to stereotypical thinking than control group students (pre-tests were conducted to determine "happy" mood students and neutral students) by being more likely to find a student named "Juan Garcia" guilty of beating up a roommate than one identified as "John Garner" (Zaraska 2012).

That happy people are more prone to stereotypic thinking was supported in research by Joe Forgas, a psychology professor at the University of New South Wales in Australia. In an experiment published in the December 2011 issue of the *European Journal of Social Psychology*, Forgas asked students to read a philosophical essay by a "Robin Taylor." Pre-tests were conducted in order to attain a control group and a "happy" group.

Participants were given a copy of the essay along with a photograph of the presumed author. In some cases, the essay was accompanied by a picture of a middle-aged, bearded man and in other cases the essay was accompanied by a picture of a young woman in a T-shirt. Even though the essays were identical, the "happy" students judged the man's work more competent than the woman's while the control group declared both essays of equal worth (Zaraska 2012).

According to psychologist Iris Mauss, the more someone pursues happiness, the more likely he or she is going to end up feeling disappointed (Zaraska 2012). In a study of 43 women published in August 2011 in the journal *Emotion*, Mauss, after conducting a pre-test, found that happier participants felt lonelier after watching Clint Eastwood's and Meryl Streep's doomed love in the film *The Bridges of Madison County* than did a control group. In an attempt to confirm this research, Mauss conducted a follow-up study utilizing diary entries. Mauss found that those who expressed "being happy was extremely important" to them felt lonelier after experiencing a stressful event than those who didn't make such a big deal of wanting to be cheerful (Zaraska 2012).

Perhaps reflecting the conclusion of all these psychologists, Gruber indicates that we need to accept realistic levels of happiness and learn to accept—as long as one is not clinically depressed—that being happy all the time, or being extremely happy, is not necessarily the goal we should strive to attain (Zaraska 2012).

The above discussion suggesting that the pursuit of happiness is over-rated might seem contradictory to the point of this chapter. However, we wanted to acknowledge that there is some risk involved when pursuing happiness but bear in mind all the research cited here was from a psychological perspective. Our discussion of happiness is far more expansive than the limited concerns of psychology. Furthermore, psychologists themselves will attest that being happy, at least to some degree, is important. While suggesting that we don't have to strive for high levels of happiness, Gruber counters that we should embrace three positive feelings (such as joy, compassion, gratitude or hope) for every one negative (e.g., disgust, embarrassment, fear, guilt, or sadness) (Zaraska 2012). One might argue that positive feelings are a characteristic of happiness. Even our curmudgeon *USA Today* writer Craig Wilson admits that he has "been known to be happy on occasion" and states that he gets miffed when people don't at least say hello back when he greets them with a "hello." (Once again, the authors agree with Wilson on that issue.) Wilson describes walking his dog and notes that set routines such as dog-walking implies encounters with other regular morning folks going about their ritualistic behavior and notes his dismay that he receives nary a return "hello."

The study of happiness has become so important to researchers that there is now a journal devoted entirely to this field: *The Journal of Happiness Studies*. Founded in 2000, it is a peer-reviewed "interdisciplinary forum on subjective well-being" which addresses such topics as life-satisfaction and enjoyment (including mood levels), job satisfaction, and the perceived meaning of life. The journal "provides a forum for two main traditions in happiness research, speculative reflection on the good life and empirical investigation of subjective well-being" (*Journal of Happiness Studies* 2013). Contributors come from such fields as anthropology, economics, philosophy, psychology, and sociology, as well as the health-related sciences.

The pursuit of happiness is therefore important, at least for most of us. We like to be happy; we feel better when we are happy. Unlike the research previously cited, there have been numerous studies conducted on the positive benefits of happiness. For example,

Proctor, Linley and Maltby (2010), who conducted research on 410 adolescents (ages 16–18) who were divided into three life satisfaction groups—very high (top 10 percent), average (middle 25 percent) and very low (lowest 10 percent)—found that "very happy" youths had significantly higher mean scores about their attitudes toward education, academic aspirations, academic achievement, life satisfaction, altruism, self-esteem, and parental relations. Relevant to our concerns, the study found the "very happy" adolescents had stronger environmental views. The very happy respondents reported that it was very important to protect the environment; they placed a greater importance on recycling and they even reported to recycle more often than those who reported to be very unhappy (Procter et al. 2012).

In the years in between the first and second editions of *Beyond Sustainability*, the authors themselves published quite extensively (two books and many articles) in the field of friendship and happiness. In our co-authored book *Friendship and Happiness* (2017, McFarland), we collected data on college students and their views of friendship and happiness as a means of examining the connection between the two topics. When it comes to the pursuit of happiness, we were inspired by the work of philosopher Bertrand Russell (1872–1970) and his 1930 publication of *The Conquest of Happiness*, a book that predates the contemporary fascination with self-help publications by decades. Russell's use of the word "conquest" in the book's title reinforces his primary contention that happiness, except in rare cases, is *not* something that simply presents itself to people but rather is something that must be achieved (conquered). Following this line of thinking, we proposed that in order to be happy one must first conquer unhappiness (Delaney and Madigan 2017). While this might initially seem like something that "Captain Obvious" would say, it's really not the case as most people try to find things to make them happy but ignore all the things that make them unhappy. Thus, all the types of happiness discussed in *Friendship and Happiness* and in *Beyond Sustainability* are essentially unachievable until we overcome the unhappiness in our lives. As the book title might imply, Delaney and Madigan (2017) found that "we are all better off if we have a number of close friends and if we can find activities that bring us happiness. Having friends is one of the most fundamental aspects of finding and achieving happiness" (p. 256). In our *A Global Perspective on Friendship & Happiness* (2018), Delaney and Madigan serve as co-editors and contributing authors and once again, demonstrate the link between the topics of friendship and happiness. The chapters found in this publication are the result of the "Happiness & Friendship" conference held June 12–14, 2017 at Mount Melleray Abbey, Waterford, Ireland. Contributing authors are from multiple and quite diverse nations around the world and they share their vision on the meaning of these two topics.

A Constitutional Right to *Pursue* Happiness

Is the pursuit of happiness ingrained in all of us? Maybe not. According to Malloch and Massey (2006), "The pursuit of happiness is a uniquely American dynamic and outlook that continues to shape our destiny and now affects people around the world" (p. 2). What we do know is that happiness is very much a philosophical and sociological fascination. The topic of happiness certainly holds socio-political interest. The "founding fathers" of the United States found the idea of happiness so important that they included this opening sentence of the second paragraph of the Declaration of Independence:

"We hold these truths to be self-evident, that all men are created equal, that they are endowed by their Creator with certain unalienable Rights, that among these are Life, Liberty, and the pursuit of Happiness" (The U.S. National Archives and Records Administration 2012).

Sociologists, philosophers, political scientists, historians, and lawyers, among others, have dissected the above sentence from the Declaration of Independence, as they have examined all aspects of the Declaration and the Constitution and, as this is not the forum to rehash past observations (such as the idea that "all men" literally meant men only and white men at that, and the presumption of "a Creator" has the same meaning to all people), we can skip ahead to the subject at hand, the right to *pursue* happiness. It is much more than a matter of semantics that the Declaration of Independence gives us the right to pursue happiness and not the right to *be* happy. Benjamin Franklin receives credit for having said, "The Constitution only gives people the right to pursue happiness. You have to catch it yourself" (Think Exist 2012). Alexander Marriott, in his website "Alexander Marriott's Wit and Wisdom," points out that this is obviously spurious, since the U.S. Constitution never uses the phrase "the pursuit of happiness" and there is no attribution of the quote documented in any of Franklin's papers (Marriott 2011). But it is not surprising that the quote has become so popular (no doubt due to Franklin's reputation for being a wise man himself), since it nicely captures the fact that happiness is both subjective and elusive.

It is interesting to ponder why the founders of the United States felt that it was important enough to mention the pursuit of happiness in the Declaration of Independence from England in the second paragraph, which followed a first paragraph that consisted of one long sentence stating the intentions of the colonists to dissolve political associations with the existing power structure. At the most basic level of analysis, it is clear that the framers of the Declaration of Independence believed the colonists were not reaching their full human potential under the rule of England because, in short, they were unhappy with the arrangement. Declaring independence from an authoritarian implies a level of unhappiness with the status quo. By using the words "pursuit of happiness" the architects of the Declaration were leaving the door open for people to figure out what makes them happiest whether it involves material or nonmaterial, spiritual or otherwise, or whatever else it might take someone to be happy so long as this pursuit for happiness was free from tyrannical rule.

The right to pursue happiness in the Declaration of Independence also reflects the notion that the colonists had not achieved happiness under British rule because they were unhappy—the point made by Russell. Furthermore, it is interesting to note that while the U.S. Constitution does not guarantee its citizens the right to live in a healthy environment, but it does provide citizens the right to pursue happiness. As we shall discuss later in this chapter, the authors promote the notion of "environmental happiness" and how this should be a source of happiness for everyone—to live in a thriving environment. Thus, we are left with this question: Would it be true that if we are not happy because of the compromised environment, has our Constitutional right to pursue happiness been violated by politicians and social policymakers?

In this chapter, we want to explore the idea of happiness in its many forms, concluding with a discussion on happiness via living in a thriving environment. We begin with a look at what Aristotle had to say about happiness, and how this relates to the concept of thrivability.

Aristotle and Happiness

What is "happiness"? For many religious believers, it is a state of perfection that can only be achieved in the afterlife, when they will receive eternal reward for their good deeds on Earth. More down-to-earth philosophers, like the English utilitarian Jeremy Bentham (1748–1832), felt that happiness is indistinguishable from physical pleasure: acts that we enjoy should be maximized, while those that cause harm should be minimized. He even developed a "hedonic calculus" to rate activities on a scale from greater to lesser pleasure. Yet the person he most influenced, John Stuart Mill (1806–1873), recognized the inadequacy of a system that focused primarily on physical gratifications, which are fleeting and quite often detrimental in the long run to one's well-being. In Mill's famous words, "It is better to be Socrates unsatisfied than a fool satisfied." While ignorance may at times be blissful, it is nothing to be proud of. The intellectual pleasures provide longer-lasting benefits. For Mill, the highest good was to pursue a life of learning, and to apply what was learned to the betterment of the human condition. As mentioned earlier in the book, the contemporary utilitarian philosopher Peter Singer has expanded this notion to include not only the human condition but the condition of all sentient creatures.

Mill's focus upon a life of learning harkens back to what is still perhaps the most relevant discussion of "happiness": Aristotle's *Nichomachean Ethics*. In this work, written as a manual for his son Nicomachus, the Ancient Greek philosopher discusses the concept of *eudaimonia*. Usually translated from the Greek as "happiness," a better translation would be "self-fulfillment through personal excellence" or "human flourishing"—i.e., thrivability. For Aristotle, the good life consisted of developing one's natural abilities through the use of reason. A virtuous life is one where proper habits are formed that allow one to reach one's full potential. The word "eudaimonia" itself comes from combining two Greek words—"eu" meaning "good" and "daimonia" meaning "soul" or "spirit." Literally, then, a happy person is one who has achieved a state of fulfillment and thus demonstrates having a "good soul."

Yet Aristotle felt that only a small number of people—excluding poor men, those without families, those who were unattractive, and all women—had the intellectual capacities and the strength of will to achieve *eudaimonia*. However, this elitist concept of happiness is not one that most people today would rightly feel comfortable espousing.

Another problem with Aristotle's description is the assumption that the pursuit of knowledge—including self-knowledge—will necessarily be beneficial. For the truth can often be painful. The desire to retreat from reality, to find refuge in comforting myths or consoling hopes, is also a basic part of human nature. This can also help explain why so many people today refuse to admit that there is an environmental crisis occurring throughout the globe.

Comfortable myths often help shield people from painful realities, such as human mortality, the lack of ultimate justice, and the ravages of an indifferent natural world. For those who believe that it is not this world but rather the afterlife which is most important, concerns about sustainability may not predominate when it comes to living a happy life. It is not surprising that several recent studies report that religious believers are happier than nonbelievers. According to happiness researchers David G. Myers and Ed Diener: "In one Gallup Poll, highly spiritual people were twice as likely as those lowest in spiritual commitment to declare themselves very happy. Other surveys find that happiness and life satisfaction rise with strength or religious affiliation and frequency of worship attendance. One statistical digest of research among the elderly found that one of the best predictors of life satisfaction is religiousness" (1997: 7).

One needs to explore in just what sense religious people claim to be happy. Clearly, they experience a sense of social support, as well as a sense of purpose and hope for the future. Within their various religious traditions, there is a commitment to something greater than themselves. But is *religion* per se necessary for such happiness, especially if one sees it to be a support system based on lies, misinformation, and indoctrination? Can one advocate the need for religion if one is not able to find its doctrines grounded in verifiable facts? In the earlier-mentioned popular book from 1930, *The Conquest of Happiness* (still one of the best manuals on how to get the most out of life), the philosopher Bertrand Russell discusses this dilemma: "The men I have known who believed that the English were the lost ten tribes were almost invariably happy, while as for those who believed that the English were only the tribes of Ephriam and Manasseh, their bliss knew no bounds. I am not suggesting that the reader should adopt this creed, since I cannot advocate any happiness based upon what seem to me to be false beliefs" (Russell 1930: 42).

Many of the world's religions are now taking sustainability and the environmental crisis to heart. Comforting beliefs are changing in light of sobering realities. Where do intellectual honesty and personal excellence enter into the composition when it comes to being happy? Should contentment alone be the goal of human beings? Or is happiness more connected with trying to overcome existing challenges and excelling at addressing problems head-on rather than ignoring them or hoping that they will go away on their own accord?

It is here that the work of University of Chicago psychologist Mihaly Csikszentmihalyi (1934–) proves helpful in understanding the contemporary meaning of "happiness." For several decades now, he has scientifically studied individuals who claim to be living rewarding existences—modern-day exponents of Aristotelian self-fulfillment. He refers to such an experience as "flow," since most people studied describe it in terms of being so involved in an activity that they lose track of everything else, even time. Yet, unlike Aristotle's conception of *eudaimonia*, flow is something that is experienced by men and women from all walks of life.

Csikszentmihalyi's concept of flow focuses not so much on happiness as a feeling, but rather the process of achievement. It is the matching of internal skills with external possibilities. "What I discovered," he writes, "was that happiness is not something that happens. It is not the result of good fortune or random chance. It is not something that money can buy or power command. It does not depend on outside events, but, rather, on how we interpret them. Happiness, in fact, is a condition that must be prepared for, cultivated, and defended privately by each person. People who learn to control inner experience will be able to determine the quality of their lives, which is as close as any of us can come to being happy" (Csikszentmihalyi 1997: 122). Csikszentmihalyi is not merely a theoretician. He is engaged in projects to help people better understand, and utilize, the concept of flow. For instance, he has joined forces with Martin E.P. Seligman (1942–), the former president of the American Psychological Association, to establish a research network to refocus the attention of the psychological community toward happiness, rather than its usual emphasis on depression, alienation, and aberrant behavior.

Seligman has written his own book, titled *Flourish: A Visionary New Understanding of Happiness and Well-being*, in which he writes: "You go into flow when your highest strengths are deployed to meet the highest challenges that come your way" (Seligman 2011: 24). This is very much in line with the approach the authors of this book call "thrivability"—to not only endure and sustain one's self but to bring out one's talents in the highest possible way.

Yet, as Csikszentmihalyi and Seligman both stress, flow is not synonymous with contentment. An element of dissatisfaction is a key part of it—one must keep reaching for goals not yet attained. Increasing complexity is a necessary component. In order to achieve this, though, people tend to need structures in which to operate, and like-minded individuals who encourage them to reach their "personal best." (It is not coincidental that Aristotle devotes a good part of the *Nichomachean Ethics* to describing the important role that friends provide in motivating one to strive for the good life.) Religion can often provide such order, especially with its rituals, as well as a circle of co-believers. But this can occur at the expense of self-development. When institutions encourage one to accept dogmatically "revealed" doctrines and time-honored practices, the chance to be creative is stifled.

The exciting challenge facing those who address the current environmental crisis is to discover a rewarding way of life without reliance upon comforting myths. We should focus on the happiness that comes from facing the truth and living with it. Intellectual integrity is an important aspect of the humanistic good life.

In his seminal essay "The Myth of Sisyphus," the French existentialist writer Albert Camus (1913–1960) discusses the Ancient Greek legend of the king who—in punishment for defying the gods—was sentenced to an eternity of pushing a large rock up a steep hill in Hades, only to have it fall back every time he reaches the top. It is hard to think of a more pointless task. Yet Camus ends the essay with the words "One must imagine Sisyphus happy."

In order to get a better idea of what Camus is driving at, we should go back a step and ask the more basic question—what is the reason for Sisyphus being condemned to such a punishment in the first place?

There are many different versions of the myth, a few of which Camus relates in his essay. But basically, Sisyphus was the King of Corinth, and considered to be one of the cleverest and wisest of men. In some versions he was also considered to be a thief and a scoundrel (no necessary contradiction, in Camus's opinion). Most of all, he loves life, and wants to do everything he can to prevents its end. Through trickery, he manages to handcuff the God of the Dead, Hades himself, and thereby prevent himself from being dragged down into the Underworld. Death takes an unexpected holiday, and while in captivity cannot unleash his powers. All things remain alive. So, as long as he has Death in his power, Sisyphus cannot die. Ultimately, he loses out, when Ares, the God of War, becomes upset that battles are being fought without any victims. While he cannot kill Sisyphus, Ares threatens to strangle him for all eternity unless he releases Hades. Begrudgingly, the Corinthian King relents. Not surprisingly, Hades is none too pleased over his humiliating captivity and sentences Sisyphus to his eternal rock and roll punishment. All things considered, though, pushing a rock eternally is probably better than being choked forever, in the grand scheme of things.

Let us then return to Camus. The rock, in one sense, represents the countless futile tasks that mark our lives. But in another sense, the rock represents that life itself. Every second of every day in which we live, we can say with certainty "I'm not dead yet." Death may have the last laugh, but while we're alive we may feel his sting but not his scythe. And we can choose to smile back at him, in contempt. Existence itself has no meaning, but we can choose to give meaning to *our* existence. That is the power of humanity, and perhaps the key message of the philosophy known as existentialism. Accept the futility and the absurdity of existence and push on. As long as we are alive, we are holding Death at bay. The meaning of life therefore is, in part, knowing that one is alive, with an awareness that all such life, including one's own, is fleeting. The smile on Sisyphus's face is due to his recollection that he once held Death in chains and, while he's pushing that rock, Death has still not conquered him. Our lives are the rocks that we push. To cease to push them is to cease to live. It is, Camus points out, our ability to think that gives life to this story. The King of Corinth on his lonely mountain is the absurd hero par excellence. He defied the gods right up until the end. He earned his rock, and he pushes it proudly. The King of Corinth on his lonely mountain is the absurd hero par excellence. Sisyphus rocks on!

The excellent life, then, is one of striving to be the best person one is capable of becoming. In Mill's example, Socrates is not dissatisfied because of his inadequacies, but rather because of his awareness that he has still so much to learn and experience in the world. His dissatisfaction spurs him on; it does not cause him despair. And the best way to do this is to use your talents to improve the world around you as best you can—not by hiding from reality but by embracing it. When it comes to the current challenges connected to environmental sustainability, there is reason to believe that the pursuit of happiness is a central component and motivating factor.

The Many Types of Happiness

Aristotle provides us with a starting point on so many topics, including happiness. Happiness researchers David Myers and Ed Diener define happiness as primarily a subjective phenomenon "for which the final judge is whoever lives inside a person's skin" (Myers and Diener 1995: 6). Aristotle, on the other hand, stressed that while there is certainly a subjective component to happiness, the external realities of the outside world are also a necessary component. For him, happiness involves a combination of developing one's skills and matching them to the needs of one's society.

As the previous discussion reveals, and as you undoubtedly already knew, happiness means different things to different people. Some people find many separate things that make them happy while other people need many things to be happy. Finding a job might make you happy, finding a better job even happier. Getting into college is reason to celebrate, earning good grades and graduating is even better. Enjoying good health, family and friends can make many people happy. Assuring that your children have good health brings happiness to parents. Traveling, seeing new sites and meeting new people makes some of us happy and if you don't enjoy exotic travel to places like Antarctica, how about a weekend road trip with friends? Going away for a couple of days, or going on vacation, or going away to school, coming back home and seeing your favorite pets can bring happiness. Having close friends and loved ones brings people happiness. Enjoying the freedoms that come with living in a democracy brings happiness to many Americans and others who enjoy living in societies that try to assure individual freedoms.

Happiness for many people means the absence of stress and negativity. One of our students, a self-admitted "neat-freak," stated that she cannot possibly be happy unless her surroundings are clean and neat or otherwise she is stressed out. "I cannot function or concentrate on anything if there is a mess," the student stated. Some people find happiness when they are not doing anything at all, that is, when they are completely relaxed, an attitude that can be attained by some simply staying at home and "chilling." Conversely, others feel as though they must have a purpose in life and must be contributing to the well-being of society in order to be happy.

And while any one of the activities previously mentioned may make some people happy, some other people need the combination of many elements to be happy. For example, let's say a student receives an A on a term paper, but she is not happy because she is worried that her boyfriend may break up with her; thus, she cannot be happy with the A paper because other concerns are negatively weighing on her. Some college students need many things at the same time to happen in order to be happy; they want good grades, good health, and healthy relationships with significant others and family members. Regardless of whether a person needs a lot of things to be happy or just one significant thing, the pursuit of happiness seems important to most of us.

Both of the authors have taught the topic of happiness, with a focus on environmental happiness, in their respective environmental courses. We had a number of categories of happiness in mind that we discussed with students and asked for their views on. From these discussions, the material previously described in this chapter, and our growing research on the topic, we propose a number of different types of happiness. We do not proclaim this to be an exhaustive list of categories of happiness and we welcome feedback from all readers on their views of happiness. As a result of our research in the area of happiness we have doubled (from 10 to 20) the number of categories of happiness since the first edition of this book. The list below is in no particular order of what we may or may not consider the most important forms of happiness. However, we do conclude with a specific category of emphasis—environmental happiness—because of its relevancy to the topic of environmental thrivability.

Religious and Spiritual Happiness

Religion is a system of beliefs and rituals than binds people together in a social group while attempting to answer the dilemmas and questions of human existence by making the world more meaningful to adherents. It is among the oldest of all social institutions of human society (Delaney 2012). It was created as a result of the human need to explain and understand life's many mysteries. Primitive humans, who were smart enough to outwit animals of lesser intelligence, were still not enlightened enough to explain, by today's standards, relatively simple natural occurrences such as lunar and solar eclipses. Lacking this intelligence, humans began to create outer-worldly explanations like angry gods, and they found comfort in such explanations, even though they were devoid of logic and rationality. Over time, many religious adherents switched from polytheism to monotheism—the belief in one true God. Even with the rise of science, rationality, and the Enlightenment era, large numbers of people still found comfort and happiness in religion, as it provides believers with a sense of hope and meaning and a community of like-minded people who reinforce collective beliefs. Religious people find happiness in many ritualistic behaviors and ceremonies, such as baptisms, bar mitzvah, the Shahada, the Sangeet, Sangha Day, First Holy Communion, and so on.

Religious happiness generally refers to a state of perfection that can only be achieved in the afterlife, when people receive eternal reward for their good deeds on

Earth. Spiritual happiness can be achieved while living a good life while humanly alive. "Spiritual happiness is characterized by such traits as serenity, kindness, humility, and joy for self and others" (Delaney and Madigan 2017: 239). Religious and spiritual people find happiness when they feel they are living a righteous life. For people who believe in an afterlife, living a "good" life can equate to eternal happiness, a very pleasant thought indeed. People who do not believe in an afterlife can find spiritual happiness on earth by leading a good life simply because it is the right thing to do. Furthermore, people who do not believe in an afterlife understandably want to be happy in their lives on earth.

Physical and Touching Happiness

As stated earlier, Jeremy Bentham put forth the notion that happiness is derived from physical pleasure, what he referred to as "hedonic calculus." Although attaining happiness through physical intimacy with a loved one is certainly treasured by most people, physical happiness is not limited to sexual satisfaction and intimacy with a loving partner; it can be attained in a number of other ways. For example, many people find happiness through a strenuous physical workout. An arduous exercise routine, such as a long and fast run, releases endorphins within the human body. Endorphins provide individuals with an adrenaline rush. Endorphins are morphine-like substances in the brain that block pain, heighten pleasure, and have been associated with some addictions (Delaney 2014a). Endorphins are neurotransmitters produced in the brain which function to transmit electrical signals within the nervous system that may lead to feelings of euphoria, modulate appetite, release sex hormones, and enhance our immune responses. Endorphin release varies among individuals and this explains why two people experiencing the same stimuli (e.g., physical exercise, or physical threat) may release different levels of endorphins and consequently lead to different reactions among participants. The "runner's high" that some people experience and others do not is explained, in part, by the varying release of endorphins among runners (Delaney 2014a).

One of the authors' students commented, "Running is the cure to everything for me. Whether I'm in a bad mood, stressed about school, or just bored, running can fix anything. Sometimes I have the best runs when I'm really angry because I feel like I can fun forever away from my problems. The feeling you get at the end of a long hard run cannot be achieved from doing anything else. It's as if you've released all the bad in your life and can have a fresh new start to the day. As long as I have my sneakers and my iPod, I feel like I can run forever." This student clearly understands the meaning of the "runner's high."

Physical touching can also bring about happiness as a result of the release of oxytocin, a bonding hormone which strengthen the relationship between people and produces such feelings as trust, contentment and sexual pleasure, as well as reduces stress, anxiety, and fear. Hugging someone (or cuddling) stimulates the production of oxytocin which makes you feel closer to the person and relaxes you. Lying down next to someone slows the heart rate, reduces cortisol levels (the stress hormone) and brings feelings of joy and happiness (Delaney and Madigan 2017; DeAngelis 2008).

The COVID-19 pandemic provided us with a stark reminder of just how important physical closeness and the touch of another person can be ranging from the simple desire of shaking someone's hand upon greeting to grandparents hugging a grandchild, sports fans high-fiving one another, and everything else that was put on hold. Will shaking hands be a thing of past now? If so, this form of happiness is in trouble.

Intellectual Happiness and Happiness of the Mind

A number of people find happiness through intellectual pursuits. They too can experience an endorphin rush via the sense of accomplishment. Authors sometimes get a "writer's high" when they accomplish a great deal of work in any given time period. Ask any prolific or first-time author to explain how they feel when they finish writing a book or see the book in publication and they may explain to you how this writer's high feels. (The authors of this text will tell you that the writer's high is very real.) Teachers and professors may experience the "teacher's high" when they realize students have grasped the material at hand. Once again, speaking from personal experience, the authors of this text can tell you one of the most rewarding aspects of being professors is hearing from students years after they graduate how much you inspired them, or how much they learned in your classes, or how they have taken the knowledge they attained in your class and have used it in their own personal, academic, or professional lives. The writer's and teacher's high certainly represent examples of intellectual happiness.

Attaining new knowledge is an ultimate form of intellectual happiness. We could call this the "researcher's high." Scientists, especially natural scientists, may experience the researcher's high when a project they have worked on for years finally pays dividends, such as when the work is published and acknowledged by the academic community or sets a new standard in the field.

John Stuart Mill introduced us to another important aspect of intellectual happiness: the sense of achievement brought about as the result of intellectual happiness lasts longer than physical pleasures. Thus, while the runner's high represents a great blast of endorphin release within the body, that sense of accomplishment will dissipate rather quickly; the intellectual achievement lasts, potentially, indefinitely. From this perspective, intellectual pleasures such as reading a good book, seeing an intriguing, thought-provoking movie, or having a deep conversation with someone can stay with you longer than a physical touch. This is what Mill meant by his statement: Better a Socrates unsatisfied than a fool satisfied, since Socrates is still engaged in intellectual pursuits. He may not have all the knowledge he desires, but his search for it is itself pleasurable and gives meaning to his life and happiness of the mind. This should make sense after all; happiness is a state of mind. Edmund Spenser so aptly stated, "It is the mind that maketh good of ill, that maketh wretch or happy, rich or poor."

Going with the Flow Happiness

The expression "going with the flow" used to really bother one of the authors of this book. It occurred to him that if someone is going with the flow, they are not the "captain" of their own ship (their life course) because they are merely going along in life with what everyone else does. If you find yourself going with the flow, you never know, that current may lead you to a cliff or waterfalls; best to not get caught up in the flow, a critic might say. However, psychologist Mihaly Csikszentmihalyi's idea of going with the flow gives the concept a more sensible perspective.

As we saw earlier, Csikszentmihalyi (1997, 2008) argues that a link exists between happiness and those heavily involved in an activity to the point that they lose track of everything else, even the time of day. This going with the flow activity can contribute to the previously mentioned "writer's high" and "runner's high" and so much more. A runner's favorite run is often the one where the runner does not need to be anywhere in particular anytime soon, they can just run for as long as they want and take the path of their choice. Anyone who works can tell you that the workday seems to go by so much faster when they are busy. This is because when we are busy, we get caught up with the task at hand; we become consumed by accomplishing all the deeds, chores and activities associated with the workday. Before we know it, the workday is over, and we are happy.

The happiness experienced by some because of a "going with the flow" attitude is certainly not limited to the work and workout environments. Spending time with loved ones and engaging in pleasurable activities to the point where you are totally consumed by what you are doing can be very rewarding. People in love go with the flow often and they certainly seem happy. But, as Csikszentmihalyi indicates (1997, 2006), the "going with the flow" concept does not rely on one being happy at the conclusion of some activity; rather, the process of achievement presents itself as a feeling of happiness to the participant. Put another way, the thrill is in the journey, not in reaching the destination.

Cultural Happiness

Nearly all people take pride in their culture and the society of their origin. This is understandable when we consider the definition of culture—the shared knowledge, values, norms, and behavioral patterns of a given society that are passed on from one generation to the next, forming a way of life for its members (Delaney 2012). Celebrating one's culture is akin to celebrating one's place in society. This helps to explain why so many people take pride in their country and love to participate in its national holidays. In the United States, the 4th of July is a favorite day of the year for millions because it represents a celebration of American culture. That fireworks, cookouts, and summer weather are also associated with the Independence Day; we can easily see why this day brings happiness to so many people.

Some people like to feel as though their lives have meaning and this meaning can be achieved by contributing, in a positive manner, to the general well-being of society, in contrast to being a drain on society. Putting in a good honest day's work, helping out others through volunteerism, cleaning local neighborhoods of trash and so on are ways that people can feel like they are contributing to society. Once they feel as though they have made a positive contribution to society, such people have achieved cultural happiness.

BIRG, CORF and BIAF Happiness

The authors have conducted a great deal of research in many areas of sports, including the concepts of BIRG, CORF, and BIAF (Delaney and Madigan 2021; Delaney and Madigan 2009; Delaney 2001). BIRG is short for "basking in reflected glory," a phenomenon that involves people connecting themselves with successful others, often using the expression "we" (as in "we won the game!") as a way to enhance their own positive sense of self. The BIRGing behavior leads to happiness. CORFing refers to "cutting off reflective failure," a technique utilized by folks who want to distance themselves from perceived losers often using the term "they" (as in "they lost the game"). Cutting ourselves off from perceived failures serves a defense mechanism. People may CORF so that they don't feel unhappy and that, in turn, may trigger happiness. BIAF is a new concept that the authors have developed and refers to "basking in another's failure," a blasting technique that is similar to the concepts of *schadenfreude* (taking joy or pleasure in the misfortune of others) and *epicaricacy* (rejoicing at or deriving pleasure from the misfortunes of others). This technique is often a diversionary tactic to draw someone's attention away from your own shortcomings, perceived or real, by pointing to another target as a scapegoat. There many be elements of amusement (finding entertainment value, humor, or satisfaction in the misfortune of others) and sadism (taking a certain sadistic pleasure in the misfortunes of others) involved with BIAF as well.

While this is perhaps most noticeable when it comes to sports, the concepts also apply to many other fields, particularly in popular culture and politics. We often identify with actors we like, politicians we admire, and specific professions we are interested in even if we ourselves are not adept in these. For instance, we may take a great interest in space exploration and closely follow the careers of astronauts, as was done by millions of people in the late 1960s and early 1970s. Only a handful of human beings ever actually walked on the moon, but countless humans identified with their achievements and took inspiration from their heroic actions—they experienced BIRG.

BIRGing is a phenomenon which gives tremendous pleasure to almost all people on the planet. In his writings on art, Aristotle referred to this as "living vicariously"—experiencing in a secondary way what others have achieved. As the old ABC TV show *Wide World of Sports* so memorably put it, we have all experienced the thrill of victory and the agony of defeat in vicarious ways.

Ignorance Is Bliss Happiness

In a negative way, some people find comfort in the ostrich approach to life, that is, if I bury my head in sand all the problems that confront and surround me are gone. Thus, being oblivious, or stupid, is, arguably, a path toward happiness. What you don't know can't hurt you. Or can it? Surely if you have a medical condition that is leading to your own demise you would be better off knowing about it and seeking treatments. If your car has a mechanical problem that might cause you to lose control and crash, you would want to know about it before you get behind the wheel and take off. Not being aware of problems certainly doesn't mean the problems aren't real, and quite often by the time one does acknowledge them it is too late to do anything about it.

It is our contention that such an issue is at hand in regard to the environment. As detailed in earlier chapters, there are a myriad of issues that all people should be made aware of, in the hopes that their collective wisdom can be channeled in order to overcome the problems and allow for human happiness to flourish. Pretending that global warming does not exist is certainly a sign of ignorance and the corresponding inaction towards this reality exacerbates the problem. The pragmatic approach takes strong objection to the "ignorance is bliss" motto and allies itself with John Stuart Mill's famous slogan: Better Socrates unsatisfied than a fool satisfied.

Economic Happiness

Everyone has heard the cliché "money can't buy you happiness." Many of us, even if we believe in the true meaning of this cliché, chuckle because we know darn well that money is important. There are many things we would like to do in life, like take trips to exotic tropical islands, or go on snowboarding trips to the mountains, but we don't for one simple reason—we don't have the money to do so.

Money is an important aspect of economics and the economy. The word "economy" is derived from the Greek *oikonomos*, meaning "one who manages a household" (*American Heritage Dictionary* 2006). Today, the term refers to the social system that coordinates a society's production, distribution, and consumption of goods and services (Delaney 2012). Within economic systems, money is attained, legally, through income and wealth. Income refers to the amount of money a person or household makes in given periods of time, such as one's reported income for a calendar year, while wealth refers to the total value of everything one (or a household) owns minus the debts.

Christopher Rugaber, in an article titled "Happiness Gauges Economic Progress," points out that research shows there is no necessary correlation between wealth and happiness, and measuring economic happiness is an inexact science at best. He adds, "Most efforts have involved surveys in which people are asked about whether they are happy and what contributed to their happiness. Those surveys have found some consistent answers: physical and mental health, the strength of family and community ties, a sense of control over one's life, and opportunities for leisure activity" (Rugaber 12: A8).

Economic success does not guarantee happiness. But economic success affords us the opportunity to take care of the essentials in life—food, clothing and shelter—and it provides us with the means to pursue our interests that lead to happiness. So while there are people who enjoy economic success and are not happy and there are happy people who do not enjoy economic success, most people benefit from economic security, if for no other reason that the realization that bills and expenses are covered. This realization makes many people happy. Once again, we can look at the 2020 global pandemic as a case in point; millions of people lost their jobs and businesses and millions more had to wait in line for food handouts. Those who protested for the states to lift quarantine orders did so essentially because of economic unhappiness. They were willing to risk their lives, and the lives of others, just so that they could get back to work and earn money. Such is the motivation of economic need. Had the economy been truly strong before the pandemic, people would have been able to survive for a few months without working, but the economy was never really that strong.

Love Happiness

Sigmund Freud is said to have held that there are two basic causes of happiness: work and love. We all desire meaningful pursuits that make our lives seem worthwhile, be it the "writer's high" mentioned previously, the joy of athletic pursuits, the intellectual pleasures of performing mathematical calculations, or the countless other ways in which people both earn a living and find reward in their pursuits. But perhaps there is nothing more identified with happiness than the concept of "love."

There is an age-old question: What is this thing called "love?" There is perhaps something a bit ridiculous about philosophers, sociologists and other theorists attempting to analyze the nature of love, and yet ultimately what could be a worthier topic? One of the people who dedicated his life's work to better understanding this topic was the late Irving Singer (1925–2015), a professor of philosophy at the Massachusetts Institute of Technology and author of the monumental three-part work *The Nature of Love*

(1984–1987). Singer attempted to provide a naturalistic outlook toward the concept of love. He gave a sweeping overview of the myriad philosophers, theologians, poets and novelists who have tackled this subject. He sought to supplement his own analytically trained explications with plentiful illustrations from the works of creative writers such as Dante, Shakespeare, Shelley, Proust, Lawrence and even the Marquis de Sade. For Singer, there are basically three broad categories of love: the love of things, the love of persons, and the love of ideals. All of these give meaning to our lives, and we are at our best when we can somehow combine these in a pluralistic approach, a kind of equilibrium (or what the Ancient Greeks would call a harmony of the soul).

Singer published a distillation of his explorations titled *Philosophy of Love: A Partial Summing Up*. In it, he wrote: "Approaching the love of persons, the love of things, and the love of ideals as I do, I perceive the danger of denigrating any one of them…⊠. For that reason, I have constantly refused to rank the types of love by means of any *a priori* hierarchy, or to diminish the value of one or another. I want to fashion pluralistic ideas of what operates in our nature through them all separately and with regard to their feasible coordination. A good life requires an awareness, and acceptance, of that diversified prospect" (Singer 2009: 110–111).

This threefold approach to understanding the nature of love is quite helpful. But the overall context in which love of things, persons and ideals takes place is also important—namely, the biosphere itself.

No matter how one tries to explain love, it is certain that two people who are truly in love with each other find happiness in such an emotion.

Relaxation Happiness

The next nine categories of happiness were not in the first edition of *Beyond Sustainability*. Perhaps one of the more obvious forms of happiness we excluded was relaxation happiness. Relaxation refers to a state of being free from tension and anxiety and a time when the mind and body refresh. This sense of calmness is enough to make someone happy. Think about how good it feels to come home from work, put on your sweats, grab a treat, sit in a comfortable couch and chill. Relaxation allows for the restoration of equilibrium following a disturbance; relief from bodily or mental work; and a diversion from sources of stress. "Relaxation happiness then refers to times when we are doing nothing in particular, we are free from commitments, and we are slacking off and enjoying life" (Delaney and Madigan 2017:245).

Gift of Time Happiness

There are few gifts more valuable than time, especially when one considers that each of us has a limited time to live. The gift of time may be quite dramatic and come in the form of an *extended lifetime* because you managed to survive a near-death experience or a dire medical condition. The gift of time may be more mundane and come in the form of *situational unexpected free time*, such as receiving an unexpected day off from work or from school, or you had a meeting cancelled, or some other unexpected scenario arises that affords you free time. The gift of time is like "bonus" time that provides an opportunity for you to pursue some form of happiness. While it is true that the majority of Americans would choose more money over more free time, choosing more time was associated with greater happiness (Hershfield, Mogilner and Barnea 2016). Research suggests that more money is associated with greater daily happiness up to an approximate annual income of USD$75,000 and with life satisfaction beyond that (Kahneman and Deaton 2010). Research also suggests that even after controlling for material affluence, having more time is associated with greater feelings of happiness and life satisfaction (Kasser and Sheldon 2009).

Work Happiness

Work represents a dichotomy when discussing happiness. On the one hand, many people are unable with their job; research reveals that as many as 80 percent of American workers are dissatisfied with their jobs and 63 percent are "not engaged" with their jobs (Delaney and Madigan 2017). On the other hand, not having a job (or a steady stream of income) is linked to unhappiness and depression. The simple truth is, many people work at a job (opposed to working for themselves) because they *have to*, they need the money.

On the other hand, a number of people are happy at work and find great meaning in their work and their identities are positively correlated with their job. Some of the specific aspects of work that make employees happy are they like their co-workers; they have freedom, autonomy and flexibility; they feel appreciated; they learn new things; they like the challenges that their job provides; they feel engaged, which fuels their desire to continue working; they are good at their jobs; they receive outside recognition; their co-workers are as passionate as they are about their jobs; they like the financial benefits (e.g., salary, health and dental benefits, and retirement benefits); and they would be bored if they weren't working.

Some people love their work so much they become workaholics. The authors are workaholics. We have full-time jobs as professors which involve teaching, holding office hours, advising students, engaging in service activities on campus and in the community, and conducting research as we gain international reputations as experts in our fields. As tenured professors, we could just show up and teach our classes, go to our respective homes and relax. Instead, we do our primary job functions, retreat to our respective homes, and work a great deal more. We do manage to find time for leisure but even those pursuits are often connected to work. Thus, as we travel to various places around the world, whether together, alone, or with others, and we are asked by a flight attendant, "Is this work or pleasure?" We generally answer, "Both!" For us, and many other workaholics, the addiction to work can lead to happiness.

Personal Happiness

We are using "personal happiness" here to refer to a number of related variations of happiness including self-happiness, being single happiness, and being proven right happiness. *Self-happiness* refers to any activity that a person participates in that brings them joy and fulfillment. Being satisfied with oneself—whether this involves personal achievements, personal appearance, or being in a happy relationship—is a rewarding type of happiness as we can reflect upon our own personal achievements (Delaney and Madigan 2017). Knowing that you are doing all you can to help the environment thrive will lead to personal happiness. Being single happiness is achieved in many ways but especially when the individual has a rewarding social life or a fulfilling job. Being single is often viewed as a gift to those who were in destructive, dysfunctional, or loveless relationships. Single people find joy in knowing that they do not have to "ask permission" or check with their significant other before they do something. The happiness found from being proven right is especially important for public figures and professionals who offer advice on a topic that they possess knowledge on. There is a certain rush to being proven right and a corresponding degree of happiness.

Humor Happiness

Perhaps one of the more obvious sources of happiness is humor. Humor comes in many forms and can be enjoyed with friends, family, significant others or by oneself. Humor is a comic, absurd, or incongruous quality that causes amusement; with such sources as talk in general, comical books, skits, plays, movies, television shows, stand-up comedians, and so-on. Humor happiness is something that should be embraced and enjoyed; whether it is enjoyed periodically or on a regular basis. Finding the humor in certain events is a way to "take the edge off" stressful and sad situations. In situations that are particularly stressful, people may employ the concept of *gallows humor*—grim and ironic humor in a desperate or hopeless situation. Gallows humor is often quite applicable when thinking about the dire consequences of global warming and a young person saying, "At least I don't have to worry about saving money for retirement."

Friendship and Family Happiness

In our *Friendship and Happiness* book, we reviewed a number of theories explaining how happiness is attained. We also presented our own theory—"The Delaney and Madigan Theory of Happiness." In this theory we emphasized the role of *engaging in pleasurable intentional activities* as a fundamental aspect in individuals' attempt to attain happiness. We also recognize that life circumstances play a significant, albeit a lesser, role than engaging in pleasurable intentional activities; and we acknowledge that genetics plays a minor role in our pursuit of happiness. Engaging in pleasurable intentional activities may have a hedonistic quality to them as the primary reason we engage in unobligated forms of behaviors—in contrast to obligated forms of behavior (e.g., work and household chores)—centers on having pleasurable experiences. Anticipation is a great motivator for happiness as it stimulates our desire to engage in certain activities over others. However, while anticipation of happiness pales in comparison to actually achieving happiness (via the completion of the steps taken to reach the climatic moment). In this regard, reaching the destination is more rewarding the journey toward happiness. As the Chevy Chase character Clark Griswold states in the 2015 adaptation of the film *National Lampoon's Vacation*, "The journey sucks; that's what makes you appreciate the destination." While it is a stretch to suggest that the journey itself, especially the anticipation aspect, "sucks" it is the pleasurable intentional activity that brings us the greatest happiness.

In our own research of college students, we asked survey participants to identify some of the activities that make them happy and the results should be predictable (especially within the context this information is being shared here). For males and females, the top response was "hanging out with friends" and the second most frequent answer was "being with family." Thus, when given a choice of what type of pleasurable intentional activities to engage in, respondents choose friends and family. In case you were wondering, other responses included: being with my significant other, watching sports, playing sports, listening to music, playing with pets, eating good food, doing schoolwork, going to classes, smoking pot, and being outdoors (Delaney and Madigan 2017).

Awe-Inspiring Happiness

When was the last time you were filled with awe? Being in awe of something involves an overwhelming feeling of reverence, admiration, and respect for something that is so grand, extremely powerful, or beyond words. Dacher Keltner, a pioneer in the study of emotions, and the person who helped Facebook create the "like" button emojis, describes awe as "the feeling of being in the presence of something vast or beyond human scale that transcends our current understanding of things" (Scott 2016). Awe-inspiring experiences are beneficial to our core being as we come to understand and appreciate that there are sights of beauty or immenseness that transcend our everyday life experiences. Feelings of awe make us stop and think about the world in such a manner that we get lost in our own thought processes. Consequently, awe-inspiring happiness is among the more intense forms of happiness.

Sources of awe-inspiring happiness may include observing human-made structures, such as skyscrapers and architecturally magnificently designed buildings such as the Sydney Opera House, an iconic building on the shores of Sydney Harbor (Australia) that is shaped like sails of a boat. More often than not, it is nature that provides us with awe-inspiring happiness. As Scott (2016) states, seventy-five percent of awe is believed to be inspired by the natural world. One can only imagine the feeling of awe astronauts in space experience when they get to look at the vastness of space but also look at the planet Earth as a distant object. We do not have to go to outer space to experience awe, we can look at the stars from earth and marvel at all the possible life forms that exist beyond our tiny little blue marble. Many people find happiness looking at volcano eruptions, lightning storms, tornadoes, snowstorms, waterfalls, deserts, exotic animals, sunsets and sunrises. Those of us who love to travel for reasons in addition to our desire to expand personal knowledge and awareness of cultural diversity do so because we love seeing new things, sights, sounds, people, and nature. Sharing these experiences with another person(s) with whom you traveled with helps to establish a lifelong bond based on awe-inspiring happiness.

The planet earth is filled with sites of beauty and opportunities for awe-inspiring experiences. Why then, are so many people willing to destroy the environment?

E-Happiness

We live in a world that has come to be dominated by electronic devices and electronic forms of communication. In 2017, the average adult spent 5.9 hours per day with digital media, up from 3 hours a day in 2009 (Marvin 2018). The upward trend with time spent on electronic devices continued to rise as we entered the '20s. In some cases, we use electronic devices for work purposes and in other cases we use electronic for pleasure and happiness. During the COVID-19 pandemic the amount of time people spent in front of their screens for work increased dramatically as a large number of people were working from home and engaging in such e-activities as Zoom meetings.

It became all the more apparent that many people could perform their job duties at home (a big plus for the environment) but they would also have to become far more tech-savvy (a number of workers initially struggled with meetings on Zoom). For many, working from home represented a type of e-happiness as they no longer had to commute to work, thus saving time, money and lowering their stress levels; and they did not have to fret over what to wear and saved money on dry cleaning and doing laundry. Conversely, many new at-home workers had to content with a slew of problems including balancing work with childcare and problems with internet connections; this resulted in e-unhappiness.

While working from home represents e-happiness for some, where we really see e-happiness is within the realm of social media. And this is where our categorization of electronic happiness primarily resides as *e-happiness* refers to the joy and self-satisfaction one feels by receiving positive feedback from others (most of whom are strangers) in the electronic world. There are many basic forms of e-happiness including having a large number of followers or friends; feeding the social media world with content that results in "likes" or "loves"; posting selfies and having people comment positively about your attractiveness; and becoming a social media influencer and getting a rush from realizing your actions or opinion on a subject somehow has meaning to others. Making a huge impact on the social world can transform an individual into someone famous, which is often the goal of people who find elation with e-happiness.

Thrivability Happiness

This type of happiness is central to our book's theme: the desire to not only survive and make do, but to embrace challenges and, in so doing, find new ways of excelling in life. Instead of trying to escape from difficulties (and take a "head in the sand" approach to life), those who are inspired by thrivability welcome the opportunity to find new ways to grow. Rather than look upon problems as something to avoid or despair about, they understand that it is the very essence of life to be ready to change one's routines if necessary, and to constantly be on the lookout for new opportunities. In the words of Michelle Holliday, "Thrivability is a growing global movement, with passionate champions around the world. We define it as the intention and practice of aligning organizations with how living systems thrive and how people thrive. It's the recognition that organizations work the same way living systems do—living systems like our bodies, rain forests and coral reefs—because they are living systems. And when we bring what we know about thriving living systems to our organizations, we get amazing results" (Holliday 2014). Thrivability happiness is connected with creativity, and an openness to paradigm shifts in how we view reality. It is striving for greatness. While realizing that the highest of goals might never be achieved, the effort put in to doing so nonetheless is invigorating and life changing. As we stated in Chapter 1, thrivability emerges from the persistent intention to create more value than one consumes.

Fear of Happiness

Some people are in fear of happiness because they fret when things are going well and assume it will balance out with something bad. In this regard, some people are afraid to admit to being happy as they believe karma, the cosmos, law of averages, or something, will make them pay for daring to say aloud that they were happy. This idea of believing in a "jinx" or a "curse" is held by quite a few people, including athletes, superstitious people and some religious adherents. In sports, if a pitcher is throwing a no-hitter late in the game it is very common for his/her teammates to stay far away from them in the dugout and never, ever speak to the pitcher about their no-hitter. Some sports fans will blame an announcer if he/she says something like "Lopez is just three outs away from a no-hitter," and then the batter gets a hit and ruins the no-hitter. Superstitious persons find casual relations between certain behaviors and outcomes where they do not really exist, such as believing that you must wear your "lucky" shirt in order for your team to win. Religious people will cringe when someone evokes the name of God with some event such as God wants me to get this job. For these reasons, and many others, people who fear happiness prefer to play life overly cautiously. They will not dare say or doing something that draws attention to the fact that they are happy about something. Instead, they will keep their excitement to levels far below others who are celebrating. In his research, Peter Lambrou (2014) found that people who suffer from depression often steer away from activities that could bring about feelings of happiness as they worry that the anticipation of happiness will inevitably lead to disappointment and emotional letdown. In this regard, the person has a fear of happiness because they think an event might lead to unhappiness.

Environmental Happiness

Perhaps in trying to understand just why "environmental happiness" is so crucial, the phrase "a breath of fresh air" should come to mind. Metaphorically, it has come to mean a pleasant change of direction or a new way of looking at a problem that allows one to come up with a better solution. But of course, there is the literal meaning as well. Without fresh air we cannot survive, nor can most of the other living things in our ecosystem. Both through our own inventions which have created tremendous pollution on the planet as well as through naturally caused poisons that can end our existence, the reality of fresh air remains constantly at issue. If we can focus attention on this and be consciously aware of the question is there fresh air to breathe or not, we would raise our consciousness of the environment and contribute to the greater happiness of humanity. Just think of what the alternative is.

The authors, like other environmentalists would love to live in a world where people valued *environmental happiness*—finding joy in a healthy and thriving environment—as much as they value other forms of happiness. We would all benefit from an environment with rights to thrive as discussed in Chapter 7. Furthermore, as we have stressed several times throughout this text, when it comes to environmental happiness, the choice is simple: thrive or die.

Gross National Happiness

We have briefly looked at 20 different categories of happiness. No doubt readers may have come up with other types of happiness as well. With such divergent categories of happiness, it is difficult to measure the degree of happiness that any one person may have in comparison to another. For example, is someone who is very economically happy more or less happy than someone who is very happy religiously speaking or very happy with love? It is relatively difficult to objectively compare people with such a subjective topic as happiness. Social scientists like to measure all sorts of things including happiness. As a result, in addition to simply asking people to report their degree of happiness there are a number of popular happiness quizzes that people may take. See Popular Culture Section 8 below for a discussion on happiness quizzes.

If measuring individual happiness is challenging, is it possible to measure the level of happiness in a given nation? One such attempt to measure national happiness comes courtesy of Bhutan's former king, Jigme Singye Wangchuck and his idea of "gross national happiness," coined in 1972 (Nelson 2011; Leigh 2006). From this concept, the "Gross National Happiness" (GNH) index was created. The GNH has four pillars:

- the promotion of equitable and sustainable socio-economic development;
- the preservation and promotion of cultural values;
- the conservation of the natural environment; and
- the establishment of good governance.

Wangchuck developed the GNH while transforming Bhutan, a landlocked mountain monarchy, into a more modern society. He wanted to promote equitable and sustainable socio-economic development but preserve and promote Bhutan's traditional Buddhist spiritual and cultural values. Among the Buddhist core spiritual tenets is the conservation of the natural environment. To do all this, Wangchuck mandated that any proposed economic and development plans must go through a GNH review. The idea of GNH review is similar to the environmental impact statement required for development in the United States. The GNH review process addresses the "establishment of good governance" pillars. The four pillars of the GNH are general enough that they can be applied to all nations, regardless of its societal core beliefs because the GNH review promotes the preservation of cultural values particular to any given society.

It was Wangchuck's hope that people would find happiness when socio-economic development was combined with the preservation of cultural values and the conservation of the natural environment under guidance of good government. Wangchuck proclaimed that Gross National Happiness is more important than the Gross National Product (de Graaf 2009).

Washington Post reporter Peter Whoriskey, in an article titled "Coming Soon: Measuring Our Gross National Happiness," points out that many American economists are taking the GNH index quite seriously. "According to proponents," he writes, "a measure of happiness could help assess the success of governmental policies. It could gauge the virtues of a health benefit or establish whether education has more value than simply higher incomes. It might also detect extremes of inequality or imbalances in how people divide time between work and leisure" (Whoriskey 2011: A-18). A panel of experts selected by the Department of Health and Human Services is looking in detail at the Bhutan plan and its relevance to the United States, and similar panels have been initiated in Britain and France as well. "When talking about measuring happiness," Whoriskey adds, "researchers make what they say is a critical distinction between two ways that people feel happy. There is 'experiential well-being,' or how people feel about their daily activities, and then there is 'life satisfaction,' or how people remember and judge their lives or parts of it" (Whoriskey 2011: A-18).

New York Times best-selling author Thomas Friedman (2011) states there will come a time when people realize "that the consumer-driven model is broken and we have to move to a more happiness-driven growth model, based on people working less and owning less" (p. A-12). On their death beds, people will not say, "I wish I had worked harder or built more shareholder value." Instead, they will say, "I wish I had gone to more ballgames, read more books to my kids, taken more walks" (Friedman 2011: A-12). Which is to say, the Gross National Happiness of a nation, and of the world itself, is no longer a frivolous concept.

The GNH scale has its detractors. For one, critics point out that Bhutan is a far cry from paradise. Bhutan is a society that has faced criticism for expelling residents it claims are illegal Nepali immigrants, but whom some human rights groups assert were citizens opposed to the monarchy (Nelson 2011). Bhutan has a culture that accepts domestic violence, and based on its National Statistics Bureau, "around 70 percent of women felt they deserved to be beaten if they refused to have sex with their husbands, argued with them, or burned the dinner" (Nelson 2011). Other critics of the GNH model point to the difficulty of measuring the concept of "happiness" (Cayo 2005).

Despite the criticisms of Bhutan's cultural norms and values and the difficulty in measuring "happiness," it is reasonable to work with the premise that happiness does exist in some form.

Happiest Nations in the World

There are a number of indexes designed to measure the happiest nations in the world including the "World Database of Happiness" that compiles an archive of research findings on subjective enjoyment of life and measurements of happiness (World Database of Happiness 2019). Among other things, the World Database of Happiness website provides data of worldwide nations on three happiness categories: "Average Happiness," subjective appreciation of life-as-a-whole (0–10 range); "Happy Life Years," subjective analysis of how many years, on average, residents reported to be happy (range of 0–100); and "Inequality of Happiness," subjective analysis of the disparity in reported happiness (0–3.5). The nation receiving the highest score for "Average Happiness" is Mexico (8.3) and lowest score was Central African Republic. The highest score for "Happy Life Years" is Switzerland (66.8) and the lowest from Central African Republic (18.8). The lowest score for "Inequality of Happiness" was Netherlands (1.41) and the highest level of inequality was Egypt (3.16). The United States received respective grades (for the 2010–2018 data bank) of 7.1; 55.9; and 2.14. Thus, the U.S. is relatively happy but there is a good deal of happiness inequality. Russia, by comparison, had scores of 5.7; 39.1; and 2.37 (World Database of Happiness). (Note: If you are interested, you can go to the World Database of Happiness website and look up any nation you want to see how they rank.)

The Happy Planet Index (HPI) measures the level of nations across the globe. The HPI is calculated by examining four elements to show efficiently residents of different countries are using environmental resources to lead long, happy lives. The formula is: "Wellbeing" (how satisfied the residents of each country say they feel with life overall on a scale from zero to ten) times "Life expectancy" (the average number of years a person is expected to live in each country) times "Inequality of outcomes" (the inequalities between people within a country in terms of wellbeing and life expectancy) divided by

"Ecological Footprint" (the average impact that each resident of country places on the environment) (Happy Planet Index 2020a). The HPI website provides a map of the world wherein visitors to the site can click on any nation to learn of their score. In 2020, the United States, for example, had a score of 20.7 ranking it 108th out of 140 countries. This score is slightly above Russia which had a score of 18.7, ranking it 116th out of 140 countries. In contrast, Mexico had a score of 40.7 and ranked 2nd of 140. The nation with the highest score was Costa Rica with a HPI score of 44.7, making it 1st out of 140 countries (Happy Planet Index 2020b).

The Paris-Based Organization for Economic Cooperation and Development (OECD) has a "Better Life Index" that consists of 11 categories: housing, income, jobs, community, education, environment, civic engagement, health, life satisfaction, safety, and work-life balance (OECD 2020). The OECD tracks forty member nations on each of these eleven categories and provides a summary report. Interestingly, the OECD does not assign rankings to countries but instead states that the Index is designed to let the user investigate and visualize how each of the 11 topics can contribute to well-being.

The next happiness index to be discussed is the London-based Legatum Institute's "Legatum Prosperity Index." This Index examines "nations around the world on their strengths and weaknesses in order to determine the economic and strategic choices that need to be made to further build inclusive societies, open economies, and empowered people to drive greater levels of prosperity" (Legatum Prosperity Index 2019). The most recent Index data available at the time of this writing was 2019 when 148 out of 167 countries measured enjoyed an improvement in their prosperity since 2009. (Note: We would speculate that many nations will take a dip in 2020 because of the economic havoc caused by the COVID-19 pandemic.) The Index utilizes 9 measurements to compile a comparative ranking; these variables are safety and security; personal freedom; governance; social capital; investment environment; enterprise conditions; market access and infrastructure; economic quality; and living conditions. The top five ranked nations in 2019 were Denmark, Norway, Switzerland, Sweden, and Finland. The bottom five ranked nations in 2019 were South Sudan (167th), Yemen, Central African Republic, Chad, and Afghanistan. The United States was ranked 18th and Russia 74th. The U.S. performs most strongly in "enterprise conditions" and "market access and infrastructure" but (no surprise here) is weak in "health" (Legatum Prosperity Index 2019).

The final happiness index we will look at is the "World Happiness Report," an annual survey conducted by the Sustainable Development Solutions Network for the United Nations. This index looks at six variables: income, freedom, trust, healthy life expectancy, social support and generosity (Bloom 2019). Based on this happiness index, the top ten happiest nations in 2019 were Finland, Denmark, Norway, Iceland, Netherlands, Switzerland, Sweden, New Zealand, Canada, and Australia. The bottom ten (saddest) nations in 2019 were South Sudan, Central African Republic, Afghanistan, Tanzania, Rwanda, Yemen, Malawi, Syria, Botswana, and Haiti. As usual, the United States is not among the top ten nations but also no where near the bottom ten. The U.S. ranked 19th on this index as it fares well with income (ranked 10th) but does poorly with freedom (61st), corruption (42nd) and social support (37th) (Bloom 2019).

If you are interested in how individual happiness is calculated take a look at the popular culture section below.

Popular Culture Section 8

Popular Happiness Quizzes: Measuring Individual Happiness

Are you happy? What makes you happy? Are people happier today than in the past? Are people in some societies happier than others? Does your significant other make you happy, and do you make your significant other happy? There are nearly as many questions about happiness as there are sources of happiness. What makes one person happy may not make another person happy. The authors of this text, for example, find happiness in spending countless hours conducting research, writing, editing, and ultimately publishing (intellectual happiness). And yet, there are many people, including students, who do not find happiness in writing. Furthermore, there are many people who would just as soon not know about all the problems with a specific topic like the environment; these people find happiness in ignorance.

Happiness as a topic of scientific inquiry is a relatively new subject area. After all, throughout most of human history, including today, most people have been too busy securing basic necessities to be concerned with such things as happiness and self-fulfillment. Happiness, if it was even described *as* happiness, in the past was generally tied to having sufficient food, shelter, clothing and other basic needs. Today, there are entire industries (e.g., entertainment, sports, spas, cruises, spirituality and meditation centers, and fantasy camps) built around providing people, who can afford such luxuries, happiness in a variety of shapes and forms. Not surprisingly, as happiness industries have gained prominence, so too have attempts to measure happiness flourished.

Among the most popular methods to measure happiness are quizzes. While doing research in this area the authors found and took a variety of happiness quizzes. Chances are, you too have taken a happiness quiz; there are certainly enough of them available. Happiness quizzes can be found in books and magazines, on television, and online where there are literally millions to choose from.

Even if this were a book specifically covering just happiness quizzes, it would be impossible to review a true representative sample of the millions of popular happiness quizzes that exist. Consequently, we will discuss just a few examples. The first happiness quiz we will look at is the "Be Happy Index" (BHI). The BHI is a happiness quiz promoted by Oprah Winfrey (on her show and her website) and used during Dr. Robert Holden's eight-week "Be Happy" program, which was tested by independent scientists for the BBC documentary *How to Be Happy* (Holden 2010). The scientists described the quiz as a fast track to happiness that both changes the way you feel and the way your brain functions (Holden 2010). As Yoda (*Star Wars*) might say, "Lofty ambitions, this quiz has."

The BHI is a simple quiz that utilizes, like many happiness quizzes, a Likert Scale for responses (ranging from "not true" to "very true") to 10 questions on such topics as self ("I know who I am and I like myself"), relationships ("My most important relationships get the most attention"), and sense of purpose ("I have a strong sense of purpose and/or I love my work"). Respondents click their responses and a final score with an interpretation of the score is provided at the conclusion of the quiz. (The score is determined by the 1–5 point value for each question multiplied by 2 to garner a possible high score of 100 and a low score of 10.) Advice (described as "scientifically proven methods for raising your happiness level") is offered to the respondent for each of the 10 topic areas. The quiz, then, measures both one's level of happiness and offers advice on how to become happier. (To take this test go to http://www.oprah.com/spirit/Take-the-Happiness-Test-Quiz.)

The next happiness quiz we will discuss is provided by *Scientific American*. This quiz has 10 "yes" or "no" questions (5 that address positive feelings and 5 that address negative feelings) and one question that asks respondents to "Rate Your Life as a Whole" with a "0" to "10" response scale (Wickelgren 2011). After answering the questions, respondents are given an analysis of their responses to the "positive feelings" and "negative feelings" categories, followed by a "feelings balance" score and analysis. The feelings balance score is based on one's positive and negative responses. A "Life Satisfaction" score is always provided along with an analysis and suggestions on how to improve your life satisfaction (if necessary). A very interesting final analysis is also provided under the heading of "Where Do You Fit In?" This category lets you compare your score to average scores (in all four categories) of people from nations around the world (provided in alphabetical order). (To take this test go to http://www.scientificamerican.com/article.cfm?id=happiness-how-happy-are-you-quiz.)

The final happiness quiz we will look at comes from the AARP. The AARP offers a "Key to Happiness Quiz" (Oberliesen 2013). This happiness quiz encompasses 8 questions wherein the answers are connected to brain activity, physical activity and physiological changes in the body. Questions include "What health benefits do you get from owning a pet?" "When you hear a favorite tune, what effect does it have on your brain?" "Having sex is linked with happiness, so how often do 'very happy' people have sex?" "What physical changes happen in the body when you laugh?" and "At what age do we tend to feel the least depressed?" Respondents are given the option of four responses to test their knowledge on the link between happiness (and depression) and physical well-being. At the conclusion of the test, one's results are tabulated, and analysis of the answers is provided. (To take this test go to http://www.aarp.org/health/healthy-living/info-03–2013/keys-to-happiness-quiz.html.)

Summary

The urge to be happy is as natural to humans as breathing and is one of the strongest motivating factors in life. While one can debate over the meaning of what constitutes happiness, surely the pursuit of it is something that cannot be taken lightly. As much as people want to be happy, governments do not provide constitutional rights to happiness; the United States Constitution, however, does provide Americans with the right to pursue happiness.

Environmental thrivability is, above all else, the drive to meet existing challenges to the biosphere and come up with solutions that will do more than merely sustain current realities. Rather, the point of thrivability is to allow people to fulfill their talents at the top of their game—to achieve what Aristotle called "eudaimonia" and Mihaly Csikszentmihalyi calls "flow." The best way to do so is by combining subjective and objective elements.

The ideas of Jeremy Bentham (*hedonic calculus*), Aristotle, and Csikszentmihalyi provide us with a starting point on so many topics, including happiness. Individual happiness is a very subjective phenomenon as it means different things to different people. As a result, the authors have categorized 20 different types of happiness including environmental happiness. It is critical for the well-being of the planet that all the people in the world learn to find happiness in helping the environment thrive. After all, when it comes to environmental happiness, the choice is simple: thrive or die.

A number of indexes designed to measure the happiest nations in the world were also discussed. These indexes included such variables as wellbeing, life expectancy, housing, income and employment, education, civic engagement, health, life satisfaction, personal freedom, social capital, and the degree to which nations are using environmental resources to lead long, happy lives. While all the happiness indexes are connected to environmental concerns, at least indirectly, the Happy Planet Index is the most relevant to our discussion on the goal of promoting a thriving environment.

CHAPTER 9

We Can Change, but Will We Change?

The authors have presented their case throughout this text that we are in the sixth mass extinction period and that this extinction process is being sped along because of human actions, especially since the time of industrialization. The two biggest culprits of humanity's contribution to the 6th ME are the reliance on burning fossil fuels and a general disregard of the environment. If humanity hopes to survive this critical century in our history, we must make many significant changes to our collective behaviors.

We Need to Change

While we believe in, and promote the idea that, we are in the 6th ME period, we have also acknowledged that mass extinction eras can last hundreds of thousands and even millions of years. We are somewhere along in the 6th ME era and because of human activities, we argue that we are likely nearing the end of this era before humanity and 70–80 percent of the rest of the species will die off by the end of this era. If we hope to have many future generations of humans and other animals and plant life, we need to drastically change our behaviors.

If there was a "range of concern" regarding the seriousness of this mass extinction period, the authors would be leaning toward a "dire concern" about the future as we fear the worst case scenario is inevitable but not necessarily in our immediate future. On the extreme ends of this "range of concern" is "no concern" at one end because some people, primarily fossil fuel proponents, do not fret at all about the 6th ME; and on the other extreme is the "end is very near" (think of the "Doomsday Clock" mentioned in Chapter 1). Count the increasing number of the world's leading climate scientists among those who believe the end is very near. The authors of a 2018 landmark report by the United Nations Intergovernmental Panel of Climate Change (IPCC) state that urgent and unprecedented changes are needed in order stave off global catastrophe by 2030 (Watts 2018). These scientists argue that 2030 is the year when we will exceed the maximum temperature increase allowable as per the 2015 Paris Agreement before humanity risks dramatic droughts, floods, extreme heat, and poverty for hundreds of millions of people and the full eradication of coral reefs due to global warming. They put forth the notion that if the goals of the Paris Agreement are not met, all their dire predictions will come true. This is the same agreement that President Donald Trump pulled the U.S. commitment from (made originally by President Obama).

Consider two of the many dire warning signals that have led to scientists sounding an alarm. The Antarctic ice loss has tripled in the last decade which raises concerns over the rising sea levels. This ice loss excludes the largest part of the continent, East Antarctica, which has remained more stable and hasn't contributed much ice to the ocean. However, in the last five years, it too has begun to lose ice, and this is very important as a controversial 2016 study by former NASA scientist James Hansen and associates found that the Earth could see sea level rise above one meter (or 3.3 feet) within 50 years if polar ice sheet loss doubles every year (Mooney 2018). This prediction "reinforces that nations have a short window to cut greenhouse gas emissions if they hope to avert some of the worst consequences of climate change" (Mooney 2018: A14). Deforestation and industrialization are such a serious concern that scientists project that if 2.2 billion acres of new trees were planted—around 500 billion saplings—they could absorb 220 gigatons of carbon once they reach maturity. Swiss scientists estimate this would be equivalent to about two-thirds of the manmade carbon emissions since the start of the Industrial Revolution (Larson 2019). This figure for the number of trees needed to offset increasing carbon emissions does not take into account the fact that even more trees would have to be planted to offset the continuing deforestation.

The authors believe that it would be ill-advised to state a specific date when the worse-case scenario will occur, but we are comfortable saying that every year of significant inaction draws us closer to the inevitable point of no-return. Let's quickly review why we *need* to change our ways of treating the environment: a number of animal and plant species have already gone extinct and the extinction rate is higher than the natural course of life; the earth's carrying capacity has been stretch to within its breaking point; human dependency on burning fossil fuels as a primary source of energy has directly led to increased levels of carbon dioxide and greenhouse gases; nature has responded to climate change in a dramatic fashion including melting glaciers, ocean acidification, increased wildfires and so on; overpopulation, which contributes to the demands placed on the carrying capacity of the earth; record heat and increased fire seasons; human contribution to the five horrorists, especially all the variations of pollution; the development and use of plastics; deforestation; and harmful agricultural practices. Clearly, humans are causing far more harm to the environment than good.

Reactions to the *Need* for Change

The pro-environment reaction to this need to change is obvious and is exemplified throughout this book. In brief, we need to believe in science and the overwhelming evidence that humanity is causing great harm to the environment and that our actions are contributing to the 6th ME. While the authors promote the idea of thrivability, other environmentalists are simply happy content with sustainability. As we stated in Chapter 1, thrivability involves incorporating a vision of the world that extends beyond the limitations of a sustainability attitude of the environment by promoting heightened levels of collaboration among individuals, groups, corporations and government. Thus, thrivability requires a major shift both in social consciousness and social action; that is, if we hope to avoid an ecocatastrophe such as those involved in earlier mass extinctions.

The anti-environment reaction is one that is short-sighted and detrimental to the environment, humanity, and all living things as the continued reliance on fossil fuels will surely assist the finalization of the 6th ME. The anti-environment movement is spearheaded by climate change deniers and especially the fossil fuel industry and the politicians who are duped by their influence via such methods as campaign contributions, the promise of unskilled job opportunities, and the voodoo trickle-down economics construct. So blatant is the disregard for the environment among certain politicians and the fossil fuel industry is that they would take advantage of a global pandemic to sneak in such measures as rolling back air quality and climate regulations and the elimination of many specific environmental protections (as detailed in earlier chapters). Is there any wonder why some people believe that the actions and lobbying efforts put forth by the fossil fuel industries (including oil, coal and natural gas companies and the shipping and trucking industries) and conservative politicians, and especially President Trump, is a sign (or evidence) that the COVID-19 pandemic was blown out of proportion by conservatives and the anti-environment industries as a means of distracting the public and the mainstream media from seeing what was going on right in front of their eyes? Such a belief and concern—while far-fetched—was based on the idea that Trump freely admits to targeting environmental and public health regulations because he values business over the environment. (Note: This reaction among some conspiracy theorists may have actually been a response to Trump's claim in early March 2020 that Democrats were blowing out of proportion the seriousness of the pandemic.) Disregarding scientific evidence that humans contribute to global warming is the flimsy manner in which fossil fools and greenhouse asses justify a continued reliance on fossil fuels.

In addition to the variations of human skepticism discussed in Chapter 5, there are other reactions to our need to change if we hope to save the environment for future generations and stave off the 6th ME for as long as possible. One such approach is "Hopeless Reaction—We're Doomed!" This mentality is highlighted by the idea that no matter what we try, there's nothing to be done. As reporter Elizabeth Weise points out, many biologists and paleontologists propose that we have already entered the sixth great mass extinction. She writes: "All of the mass extinctions of the past also transpired over many centuries or longer. Biologists today estimate that within the last 500 years, at least 80 mammal species have gone extinct out of beginning total of 5,570 known species.... Questions still to be answered are whether currently threatened, endangered and critically endangered species will go extinct" (Weise 2011).

In 2002, a group of scientists from 193 countries chose the year 2010 as the year in which a "significant reduction of the current rate of biodiversity loss at the global, regional and national level" would be achieved (Milius 2010:20). Did it happen? "Fat chance. Conservationists and trend watchers predict at best a few bright points among worsening losses" (Milius 2010: 20). This leads doomsayers to ponder, "Why bother trying? We're already past the deadline." The clock is ticking and the second hand points ever closer to doomsday. In fact, as we stated in Chapter 1, the "Doomsday Clock" is just 100 seconds away from midnight—the moment of extinction.

Some might even exult in this hopeless scenario and say that the extinction of one particular species, namely *Homo sapiens*, would be a blessing. But, as we pointed out, that is a rather peculiar stance for members of that species to take, especially since Homo sapiens means "thinking human"—our most outstanding trait is our ability to think and use our intelligence to solve problems. Dante famously wrote that the Gates of Hell have above them the sign "Abandon All Hope, All Ye Who Enter Here"—but we are not in Hell, no matter how hot things may be getting on Earth, and as long as we can conceptualize our problems there remains the chance that we can overcome them.

Another strategy to take is the "Head in the Sand" approach. We have already detailed the viewpoint of global warming deniers and others who stoutly insist that we are in no imminent danger. If nothing else, such a view seems both short-sighted and unethical. A *Peanuts* classic cartoon from 1965 (before the Environmental Movement itself got off the ground) shows Lucy and Linus contemplating their world. "Our generation has been given the works," Lucy bemoans. "All of the world's problems are being shoved at us." When Linus—the most philosophical of Charles Schultz's creations—wisely asks her, "What do you think we should do?" she grins wildly and, shaking her fist, intones, "Stick the next generation." While Lucy may not mind passing the buck in this way, we must ask the question whether we can continue policies that will knowingly cause harm to future generations.

The noted ecologist Garrett Hardin (1915–2003) in 1968 came up with the concept of "The Tragedy of the Commons." He imagined an agricultural society where all the citizens graze their livestock on a commonly owned field which can only support a limited number of animals. He pointed out that the rational response for most people would be to continue to graze their animals, since each individual would feel he or she was still prospering, even while the field itself was being destroyed—an ultimately irrational reality. His point was that, while no one wanted to see the common field completely degraded, it was hard to conceive that each person's actions was collectively helping to cause that very event. It is significant that he postulated this concept shortly before—inspired by the *Sputnik* challenge—America's space program did achieve its goal of putting humans (at least briefly) on the moon. The famous photographs taken of the Earth from space made it vitally clear what Hardin was getting at: it is the Earth itself that is the commons on which we all must live, until such a time as we truly are able to venture out into outer space.

In 2001 Cambridge University Press published a controversial book by a Danish statistician named Bjørn Lomborg, called *The Skeptical Environmentalist: Measuring the Real State of the World.* Taking umbrage with Hardin and others like him, Lomborg argued that in fact the commons was not being decimated at all, and that the world was becoming a much better place to live for humans. The ecosystem was not nearly as imperiled as environmentalists claimed it to be, and even if it were, humans would necessarily come up with answers to the problems using our intellectual skills. In a newspaper debate with environmental philosopher Peter Singer, Lomborg raises concerns about those he calls "Green Extremists," arguing that "Mr. Singer … evades the awkward point that an excessively green approach can actually make the environment more imperiled. As we get richer and more environmentally conscious, our growing passion for organic farming and antipathy to genetically modified crops inevitably leads us to accept decreased agricultural yields. An obvious consequence is that we end up converting more wilderness to agricultural use" (Lomborg 2011: C1). For him, radical environmental changes will in fact create problems that do not presently arise, following the law of unanticipated consequences.

Such skepticism, when applied properly, can be healthy, in that it can prevent us from acting rashly—but it can also lead to a sense of complacency. Singer responds in a telling fashion: "Mr. Lomborg is a technological optimist but an ethical pessimist. I'm all for sustainable technology and economic growth, but I also think we should do what we can to encourage people to take a more ethical approach to global issues" (Singer 2011: C2). Which is to say, as Singer has argued in many other contexts, a new global ethics is essential for addressing the changing needs of an ever-expanding human population.

Somehow hoping that all will be well without first asking what negative impact our technology may be having on the environment can be reckless rather than helpful.

As Mark Bittman points out, following along the lines of Singer's position, "For the first 200 years of the fossil fuel age, we could claim ignorance of its lasting harm; we cannot do that now. With knowledge comes responsibility, and with that responsibility must come action. As the earth's stewards, our individual changes are important, but this is a bigger deal than replacing light bulbs or riding a bike. Let's make working to turn emissions around a litmus test for every politician who asks for our votes" (Bittman 2013: A21). In other words, to truly deal with the crises at hand we must mobilize a political and social sea change akin to that which marked the beginning of the Space Age.

This change in attitude can be called the "sustainability approach." But, as we have argued throughout the text, "sustainability" as a term lacks vitality. It often emphasizes, as shown in the quote above from Carl McDaniel, limitation, sacrifice, and trying to hold on to what remains. It is perhaps no wonder that writers like Lomborg, who emphasize creativity, new possibilities, and limitless horizons, prove to be more popular. But such popularity comes at a cost—perhaps the cost of our own survival.

Is sustainability too gloomy or uninspiring a concept? People of a certain age may remember President Jimmy Carter, sitting by a White House fire in his cardigan sweater, urging all Americans to lower their thermostats and start conserving energy. If he had been listened to in 1979 what might the world look like now? But alas, he came across as a scold and party-pooper, whereas his political opponent, Ronald Reagan, with his talk of "Morning in America" and his disavowal of conservation measures, was swept into office in 1980 on a wave of optimism. If "sustainability" remains connected to people's minds with such gloomy measures, it is hard to see how it can ever spark a mass movement towards sounder environmental policies.

Also, can we really sustain life in an ever-changing environment? While being concerned about the eradication of species, there will always be an ongoing process of adaptability. Not all species will be able to survive no matter what measures may be taken to prevent their disappearance. In addition, even after all previous mass extinctions, some life forms not only survived but were able to thrive in their new environments. As the paleontologist Michael Benton points out in an article in *The New Scientist*, "Mass extinctions ... have a creative side. Marginal groups sometimes get a chance to expand and become dominant. Most famously, mammals benefited from the demise of the dinosaurs" (Benton 2011: 54). Of course, he is quick to add that after most previous mass extinctions, the rate of recovery for living things averaged a few million years. We don't have such a luxury of time if we want to see the human species avoid extinction.

What can be done? Can one perhaps combine the "technological optimism" of Lomborg with the "ethical optimism" of Singer? That is what we argued for earlier, taking a Green Pragmatic viewpoint in the spirit of John Dewey—to emphasize not merely sustaining what is, but coming up with new ways of living that are inspiring and draw upon all of our abilities. This relates to the overwhelming human drive for happiness.

We Are Capable of Change

Despite the ignorance of many and the greed of the rich and powerful, there is a reason humans were able to climb to the top of food chain—our intellect. Not all members of our species possess the intellect to see a problem and find a way to combat and conqueror it. But thankfully, innovative and/or scientifically driven persons have managed to take our species to unforeseen heights even while others try to tie anchors on progressive thought and action.

Below, we will discuss two specific examples of the human capability to change. The first is a story about the ability of humankind to achieve something seemingly impossible prior to the 1950s, the idea of space travel and putting a human on the moon. This change occurred because of a group of scientists and progressive-thinking political leaders. The second story describes how we can change if we are, essentially, forced to do so.

The *Sputnik* Story

Perhaps in order to "save the Earth" it is important to better understand it. But, as the saying goes, seeing is believing. To truly understand the nature of the problem, and to appreciate more the beauty of the world in which we live, it is important to see as much of it as possible.

On October 4, 1957, the world was shocked to learn that the Soviet Union had launched a satellite into the sky. Americans, sure of their technological supremacy, were caught off guard. They had been certain that if anyone was going to put a satellite into orbit, it would be the United States. As Paul Dickson relates in his book *Sputnik: The Shock of the Century*, most Americans' initial reaction was disbelief, followed soon by fear. Just what was the Soviet Union's purpose for putting this sinister object in the sky? Cold War realities collided with the beginning of the Space Age. "There was a sudden crisis of confidence in American technology, values, politics and the military" (Dickson 2001: 4).

But the launch of *Sputnik* ended up spurring fundamental changes in the United States, as the urge to beat the Soviets became a predominant cause, uniting differing political and social factions. Dickson adds: "Science, technology, and engineering were totally reworked and massively funded in the shadow of *Sputnik*." The challenge was accepted, and America rose to the occasion. In 1961 President John F. Kennedy proudly proclaimed that America would put a man on the moon by the end of the decade. On July 20, 1969, the United States achieved the goal set by President Kennedy as American astronauts reached the Moon, landed, and walked on the surface, and made it back to their orbiting space craft marking, perhaps, the most epic moment in civilization. Something that a few years before would have been deemed a mad quest or a scenario purely out of a science fiction novel now became a reality. "The Russian satellite essentially forced the United States to place a new national priority on research science, which led to the development of microelectronics—the technology used in today's laptop, personal, and handheld computers. Many essential technologies of modern life, including the Internet, owe their early development to the accelerated pace of applied research triggered by *Sputnik*" (Dickson 2001: 4). Ever since, we've become used to the saying, "We can put a man on the moon—why can't we solve our other problems?" This is precisely the question we have been asking throughout this book regarding environmental crises.

Just as Americans in 1957 came to a rude awakening when they saw *Sputnik* in the sky, so those of us in the early 21st century now face a similar wake up call. Can our species escape destruction? Will we be able to overcome the ways of acting that have to a large extent gotten us into dire environmental crises? Previous chapters in this book have shown the very real problems faced by all life on earth, not merely the human species.

The COVID-19 Pandemic

The year 2020 will forever be linked to a global pandemic. With origins in Wuhan, China, the novel Coronavirus would first appear in late 2019 and quickly spread worldwide in the early months of 2020. Known as COVID-19, this disease from the family of Coronaviruses that causes respiratory illness would infect tens of millions of people globally and kill more than one million people. The planet was not prepared for the microbes associated with COVID-19 and the lack of leadership from the White House contributed to spread of the disease in the U.S. and a dramatic increase in the number of deaths. On May 20, 2020, *The New York Times* reported that lockdown delays in early March cost at least 36,000 American lives (Glanz and Robertson 2020). "And if the country had begun locking down cities and limiting social contact on March 1, two weeks earlier than most people started staying home, the vast majority of the nation's deaths—about 83 percent—would have been avoided [Columbia University] researchers estimated" (Glanz and Robertson 2020). On March 1, 2020, Trump held one of his infamous self-congratulatory rallies at the North Charleston Coliseum in Charleston, South Carolina, and proclaimed to his adoring crowd that the Democrats want to politicize the coronavirus. Trump pauses so that the crowd can boo. And then continues to state, "This is their new hoax." Instead of taking action, Trump blamed the Democrats for a "global hoax." On May 19, *Yahoo Finance*, citing data from the Penn Wharton Budget Model (developed at the Wharton School of Business, Trump's alma mater), reported that if states fully reopen with no social distancing rules in place that as many as 5.4 million Americans could test positive for coronavirus, and if states reopen while still practicing measures of social distancing, nearly 4.3 Americans were projected to be diagnosed with COVID-19 by the third week of July (Myers 2020). By mid-February, more than 28 million Americans had contracted the disease and nearly a half million had died.

Americans had no idea that Trump actually knew about the dangers of the virus but instead chose to lie to the public and Congress. In September 2020, recordings of Trump during interviews with journalist Bob Woodward (for a forthcoming book, *Rage*) were made public. On February 7, Trump told Woodward (about the coronavirus), "You just breathe the air and that's how it's passed. And so that's a very tricky one. That's a very delicate one. It's also more deadly than even your strenuous flues. This is deadly stuff" (*Associated Press* 2020a). On March 19, Trump told Woodward he deliberately minimized the danger, "I wanted to always play it down. I still like playing it down because I don't want to create a panic" (*Associated Press* 2020d). Thus, while Trump lied to the public that the coronavirus was nothing to be concerned about, was a Democratic hoax, and a media hoax, Americans were dying because they did not have a chance to isolate themselves. When most Americans began to realize the seriousness of the COVID-19 pandemic they did panic, they emptied shelves of toilet paper, cleaning supplies, protective gear and so on. When the recordings were released Trump tried to justify his actions by saying he is "a cheerleader for the country" (*Associated Press* 2020d). Trump certainly wasn't a leader and the American voters do not vote for a cheerleader, they vote for someone to lead the country, in good times and bad times.

By early March 2020 it became apparent to many state Governors, especially in those states (i.e., New York and California) that were recording high numbers of COVID-19 hospitalizations and deaths that they would have to take charge if there were to be any hope of containing the virus. Most states enacted stay-at-home orders, which included requiring most businesses having their employees work from home; banning large gatherings (e.g., at beaches and church services); requiring people to wear protective masks in public; mandating physical distancing of 6 feet in public areas; and closing all but essential businesses (e.g., grocery stores, pharmacies, and liquor stores). In the U.S. the White House would, of course, eventually get involved in an attempt to curtail the spread and to "flatten the curve." In most countries around the world similar restrictions were put in place.

The possible repercussions of ignoring the dangers of this pandemic would be catastrophic unless dire actions were taken. For a month or two, depending on the state, and the country, people abided by these restrictions and why not, this was a virus with no known cure. Slogans such as "We are all in this together" were embraced, until they weren't. Before too long, people grew tired of staying at home. For some, the pandemic was "fake news" created by the media. Others, who were out of work and legitimately needed to get back to work because they desperately needed the income to support themselves and their families started to defy the orders. Some people protested. Others continued to abide by the regulations; after all, we're all in this together—although, we know that was a sentiment not actually shared by all.

This most relevant occurrence related to the COVID-19 pandemic to our concerns here was the short-term effect it had on the environment. Because most people were working at home there were far fewer people on the roads and outdoors. Countless ecosystems benefited from this. As people stayed home, the Earth turned much cleaner. As we mentioned in Chapter 7, smog cleared up in New Delhi; nitrogen dioxide pollution in the northeastern United States was down 30 percent compared to normal in April; air pollution in Rome from mid–March to mid–April was down 49 percent from the previous year; and wild animals roamed in areas normally occupied by humans. Additionally, the air from Boston to Washington was the cleanest since a NASA satellite started measuring nitrogen dioxide, in 2016; air pollution was down 46 percent in Paris, 35 percent in Bengaluru, India, 38 percent in Sydney, 29 percent in Los Angeles, 26 percent in Rio de Janeiro, and 9 percent in Durban, South Africa, NASA measurements indicate; many other areas in India as well as in China, had the cleanest air in decades; more than 70,000 olive ridley sea turtles were seen nesting on the beaches of eastern Indian state of Odisha, and sea turtles in Costa Rica and Mexico were also able to safely lay eggs on empty beaches as humans were quarantined, leaving the turtles undisturbed; and in Venice, the water in the famed canals became clearer than they had in recent memory and even fish were visibly swimming around in them (Borenstein 2020b; Johnson 2020; Guy and Di Donato 2020). As for carbon pollution in sum, the world cut its daily carbon dioxide emissions by 17 percent at the peak of the pandemic shutdown in April (2020) (Borenstein 2020c).

We agree with the sentiment of conservation biologist Stuart Pimm (whom we mentioned in Chapter 1) who said of these extraordinary circumstances (COVID-19), "This is giving us an opportunity to magically see how much better it [earth] can be" (Borenstein 2020b). Just this small amount of time when humans had less of a negative impact on the environment and immediately the Earth got cleaner and many species found in the natural world regained a sense of balance with their ecosystems. This little social experiment of most humans staying at home opposed to their normal activities of driving automobiles to work and play and tourists being taxied around the canals of Venice provided clear evidence of how much harm we cause the planet. Proponents of fossil fuels and climate change deniers were provided with undisputed evidence of the danger human action poses to the natural environment.

The environmental lesson learned from the COVID-19 pandemic is clear, if humans act together, we are capable of change, and this change could, perhaps, be significant enough to help the environment sustain itself. Ranjit Deshmukh, an assistant professor in UC Santa Barbara's environmental studies program states, "If we have to draw lessons from these exceptional circumstances, then those lessons would be the importance of global cooperation in tackling global problems" (Tasoff 2020).

So there is at least a glimmer of hope that when humans are pushed to the brink, perhaps they will act in a manner that is best for the environment if our survival is at stake. And this despite the fact that a number of Americans struggled and eventually failed at heeding the advice of medical experts and those who warned of the dangers of COVID-19. We do see some other efforts among collectivities when it comes to doing the "green thing." The University of California system, for example, announced on May 19, 2020, that it had fully divested from all fossil fuels in their effort to fight climate change through investment strategies (Watanabe 2020). In June 2020, Cornell University announced a moratorium on fossil fuel investments (Steecker 2020). We have not forgotten about the ability of social movements to have an impact on our ability to change. Recall the activism of Greta Thunberg and ability to motivate the younger generation into action. This generation recognizes just as the 1970s generation realized, that changes must be made if there is any hope for humanity on this planet. Students across the globe are involved in a "generational strike" social movement with such goals as saving nature, saving the earth, and saving the future all the while hoping to put pressure on politicians and corporate leaders.

All this begs the ultimate question, we can change, but will we? This question mirrors the one put forth by Carl McDaniel in the final chapter title of his 2005 book *Wisdom for a Livable Planet*, "Can We Change, Will We Change?" He points out that "we are constrained by a finite Earth overflowing with humans and their artifacts. A successful cultural shift requires that we accept the limits of Earth and its resources, an idea that contradicts that which delivered the present bounty to so many of us. We require a belief system that embraces joy and celebrates humanity's limitless capacity for spiritual and intellectual growth while at the same time accepting and appreciating the physical laws and limits governing the biosphere. Can we shift from an economically centered pattern of living to adopt lifestyles compatible with physical and biological limits?" (McDaniel 2005: 225).

In our view, we can change our present course, just as Americans more than 60 years ago changed theirs in order to adapt to the realities of the Space Age. But the realities of the "Environmental Age" require that such change occur on a global scale, include a commitment from both the people and the collective energy industry, and new ways of thinking must be adopted on a hitherto unprecedented scale.

Change to Green

In addition to the two specific case studies discussed above, we have witnessed pro-environment social movements that have enjoyed certain levels of success despite those who attempt to crush them. Currently, there is social movement that we refer to as the "Change to Green." The "Change to Green" mantra incorporates the "going green" movement, green tourism, and green employment, all the while of warning against greenwashing. Let's take a closer look at this "Change to Green" movement.

Going Green: Pragmatic Approaches

The "Change to Green" concept refers to switching from using products that are harmful to the environment (e.g., fossil fuels for energy and meat for food consumption) to those that are more eco-friendly (e.g., renewable forms of energy and vegetables for food consumption). In Chapter 4, we discussed many human behaviors (e.g., fossil fuel extraction and consumption, the use plastics, deforestation, food waste, and marine waste) that contribute to the 6th ME. If we want to embrace the idea of thrivability we will have to change all of these behaviors and many other harmful practices. We discussed the need for renewable energy in Chapter 2; here, however, we will discuss a few specific examples of pragmatic approaches to going green.

Scientists have proposed clustering together millions of solar panels to form an island that could convert carbon dioxide in seawater into methanol, which can fuel airplanes and trucks (Patterson et al 2019; Davidson 2019). The scientists argue, "Humankind must cease CO_2 emissions from fossil fuel burning if dangerous climate change is to be avoided.... The recycling of atmospheric CO_2 into synthetic fuels, using renewable energy, offers an energy concept with no net CO_2 emission" (Patterson et al 2019: 12212). The scientists argue that the technology exists to build floating methanol islands on a large scale in areas of the ocean free from large waves and extreme weather (i.e., off the coasts of South America, North Australia, the Arabian Gulf and Southeast Asia) (Davidson 2019).

One of the most difficult factors in "going green" is how to encourage people to change their ways. In Los Angeles, a sprawling city long identified with energy consumption and pollution, people are getting paid to go green. A groundbreaking energy program called "Clean LA Solar" supplies zero-carbon, renewable energy for Los Angeles while creating jobs and fueling private investment in the city (Clean L.A. Solar 2020). The program allows the Los Angeles Department of Water and Power to pay customers to "generate solar power across the city's vast expanse of flat roof space" (Green 2013: B5). The State of California has mandated that 33 percent of electricity must be generated by renewable sources by 2020. In addition to increasing the amount of energy being produced by solar power, this program has had a beneficial impact on the economy. It is helping with the unemployment problem—over 1300 "solar installers" have been trained for the new green job opportunity.

In New York State, Governor Andrew Cuomo (D) signed into law (2019) the nation's most aggressive targets for reducing carbon emissions and is intended to drive dramatic changes over the next 30 years. The law "calls for all the state's electricity to come from renewable, carbon-free sources such as solar, wind and hydropower. Transportation and building heating systems would also run on clean electricity rather than oil and gas"

(*Associated Press* 2019c: A15). Among the possible methods of reaching this goal would be solar panels on every roof; parking meters that double as car chargers; wind turbines towering above farm fields and ocean waves; and cars and home furnaces and factories converted to run on electricity from renewable sources (*Associated Press* 2019c). New York's attempt to fund a variety of other projects—including investing in fish hatcheries, restoring riparian buffers to reduce nutrient runoff, removing obsolete and hazardous dams, and renovating hazardous dams, and renovating infrastructure to better withstand flooding—where put on hold as Governor Andrew Cuomo's plan to place a bond measure on the state's 2020 election ballot to fund the projects were deemed ill-advisable in light of the state's decimated economy as a result of the pandemic (*Associated Press* 2020b).

In 2019, soon after Governor Janet Mills of Maine (D) took office she announced renovation plans for the Maine's governor's mansion to add solar panels. This move was seen as a "rebuke to her predecessor, Republican Gov. Paul LePage, whose administration put a moratorium on new wind turbines and enacted policies that critics say stymied solar energy in the sate" (*Associated Press* 2019d). According to the National Conference of State Legislatures, Maine was among 11 states that either flipped the governor's seat from Republican to Democratic or saw Democrats win newfound control over the legislature in the 2018 midterm election. Not coincidently, all of these states have passed or are weight legislation that would expand renewable energy in their states (*Associated Press* 2019d).

As demonstrated a number of times in this text, the Democratic Party has long been generally known as the pro-environment political party while the Republican Party routinely opposes efforts to sustain the environment. This reality dates back to 1979 when President Jimmy Carter installed 32 solar panels on the roof of the White House long before many Americans even knew what solar power was. Carter's effort to educate the nation on the benefits of alternative energy was short-lived as President Reagan had the solar panels removed when he became president in January 1981 (Ward 2016). It is worth noting, however, that President George W. Bush (R) installed solar panels at the White House in 2003. Carter, the president who has done more humanitarian work than any other former president, did not give up on solar energy. Instead, he leased 10 acres (four hectares) of his farmland to SolAmerica Energy for the installation of more than 3,800 panels that rotate to follow the sun and will provide energy to more than half of his hometown of Plains, Georgia, which has fewer than 700 residents (Wulfhorst 2017).

If people continue to create pragmatic alternative forms of energy we can lessen our dependency of fossil fuels and save the environment. And if that happens, we could see the price of crude oil drop again, as it did during the COVID-19 pandemic, when the price fell to a negative $37.63 per barrel (Roth 2020c) as sellers had no markets to send their fuel. However, as realists, we understand that it will be a long time before people are not relying on fossil fuels as their primary source of energy. Then again, as the events of the COVID-19 pandemic have taught us all to be better prepared for the future. Let's hope politicians and Big Business also come to the realization that renewable sources of energy need to be developed on a large-scale basis, immediately.

One of the greatest ecological issues is the disposal of waste. This is especially true at big sporting events. The Carrier Dome at Syracuse University is dealing with this problem in an innovative way. It has installed a new waste disposal system to collect the rainwater that runs off the fabric roof of the dome, home to the 49,262-seat arena where the SU basketball, football, lacrosse and other teams play. "Approximately 880,000 of the 6.6

million gallons of water that pours off the Dome's 7-acre roof each year will be captured by the system and stored in tanks hung from the bottom of the arena's upper bleachers. During events at the Dome, the water will be used to flush the toilets and urinals in the building's 16 public restrooms" (Moriarty 2013: A8). The system consists of two exterior 25,000-gallon bulk storage tanks, and two interior 4,500-gallon tanks that hold rainwater captured from the roof. The tanks hold enough water to flush half the Dome's toilets and urinals during major sporting events with blue-dyed water. Building codes require the water to be dyed to avoid confusion with drinking water, even though the water will only be used in toilets and urinals (Blalock 2019).

Earlier in this chapter we cited the concern among scientists for the need of billions of new trees to be planted. It's not as if the earth does not have trees as it is estimated that there are 3 trillion trees growing on the planet, but it is estimated that the earth once had seven times that number prior to human civilization (Borenstein 2015). Yale forestry researcher Thomas Crowther found that 15 billion trees are cut down each year by people, with another 5 billion trees replanted. That's a net loss of 10 billion trees a year (Borenstein 2015). The math is clear; we need to plant far more trees. There are efforts being made. In Ireland, which has enough trees to offset its own carbon emissions, government officials announced in 2019 its plan to grow an additional 440 million trees by planting 22 million trees for the next 20 years (Krstic 2019). In Morocco, there are attempts of reforestation. The Reforest Action sustainability plan is designed to plant a wide variety of trees including those with agricultural benefits and those that are not invasive to local fauna and flora populations. Among the expected benefits are savings in energy for villages, development of the local economy thanks to the growing of biological fruits, development of ecotourism in the region, and raising awareness of the inhabitants on sustainable development (Reforest Action 2020). The Reforest Action website states that over 6.7 million trees have been planted so far. The push to grow trees in Morocco is a part of kingdom's effort to reduce its fossil fuel dependency—at 89.33 percent in 2017—down to 60 percent (World Atlas 2020a). In Syracuse, the closest major city to one of the authors, there are lots of trees (pending their potential destruction as a result of invasive species) and the plan to plant an additional 70,000 trees over 20 years would increase the land area coved by tree leaves to one-third the size of the city (Coin 2020). While Syracuse was already planting approximately 350 trees a year and an additional 1,000 trees with help from the Save the Rain program, the city loses about 800 trees per year because of disease (often caused by invasive species) (Coin 2020). Chances are, the city closest to you is also doing something about increasing the number of trees, why not check and see? And perhaps you could participate in the planting of trees in order to reduce the negative impact of carbon emissions and lower your own carbon footprint.

Green Tourism

Perhaps in order to "save the Earth" it is important to better understand it. After all, as the saying goes, seeing is believing. To truly understand the nature of the problem, and to appreciate more the beauty of the world in which we live, it is important to see as much of it as possible. At least that's the idea behind ecotourism. *Ecotourism* refers to encouraging people to travel to exotic, often threatened, natural environments throughout the world so that they can experience its wonders, interact with local populations and support conversation efforts at the destination site. The important point here is that, as much as possible, one must not endanger, degrade, or exploit local resources, such as animals, trees, or soil, but rather try to understand them. Ecotourism provides educational opportunities for travelers and economic opportunities for locals. It can be a prime means of consciousness-raising. Ecotourism is promoted in many developing countries because it can diversify local economies and the influx of foreign currency is generally very valuable. Destination sites are mostly conscious about proper waste and recycling efforts as well (Furqan, Som and Hussin 2010).

However, as we alluded to in Chapter 7, the economic benefits of ecotourism can easily be overridden by environmental upheavals. There is the ever-present danger of exploitation, the possibility of overcrowding on lands previously with little human activity, or the likelihood of harmful impacts. The potential degrading effects of ecotourism is something that must be addressed as it has become almost universally accepted as a desirable and politically appropriate approach to tourism development and sustainability awareness (Zolfani, Sedaghat, Maknoon and Zavadskas 2015).

Even remote parts of the world are now being explored. In an article titled "Antarctica Concerns Grow as Tourism Numbers Rise," science writer Rod McGuirk reports that Antarctic tourism has grown from fewer than 2,000 visitors a year in the 1980s to more than 46,000 in recent years (McGuirk 2013). "There are fears that habitat will be trampled, that tourists will introduce exotic species or microbes or will transfer native flora and fauna to parts of the continent where they have never before existed" (McGuirk 2013). Twenty-eight countries comprise the Antarctic Treaty Consultative Committee and are monitoring the various travel agencies venturing into this territory. Still, not everyone is unhappy with the increasing numbers of visitors. "Australia-based adventurer Tim Jarvis sees Antarctic tourists not as a problem, but as a part of the solution for a frozen continent where the ice is rapidly retreating. If more tourists see its wonders and the impacts of climate change, particularly on the Antarctic Peninsula, Jarvis said, the world will become more inclined to protect the continent" (McGuirk 2013). This is a key point—perhaps the best way to understand the fragility and interconnectedness of life is to experience it firsthand, but hopefully this can be done in a manner that does not compromise the terrain. In a particularly evocative sort of conscious-raising, the winter 2013 "swimsuit issue" of *Sports Illustrated* featured model Kate Upton in a bikini posing in front of the ice caps of Antarctica—while this may perhaps have spurred a certain amount of new tourism to that region, the tourists are unlikely to see any bikini-clad models there. But they will certainly learn a great deal about a previously little-known part of the world.

That is the challenge—can there be a thrivable ecotourism industry? In order to help make this a reality, a group of tourist companies, non-governmental organizations and nonprofit organizations joined together to create the Global Sustainable Tourism Council. Its purpose is to encourage the tourist industry to continue to drive conservation and poverty alleviation, as well as promote higher standards of respect among world travelers. This is particularly important as more advanced means of transportation are allowing areas of the Earth previously unreachable to be more easily accessed.

Green Employment

Undoubtedly, as you have read this sampling of green activities, it has dawned on you that there are plenty of business opportunities in the "Change to Green" wave of pro-environmentalism. The U.S. clean energy sector employed nearly 3.4 million people before the COVID-19 pandemic—three times the workforce of the fossil fuel industry according to Environmental Entrepreneurs, known as E2 (Roth 2020d). The number of clean energy jobs were increasing quickly and the Bureau of Labor Statistics projected in 2019 that the two fastest-growing jobs in America over the next decade (2020s) would be solar panel installer and wind turbine technicians (Roth 2020d).

However, the COVID-19 pandemic cost millions of people their jobs—jobless claims increased to more than 26 million people in just five weeks (mid-March to the third week of May)—in the United States alone, and a significant number of these jobs were in the clean energy sector. A report from Environmental Entrepreneurs, a clean energy advocacy group identified more than 106,000 clean energy workers filed for unemployment benefits in March—a number that most certainly grew over in the weeks since then (Roth 2020d). And this was the case as more than half a million clean energy jobs were lost in March and April alone, "reversing years of growth in an industry that has helped reduce lung-damaging air pollution and the emissions responsible for climate change" (Roth 2020e: A8). Once the pandemic is over, these jobs are expected to return, and the upward trend of green employment should resume.

Green employment wasn't the only energy industry affected by the pandemic as BP announced in June 2020 that it would cut 10,000 jobs accelerating the company's move to slim down for their green energy transition (*Los Angeles Times* 2020c). BP's attempt to embrace green employment remains commendable, if not necessary for their survival.

Greenwashing

There are times when claims of "going green" by businesses may actually be "greenwashing" in disguised. Greenwashing refers to the "token environmentally friendly initiatives as a way of hiding or deflecting criticism about existing environmentally destructive practices" (Schendler 2009: 225). Greenwashing, then, is a term used to describe the deceitful practices of some companies that try to hide what they are really doing to the environment or try to deflect critics by pointing them to the direction of a token pro-environmental initiative they have taken. Greenwashing is a marketing strategy used by deceitful companies to dupe the public about the true nature of their company's operations (Hoffman and Hoffman 2009).

For example, in 2001, Shell Oil launched a greenwashing campaign that indicated they were leaving large tracts of green forests intact and working diligently on renewable energy sources while in reality, they were spending just 0.6 percent of their annual investments on renewable energy (Schendler 2009). In 2019, China, Japan, and South Korea joined other industrialized nations in promising to reduce their use of fossil fuels in an attempt to battle climate change. However, as the fossil fuel industries in these three nations took steps to promote renewable energy at home, they were also financing dozens of new coal-fired power plants in other nations (Bengali 2019). Most of these plants are being built in Southeast Asia and Africa where emerging economies face a growing demand for cheap, reliable electricity and that need is most easily met by coal, the single largest source of the greenhouse gas emissions blamed for warming the planet (Bengali 2019). As a result, "even as demand for coal has flattened in China and decline in industrialized Western nations, its rise in the rest of the Asia has helped push carbon emissions up [in 2018] by 2% to the highest level recorded, according to the International Energy Agency" (Bengali 2019: A1).

Some companies that are doing their best to be pro-environment worry about publicizing the good things they are doing out of fear that the public may believe they are greenwashing. There are websites that monitor and publicize companies attempting to greenwash the public. One such site is the Greenwashing Index, a website that has a forum for customers to post about companies' green claims (O'Connor 2008). Greenwashing has become such an issue that the Federal Trade Commission has been reworking their environmental "Green Guides." "The Green Guides were first issued in 1992 and were revised in 1996, 1998, and 2012. The guidance they provide includes: (1) general principles that apply to all environmental marketing claims; (2) how customers are likely to interpret particular claims and how marketers can substantiate these claims; and (3) how marketers can qualify their claims to avoid deceiving consumers" (Federal Trade Commission 2020).

The Environmental Movement and Thrivability

We saw in Chapter 7 how the initial impetus for Earth Day came from United States Senator Gaylord Nelson (D–Wisconsin). He had become worried about the impact of pollution on the fishing, canoeing and other outdoor recreational activities in his native state, and had even persuaded President John F. Kennedy to give a talk about conservation in 1963, shortly before his assassination. The first Earth Day, on April 22, 1970, was a huge success, creating a firestorm of activities involving millions of Americans, including—amazingly enough—over two-thirds of the members of Congress (Lemann 2013: 73). How strange that seems today, when many political leaders seem at best distant from, if not outright hostile to, conservation efforts.

What might be done to reenergize the mass movement that Nelson and his fellow organizers helped to launch so many decades previously? How can we expand on the global recognition of Earth Day to a global acceptance of "going green"? And then, how can we get politicians, businesses, and the mass populace to join in on efforts to help the environment thrive? First of all, as we have argued throughout this book, a nationwide (and ultimately international) focus on understanding the environment and the many localized ecosystems must occur. As happened with the *Sputnik* phenomenon, everyone must be aware that the times are a changing when it comes to the environment. Secondly, as Garrett Hardin warned, we must stop thinking of ourselves as individuals concerned only, or at least primarily, with grazing our own herds, as it were, and rather understand that the commons on which we live—the Earth—is a vastly interconnected environment of which we are but a part. But third, as this chapter has attempted to show, we must also do all we can to motivate both ourselves and our fellow citizens to understand that tackling the very real problems caused in part by our own activities and in part by the very planet on which we live and the universe in which it resides can be a source of happiness.

Most environmentalists and college campuses promote the idea of sustainability—the ability of the environment to hold, endure, or bear the weight of a wide variety of social and natural forces that may compromise its functionality—but we have consistently promoted the idea of thrivability—a cycle of actions which reinvest energy for future use and stretch resources further, thus going "beyond sustainability." Thank about it this way, if you are economically poor and struggling to get by day-after-day, do you want to sustain that lifestyle? No, you want something better, you want to thrive. You want to, at the least, become financially secured enough to pay for all the essentials of life but also to the point where you can do some additional things like take a vacation, help members of your family meet their needs, donate to a favorite charity, and so on. Clearly, thrivability is preferred over sustainability. Remember the ominous proclamation of Lucy in the *Peanuts* strip ("Stick the next generation") and don't postpone your actions or hope that future generations will somehow magically solve them. As the saying goes, if we continue to follow our current practices, our children's children will curse us.

The younger generation already blames the Baby Boomer generation for the environmental nightmare that has been left to them to clean up. However, as we know, and have demonstrated in this book, the Baby Boomers are the generation that started the environmental movement so the blame should not be thrust upon them. The blame for the deterioration of the environment lies with Big Oil, Big Business, Big Agriculture, Big Pharma, and the politicians who are "in bed" with these fossil fools and greenhouse asses. Every time a pro-environment "baby boomer" hears the derogatory term of "OK, boomer" you can expect a retort of "OK, Millie." Meanwhile, the real human destroyers of the environment continue their assault and look on with great amusement watching millennials fight with baby boomers over "save the earth" issues. The people, regardless of their generation, must unite and require that the entities that are devastating the environment stop their harmful activities. Furthermore, seeking a sustainable environment is not good enough; we must demand an environment that thrives.

One simple thing that everyone can do is to join the Earth Day Network (see http://www.earthday.org/) and help to keep Senator Nelson's vision alive. Many people have already taken this straightforward first step as evidenced by the 50-year anniversary of Earth Day statistics that reveal one billion individuals have joined the network and over 75,000 partners in over 190 countries that are working to drive positive action (Earth Day Network 2020). The Earth Day Network's (2020) homepage states as its mission: "To build the world's largest environmental movement to drive transformative change for people and planet." The vision statement of the Earth Day Network reiterates the point that the world needs transformational change and that it is time for the world to hold sectors accountable for their role in our environmental crisis while also calling for bold, creative, and innovative solutions.

Getting involved with local organizations that are promoting conservation and sustainable efforts is a good second step. Every community has multiple organizations that people can join and, if your community does not, you can be the person who starts one. Surely someone in the community will join your efforts to help the environment thrive. In most cases, however, there are plenty of groups you can join. Local grassroots social movements can be very effective in making change happen. Also, local conservation organizations can join with larger, national, or global organizations. The beauty of merging with larger pro-environmental organizations is the realization that there is strength in numbers. If you're not sure how to get started, simply search for "local conservation and sustainability organizations" and all sorts of options appear. You are likely to find conversations groups in many areas, including national parks, land preservation, wilderness conversation, marine conservation, healthy rivers and water supply, climate change, outdoor access, recreation and trail, and local environmental groups. Good stewardship of local ecosystems contributes to helping the overall environment thrive.

Ultimately, when it comes to a commitment to change one's lifestyle choices and becoming more in tune with the environment, Edward O. Wilson's concept of "Biophilia" comes to the fore. If one truly holds that there is an instinctive bond between human beings and other living systems, and a real inborn sense that the Earth is our home, one can better feel the need to implement different strategies for survival. Being connected with nature is an essential aspect of well-being and healthy living. Architects are coming to a better understanding of this, as can be witnessed from the following quote from the *Green Building Manual*: "Biophilic design elements are sometimes considered an unnecessary expense because it is difficult to attach a monetary value to their benefits. As studies move forward and evidence mounts about the health and psychological benefits of incorporating these ideas into the built environment, the value of biophilic design is gaining acceptance" ("Whole Building Design Guide: Psychosocial Value of Space").

Wilson himself notes that humans can learn valuable lessons from studying other living creatures. He has traveled most of the known world (usually in search of new species of his own favorite creature, the ant) and he has made saving the natural world his highest priority. As a recent *New York Times* article pointed out, "When it comes to nature.... Dr. Wilson remains easily and infectiously wonder-struck. Before heading out of Central Park he paused by the statue of Balto, the sled dog who became a national hero after carrying medicine to diphtheria-stricken Nome, Alaska, in 1925. 'It's good to see a monument to an animal,' he said, looking up and smiling" (Schuessler 2012: C5). An animal *other* than a human, one might add.

Thrivability—A New Approach

Following in Dr. Wilson's footsteps, as it were, we argue that, in order to re-inspire and rejuvenate the environmental movement, there is a need for a new way of thinking. It involves accepting evolution as a basic fact of existence and coming to grips with the role of humanity in an ever-evolving ecological system, coupled with utilizing our problem-solving intellectual abilities to their highest level in order to adapt ourselves in the most efficient and beneficial way possible. That is the thrivability challenge. Humanity itself—through its sheer magnitude, if not bulk—has become a genuine force of nature. We are living in an era of accelerated growth and change, powered in large part by the movement of the human species across the globe. "Scientists are increasingly using a new name for this period. Rather than placing us still in the Holocene, a peculiarly stable era that began only around 10,000 years ago, the geologists say we are already living in the Anthropocene: the age of man" (*The Economist* 2011: 11). The Anthropocene era refers the current geological age, an era in which human activity has been the dominant influence on climate and the environment.

In addition to the Anthropocene era, another term being used by scientists to describe our present era is "the Plasticsphere." *The Economist* also notes: "Since 2008 geologists have been mulling over the idea of the Anthropocene, a proposed new epoch in the history of the earth that would encompass the years in which people have had profound effects on the planet's workings. Most often, discussion of the Anthropocene revolves around how atmospheric chemistry has changed since the beginning of the industrial revolution. Sometimes the effects of new terrestrial ecosystems, in the forms of fields, pastures and plantations, are also considered. To date, though, how the Anthropocene has created new ecosystems in the oceans as well as on land has not been much examined" (*The Economist* 2013: 70).

And just what exactly might these new ecosystems consist of? You guessed it—plastic. The debilitating role of plastic debris on the environment—including the existence of the Garbage Island—has previously been discussed in this book, but scientists are also now examining possible beneficial aspects of plastics on various life forms which either consume them or ride upon them. "Plastics are energy-rich substances, which is why many of them burn so readily. Any organism that could unlock and use that energy would do well in the Anthropocene. Terrestrial bacteria and fungi which can manage this trick are already familiar to experts in the field.... Conservationists intent on preserving charismatic megafauna have reason to lament the spread of plastics through the ocean. But those interested in smaller critters have been given a whole new sphere—the plasticsphere—to study" (*The Economist* 2013: 70).

Let's hope that (as George Carlin humorously said) the creation of plastic is not the only reason humans came on the scene. Whether or not we will ever answer that age-old philosophical question "why do we exist?" we can certainly use our "environmental imaginations" to try to better understand our current role in the universe and our potential possibilities in the changing Anthropocene era. Both environmental sociology and environmental philosophy are aiding us in our efforts to examine the natural and social forces which shape our existence.

Thriving within the Anthropocene era is thus our current challenge. Ever aware of the real possibilities of catastrophes, both human-caused and otherwise, we can be on our guard in trying to anticipate changes and coming up with strategies to flourish within them. George Carlin gleefully predicted that, now that our job of creating plastic is done, we can be phased out, "and I think that's really started already, don't you?" Plastic may be here to stay, but so, one can hope, are we. Perhaps, if we truly come to grips with the challenges facing us, thrivability too may be coming to the U.S. and beyond.

In the concluding chapter we will examine the importance of education for environmental thrivability, from grade school to college, as well as for individuals. Like the first sighting of *Sputnik* so many decades ago, this is a learning moment—we can change, but it is in our own hands whether we do so or not.

Let's not be like the dinosaurs. They had no chance—we do.

Popular Culture Section 9

"Soylent Green" Is Prescient

Charlton Heston's classic film *Soylent Green* is best remembered for its (spoiler alert) closing line: "Soylent Green is people!" But it is also a powerful cautionary lesson about environmental disaster caused by overpopulation, human-generated pollutants, human-caused famine, the contamination of the world's water supplies, and the negative impact of the Greenhouse Effect. Set in the year 2022 and made during the beginning of the Environmental Movement of the 1970s, the film was far ahead of its time in warning people that their increasing disregard for human beings' impact upon the overall environment could have catastrophic repercussions for future generations. Now, with the period of the film less than 2 years in the future, is a good time to re-examine *Soylent Green* to see how accurate its predictions were, and what lessons can be learned from it.

Heston in the 1960s and 1970s was the go-to guy for science fiction films set in the future with twist endings (other examples were the original *Planet of the Apes* and *The Omega Man*). *Soylent Green* was directed by Richard Fleischer, and in addition to Heston starred such well-known actors as Edward G. Robinson, Joseph Cotten, and Chuck Connors. It was filmed in 1973, a year of great upheavals in the United States, witnessing the Arab oil embargo, rampant pollution, population concerns, and the revelations of the Watergate cover-up—all of which were dealt with in various ways in the film.

Soylent Green is by no means a great film. The performances are clunky, both over-acted (Heston) and under-acted (his movie love interest Leigh Taylor-Young). Edward G. Robinson, in his final film role, hams it up as Sol Roth, investigative agent, but that adds some needed spark to what is otherwise a rather dour scenario.

The movie is set in New York City in the year 2022. Forty million people are living there under a totalitarian "Big Brother" type government (redolent of the then Nixon Administration). The Soylent Corporation, which is hand-in-glove with the government, controls the world's food supply, by creating an artificial food supposedly made from soy and lentils—"soylent." It comes in different colors, and a new type—Soylent Green—has just been introduced (obviously a rather ironic name, since green implies natural, which the food, as we learn, most definitely is not).

Heston plays a corrupt cop named Thorn (another interesting name, given that he is a thorn in the side of the government) investigating the murder of a Soylent executive played by Joseph Cotten. The murdered executive lived in a fancy townhouse, far different from the cramped and seedy apartment that Thorn shares with his friend and fellow investigator Sol Roth. In addition to having an affair with the executive's mistress, Thorn steals "real food" from the murdered man's refrigerator and brings it home. There is a poignant and funny scene between Heston and Robinson, where the two men taste real meat and vegetables—in Roth's case, after many years of surviving on artificial food, and in Thorn's case, for the first time ever.

Outside the apartment, humans start to riot when the supply of Soylent Green runs out. They are treated like garbage—literally, for giant garbage trucks scoop up the rioters and, as we learn, take them in the trucks to the Soylent factory. We also learn that the reason the executive was killed was to prevent him from revealing the shocking truth that the food supply is running out. The oceans are dying—the plankton is gone. The interconnectedness of life is a theme running throughout the film, as is the possibility of humans destroying this connection through their greed and corrupt business practices, exemplified by the sinister Soylent Corporation.

The shock ending of the movie, where Thorn (who has hidden on top of one of the garbage trucks and gone into the factory) learns the bitter truth is perhaps not so shocking now—it has been parodied many times on shows like *The Simpsons* and *The Family Guy*. *Soylent Green*, Thorn discovers to his horror, is made not from soy and lentils but from human flesh. After being shot, he is rescued by one of his colleagues, to whom he reveals the immortal last words "Soylent Green is *people*!"

While the ending may no longer be shocking, *Soylent Green* is a good film to ponder, for its historical significance (while set in the future, it is very much a picture of life in the early 1970s) and for its depiction of a dystopian world no one would wish to live in. This is best exemplified by the way in which Sol Roth, upon learning the truth about the Soylent

Corporation, decides to "go home." He gives up on life and enters a Euthanasia Center, where he asks to be put to death.

This is a very moving scene. The center is an air conditioned, pleasant environment, not crowded at all, with friendly "caretakers." This is how life should be, it seems, but it is only experienced by people at the very end of their lives. With classical music playing, Sol watches a film showing how the earth used to look before the days of pollution and over-crowding—beautiful wildlife, flowers, sunsets. Thorn arrives too late to save his friend but does get to witness the film—Roth tells him that he had known such a world in his youth, but Thorn had never seen anything like it before.

Richard Fleischer, the movie's director, mentions on the voiceover on the DVD version that this was not only the last scene of the movie for Robinson, but the last scene he was ever in. Only a handful of people, including Heston, knew that he was dying from bladder cancer and would never make another movie. Heston and Robinson were not just friends in the movie, they were very close personally, having appeared together years earlier in the classic film *The 10 Commandments*. Fleischer points out that those are genuine tears in Heston's eyes as he watches his old friend's final scene. For once, he doesn't overact.

That particular scene from *Soylent Green* still packs a punch. Like the ending of Heston's *Planet of the Apes*, the message is a simple but powerful one—we blew it. When it comes to showing us the likely consequences of our actions if we continue on our present course, and the horrible future we would thereby bestow upon our descendants, *Soylent Green* is prescient.

Summary

The case has been made throughout this book that humans need to change their behaviors dramatically if we hope to slow the process of the sixth mass extinction. To that point, in this chapter we propose that humans *can* change their behaviors but ask the ultimate question, "*Will* we change?"

It was proposed that we do, in fact, *need* to change as the dire warning signals of a compromised environment exist everywhere. To address the question of whether or not we will change this chapter describes a number of reactions to the need for change: environmentalists want immediate change; fossil fuel proponents want little or no change; some believe its too late and we are already doomed; others exult humanity's extinction; some try to hide from reality by burying their heads in the sand; and many tragically fail to act in their own best interests.

Humanity is absolutely *capable* of dramatic change as evidenced by the Sputnik story and COVID-19 pandemic case study. So, there is at least a glimmer of hope that when humans are pushed to the brink, perhaps they will act in a manner that is best for the environment if our survival is at stake. Our capability for change is spearheaded by the "change to green" mantra that incorporates the "going green" movement, green tourism, and green employment, all the while warning against greenwashing.

The change to green and joining in the environmental movement and promoting thrivability are how we can and will change. If humanity does not radically change its harmful environmental behaviors, we will meet our destiny of being like the dinosaurs long before our presence on earth was supposed to cease.

CHAPTER 10

Educating Thrivability

Several years ago, the authors of this book were on a famous university campus in another country (both of which will remain unnamed) to give lectures at a conference. While they were excited by the intellectual caliber of the proceedings, they couldn't help but notice how run-down and garbage-ridden the campus was. After walking past the same buildings for several days in a row, it was clear that the garbage on the ground was a permanent feature and was in fact being added to by passersby on a daily basis. No one, it seemed, felt it was their responsibility to pick it up. The restrooms were also, to put it mildly, horrifying, and in an act of international kindness several students at the campus, even though unable to speak our language, made it clear through gestures that we should hesitate before entering such places if we knew what was good for us.

We have frequently mentioned this to the students at our own campuses, the State University of New York at Oswego and St. John Fisher College, not to belittle the other campus but to make it clear how important it is to maintain a proper upkeep and love for one's environment. When it comes to educating about environmental thrivability, having a sense of place is a good way to start. In order to flourish, it is important to have respect for one's environment.

Reputable Environmental Organizations

One good way to learn about environmental thrivability is to connect with reputable environmental organizations that have proven their credibility. Here is a short list of some of the organizations we recommend.

Earth Day Network. As mentioned, the Earth Day Movement began in 1970 and is still celebrated across the globe. Its website gives detailed information on the activities held each year and environmentally beneficial efforts promoted worldwide. Of particular note is the "Greening Schools and Promoting Environmental Education" project.

The Nature Conservancy. Founded in 1951, this organization's mission is to "preserve the land and waters on which all life depends." Working in more than 30 countries and with over a million members, it puts its money where its mouth is, as it were, by buying millions of acres of land and protecting thousands of miles of rivers across the globe, as well as operating hundreds of conservation projects throughout the world. It is one of the most trusted environmental agencies, with a proven track record of commitment to conservation. One of its best-known efforts is the Plant a Billion Trees campaign, which is attempting to restore by the year 2015 much of the Atlantic Forest of Brazil.

World Wildlife Fund (WWF). Founded more than 60 years ago, the WWF is a leading conservation organization that works in more than 100 countries and is committed to reversing the degradation of our planet's natural environment.

Greenpeace International Organization. Founded in 1971 in British Columbia to oppose U.S. nuclear testing at Amchitka Island in Alaska, this organization is dedicated to preserving endangered species of animals, preventing environmental abuses, and heightening environmental awareness through direct confrontations with polluting corporations and governmental authorities.

National Geographic Society (NGS). Headquartered in Washington, D.C., this non-profit organization was founded in 1888 for the purpose of promoting environmental and historical conservation and the study of world culture and history.

Planeterra Foundation. Founded in 2003, Planeterra promotes responsible ecotourism via a global network of travel industry partners by providing support for local communities and encouraging travelers to respect and appreciate the people and places they venture to: http://www.planeterra.org/.

These are just a few of the many worthy organizations dedicated to the cause of environmental protection and ethically responsible living.

Efforts by Colleges

Since the authors are college professors, they are especially committed to the cause of making college campuses more attuned to learning about environmental issues in both an academic and personal way. Students are by their nature enthusiastic learners and are the key to making a beneficial impact. One of the best ways to begin to make a difference is to become active in the various "go green" movements on campuses.

In Canada, the university setting has become a place of educating students on issues related to sustainability as they recognize that environmental issues have become an increasingly important aspect in people's lives. Schools are trying to keep up with the public demand for environmentally friendly and sustainable campuses. Students and professors want their schools to adopt a "green" approach to running their operations. When it comes to the green practices of colleges the group people often responsible for these actions is Facilities Management. Some campuses will have a research center devoted to environmental research and sustainability. Many campuses have student groups that are devoted to green efforts such as recycling, organizing garbage cleanup days, provide information sessions, and other initiatives. Pretty much all schools encourage some sort of recycling and composting, especially around the food court areas of campus. Most new buildings being constructed on campuses are incorporating sustainable technology (e.g., water-reduction systems, green roofs, and solar cells on roofs). Edible landscaping such as fruit trees and vegetable gardens are being constructed for use in local food places on campus. Pedestrian and bike paths are becoming increasingly common on campuses, especially the larger ones, to encourage people to use their vehicles less often and take public transit or bikes (Canadian-Universities.net 2019).

The United States has far more campuses than Canada, and it takes a more case-by-case approach, but the green movement is certainly going strong on many of its campuses as well. The Association for the Advancement of Sustainability in Higher Education (AASHE) was established in 2005 to "help coordinate and strengthen campus sustainability efforts at regional and national levels and to serve as the first North

American professional association for those interested in advancing campus sustainability" (BestColleges.com 2020). AASHE (2020) claims to be the leading association for the advancement of sustainability in higher education serving faculty, administrators, staff and students who are change agents and drivers of sustainability innovation. AASHE is comprised of over 900 members across 48 U.S. states, 1 U.S. territory, 9 Canadian provinces, and 20 countries (AASHE 2020). AASHE recognizes colleges and universities that have achieved the greatest level of success with green initiatives on-campus and within their surrounding communities. The Association gauges these efforts using the Sustainability Tracking, Assessment & Rating System (STARS), a voluntary system that allows different colleges and universities to report trends and track their sustainability efforts. There are four different STARS ratings—platinum, gold, silver, and bronze—that may be awarded to various establishments (BestColleges.com 2020). These rankings are made public so that colleges and universities may advertise their achievements as a mechanism to attract potential in-coming students. Students and parents may want to look at the AASHE rankings to see if their school made the Top Fifteen. At the very least, you may want to see what your college's STARS rating is and take action to get your school ranked higher in this environmental ranking.

College Courses and Sustainability Studies Programs

Another effective way to promote thrivability efforts on campuses is to develop and teach courses specifically devoted to the issue. The authors of this book have each done so, through devising and offering classes, and through their involvement with the Sustainability Committees at their respective colleges. The following information may prove helpful for those who might wish to be involved in similar courses.

St. John Fisher College—Environmental Ethics

Tim Madigan has taught a course titled "Environmental Ethics" for many years. It is interdisciplinary in its focus and is open to all students regardless of their majors. The course focuses on such concepts as anthropocentrism versus biocentrism, invasive species, the Gaia theory, the Anthropocene era, the tragedy of the commons, stewardship, green pragmatism, and biophilia, all of which are examined in this book (which is used as the text for the course).

On the first day of class, each student is asked to reflect on the question: Why should we be concerned about the future of the planet? Using George Carlin's routine (from his 1992 album "Jammin' in New York") on "the Earth will be fine—it's humans that won't make it," they then discuss the question, "What would be so bad about the human species becoming extinct?" Their first written assignment is to research the term "environmental sustainability." At the end of the semester, after having examined several different ethical theories, they are asked to return to their original paper and analyze it in light of the information they have learned (especially after discussing the concept of "thrivability") and write another short paper describing whether their views and have changed and, if so, why.

While there is a thorough examination of the history of the Environmental Movement and the various ethical theories connected to it, this is more than just an academic exercise. The emphasis is on how each student can make a difference, especially through learning more about the campus itself. Campus field trips include a visit to the college cafeteria, where students learn about food waste, how it is being dealt with, and other food management policies. These include the use of local food, which reduces the number of deliveries and the carbon footprint caused by this, as well as the use of reusable mugs and recyclable containers, utensils and napkins. The cafeteria has reduced serving sizes in order to better control waste and has started a composting station. It is an eye-opening experience for the students to learn about the waste of food that occurs on a daily basis.

A tour of the various wildlife outside the classroom is led by a professor of biology whose project is to catalogue all the trees on campus. The students participate in this project, and thereby make a contribution to preservation efforts. They also meet with representatives from facilities who talk about how the campus deals with waste management and get a chance to interact with the cleaning crew—vital members of the campus whose work is too little appreciated. The students are asked to keep records of their own garbage output for one week. They also are asked to do a walking tour on their own of the campus and keep track of efforts to keep it a clean and nature-friendly environment—unlike the example mentioned at the beginning of this chapter. The students then choose a research topic of their own, write an extended paper on it, and give a presentation about the topic to the entire class.

St. John Fisher College founded a Center for Sustainability in 2017, which offers both bachelor's and master's degrees to majors, as well as a minor in Sustainability (see the Center's website for further details). Madigan is on its Executive Board, and has helped design several of the required courses, including Environmental Ethics.

State University of New York (SUNY) at Oswego—Environmental Sociology

Tim Delaney created the course "Environmental Sociology" after first teaching a course titled "Environmental Sustainability." Initially, this interdisciplinary course had as its focus the idea of "sustainability" as it embraces the current of the time. However, it became very clear to Delaney and his students that with the Earth as deeply compromised as it now is, why would anyone promote the idea of sustainability? After all, one no more wishes to sustain a compromised environment than they would sustain unemployment or poverty. It is clear that we have to help the Earth thrive, much as each of us hopes to thrive in life.

As with any course, we begin by exploring and defining key terms, in this case, such concepts as the environment, sustainability, and thrivability. This course examines humanity's role in compromising the Earth's carrying capacity and ways in which humans can prevent the environment from being permanently sullied. Over the generations, but especially recently, the Earth's carrying capacity has been stretched to its limit due to a number of threats to the environment. These threats include the spread of deserts, the destruction of forests by acid rain, deforestation, the stripping of large tracts of land for fuel, radiation fallout, and the many areas where the population is exceeding the capacity of local agriculture.

The role of nature as a negative force on the environment is also addressed as it directly ties into the central course premise of an impending mass extinction. Coupled with the fact that each of the previous five mass extinctions occurred before the presence of humans, it is clear that nature, including forces outside our planet, is a potentially devastating source of destruction. It is further argued that the sixth mass extinction is not only inevitable, but the process of this extinction is also being sped by human activities. The course does not take a completely negative outlook on the future of humanity, as it is argued that through social movements and individual human behaviors, we can indeed help the planet thrive and postpone, perhaps for thousands of years, the next mass extinction.

Students are strongly encouraged to share their thoughts and ideas on any environmental topic both in classroom discussions and via student research term papers. It is clear to this professor that students are concerned about the future of this planet and their role in helping the Earth thrive.

At SUNY Oswego, Delaney serves on Sustainability Studies Committee and is a faculty member of the Sustainability Studies minor. The minor requires 21 total credit hours (7 courses), with 9 credit hours for the core and 12 credit hours of electives, of which "Environmental Sociology" (SOC 369) is one of the "social equity and policy course" options.

The authors would like to point out that while we are focusing on the college setting, environmental activism is something that goes beyond the classroom and the campus. Every community has local environmental organizations in need of support, which can be easily found on the Internet. But above and beyond joining particular groups, we want to stress that each of us can take personal responsibility through our private actions. You can make a difference.

An Environmental Thrivability Checklist

This book has argued in favor of raising people's consciousness regarding why environmental thrivability is of paramount importance for all of us. In closing, we'd like to offer some practical guidelines which everyone can follow. Below is an "Environmental Thrivability Checklist" that includes ideas for the home, school or work, in the yard, on vacation, in your car, and some general things that all conscientious citizens can do. Many of these suggestions came from the "Save the Rainforest" organization and others from the thoughts and ideas of the authors, their students, colleagues and friends. Readers may want to make a check-mark next to all the things they already do and make note of some of the new things they can do.

In the Home

- Reduce energy consumption by lowering the heat in the winter and raising the cooling temperature in the summer. Use insulation made from recycled paper, glass and other materials.
- Turn off home computers when not in use, which could cut CO_2 impact by 8.3 million tons a year (Miller 2013).
- Preserve water in the morning by taking shorter showers and/or use a low flow showerhead.
- Use low flow toilets.

- Turn off the water when brushing your teeth (as a general rule, don't leave water running needlessly).
- Check for water leaks. Repair leaky faucets as soon as possible.
- Turn off the lights in the rooms you are not occupying.
- Switch to energy-saving light bulbs.
- Recycle old appliances, computers, and clothing.
- Recycle cans, bottles, plastics, newspapers and other paper products, motor oil, scrap metal, and so on. It is best to completely avoid single-use plastics. Do not use balloons! They are made of plastics and may cause great harm, including death, to birds.
- Investigate local recycling centers that take items your garbage hauler does not.
- Shop at farmer's markets to support local growers. Incorporate a vegan diet and eat less meat.
- Eat foods that are grown or raised locally.
- Get rid of high-energy appliances, such as old refrigerators and water heaters, blow dryers and lawnmowers.
- Set your water heater at 130 degrees or use a tank-less water heater.
- Have your water heater insulated free of charge by your utility company.
- Use green cleaning products (e.g., phosphate-free laundry and dish soaps).
- Use the washer and dryer only for full loads and not single, or a few, item(s).
- Use cold water in the washer whenever possible.
- Re-use brown bags to line your trash can instead of plastic liners. Re-use bread bags, butter tubs, etc.
- Use a dryer rack or clothesline, if possible, to dry clothes.
- Pay bills online to help eliminate paper usage.
- Paint with brushes or rollers instead of spray paint. Use latex rather than oil-based paint (oil-based is highly toxic).
- Buy and use rechargeable batteries.
- Think twice about buying "disposable" products (they really aren't disposable and are extravagant wastes of the world's resources).
- Save wire coat hangers and return them to the dry cleaners.
- Use reusable plates and utensils rather than disposable ones. Also, use reusable containers to store food rather than aluminum foil or plastic wrap.
- Get a free energy audit from your utility company.
- Burn only seasoned wood in your woodstove or fireplace and don't light them as often.
- If you use a natural tree for Christmas, be sure to recycle it by mulching. Most municipalities offer a curbside pick-up service.

In the Yard

- Start a compost pile. Do compost items such as fruits and vegetables, eggshells, coffee grounds and filters, tea bags, nut shells, shredded newspaper, cardboard, paper, yard trimmings, grass clippings, houseplants, hay and straw, leaves, sawdust, wood chips, cotton and wool rags, dryer and vacuum cleaner lint, hair and fur and fireplace ashes (Roth 2018). Do not compost black walnut tree leaves or twigs, coal or charcoal ash, dairy products, pet wastes, diseased or insect-

ridden plants, fats, grease, lard or oils, meat or fish, and yard trimmings treated with chemical pesticides (Roth 2018).
- Compost your leaves and yard debris or take them to a yard debris recycler. (Burning them creates air pollution and putting them out with the trash wastes landfill space.)
- Put up birdfeeders, birdhouses, and birdbaths.
- Pull weeds instead of using herbicides.
- Use only organic fertilizers.
- Use mulch to conserve water in your garden.
- When doing gardening, if there are any extra plastic or rubber pots, take them back to the nursery.
- Plant short, dense shrubs close to your home's foundation to help insulate your home against the cold.

At School or at Work

- Recycle, and that means recycling all the same things as recommended in the home, but especially computer paper and cardboard.
- Use scrap paper for informal notes to yourself and others.
- Print things like in-house memo pads, etc., on recycled paper. Or, whenever possible, do not print hard copies at all, save them to your hard drive or email them to yourself to save paper (Roth 2018).
- Re-use manila envelopes and file folders.
- Walk, ride a bicycle or carpool if you can.
- Turn off your computer at the end of the day.
- Use reusable water bottles rather than disposable plastic ones or re-use coffee mugs.
- If you are healthy enough, use the stairs instead of the elevator on trips less than three floors. It is good exercise too!
- Close off certain rooms from heat in the winter and air conditioning in the summer.
- Cut down on paper usage by printing on both sides of paper.
- Send as much information electronically as possible (especially in place of paper).
- Reduce ink by using high efficiency printers. Send e-mails and texts rather than letters or faxes as much as possible.
- Check the energy rating of major appliances purchased. Buy only the most energy-efficient models.
- Get involved with sustainability committees on your campus or workplace.
- As a student, take the Graduation Pledge of Social and Environmental Responsibility: "I pledge to explore and take into account the social and environmental consequences of any job I consider and will try to improve these aspects of any organizations for which I work." (See http://www.graduationpledge.org/.)

While on Vacation

- Turn down the heat and turn off the water heater at home before you leave.
- Carry reusable cups, dishes, and flatware.

- Make sure your trash doesn't end up in the ocean and don't litter the beaches.
- Make sure your trash doesn't end up on land in our parks and reserves.
- Don't pick flowers or keep wild creatures for pets. Leave plants and animals where you find them.
- Don't buy souvenirs made from wild or endangered animals.
- Watch out for wildlife ... give consideration to all living things you see crossing the road. Deer and other animals have their own idea about "right-of-ways" as they travel.
- If camping, build smaller campfires and make sure they're completely out before leaving the site.
- If hiking, stay on trails—don't trample fragile undergrowth.

In the Car

- Drive fuel-efficient automobiles. If we improved our cars' gas mileage by 5 miles a gallon, we could cut their CO_2 emissions by 239 million tons each year, a 20 percent decrease (Miller 2013).
- When it comes time to buy a new automobile, buy a more efficient one than your old car.
- Keep your car tuned properly.
- Carpool whenever possible.
- Use public transit instead of driving whenever possible.
- Minimize the total number of times you use your car, especially on the weekends, by bike riding or walking instead. At the very least, do all your errands on the same day and in an orderly, logical fashion.
- Recycle your engine oil.
- Keep your tires properly inflated to save gas.
- Keep your wheels properly aligned to save on your tires.
- Don't litter the roads and highways—this includes cigarette butts!—save the trash and dispose of it properly at a rest stop or bring it back home with you.

Green RVing

- Keep your RV on roads that are equipped to handle larger vehicles.
- Keep RV and tow vehicle engines well-tuned to conserve energy and reduce emissions.
- Recycle as you travel and pay particular attention to campground recycling guidelines.
- Minimize the use of disposables (e.g., assign a mug for each traveler instead of using paper or plastic cups).
- Purchase newer RVs that come equipped with solar panels.
- Keep campfires small to minimize the amount of ash and pollution and don't put anything into the fire pit that will not burn, such as plastics, folks, and metals. Observe fire rules and pay attention to weather conditions.
- Use nontoxic cleaning supplies and tank additives.
- Keep your music at a respectable volume, as your favorite music may be views as noise pollution by your neighbors.

- Work with nature; that is, in hot weather, use natural shade, awnings and canvas covers, in cold weather, park where the RV will be protected from cold winds.
- At the end of your trip, dispose of all trash properly (*The Post-Standard* 2011b).

No doubt you can think of many other ways to be socially responsible and environmentally aware. The point is, each and every one of us can make a difference. Changing our practices to be more environmentally aware is a good way to show one's commitment to dealing with the myriad issues addressed in this book in a constructive and hopefully happy way. "Going green" is the key to a flourishing lifestyle.

Popular Culture Section 10

Garbology: What the Study of Garbage Teaches Us

We have addressed the need for college courses on environmental issues. Such courses can focus on topics like understanding the biosphere, the environmental imagination, or taking an evolutionary perspective. But what about a course devoted to the topic of *garbage*? Does such a curriculum exist?

In fact, it does. Bill Rathje, a University of Arizona at Tucson professor, was a Harvard-educated archeologist who studied not ancient civilizations but rather our own contemporary world. Called the world's first "garbologist," he founded the Garbage Project on his campus as a way of using archeological skills to study landfills and thereby learn about current consumption habits. Students would unearth the waste materials of the past few decades to extrapolate data on economic, social, and cultural realities of the times. As Rathje often pointed out, it's not "we are what we eat" but rather "we are what we throw out." And, as we wrote in Chapter 3, Americans generate a great deal of waste, the equivalent of 4.51 pounds per person per day (EPA 2020c).

You can learn about Rathje and his work in the book *Garbology: Our Dirty Love Affair with Trash* by the Pulitzer Prize winning journalist Edward Humes. Much of *Garbology* addresses what Humes calls "the plasticization of America." He writes, "Municipal waste was only .4 percent plastic by weight in 1960. Our trash cans had almost no plastic inside them back then, but that began to change rapidly. By the end of the sixties, plastic trash had increased sevenfold. By the year 2000, American households were throwing away sixty-three times as much plastic as in 1960" (Humes 2013: 73). According to Humes, if every country consumed and threw away as much as Americans do, and at the same rate, it would take the resources of *five* planet Earths to meet the demand.

Rathje (who died in 2011) made a good point about the meaning of the word "waste." We use it to mean that which we throw out, but it originally meant "to waste," which is to foolishly misuse something of value. "Why are we wasting stuff that we pay for as product or packaging, then pay again for as trash to be hauled away? Now it's no longer the waste itself that's negative, but the act of creating it that's at issue" (Humes 2013). For all its gloomy facts about the sheer amount of waste produced in America on a daily basis, the Humes (2013) concludes that garbage is an environmental problem over which ordinary individuals can exert control.

When it comes to asking what each and every one of us can do to make an environmental difference, we should all be talking trash!

Parting Thoughts

As good citizens of the planet there are a number of things we can do to help preserve and protect the environment. We can join conservation organizations and volunteer our time to conservation projects. If you are strapped for time, donate money to conservation projects. Get involved in any matter possible, even if it takes contacting your representatives in political office or writing letters to the editors of local newspapers. Many people have blogs and Facebook pages dedicated to sustaining the environment. Environmental social influencers are also coming to the forefront in the world of social media. Further, all of us can join the conversation in an attempt to make a positive change. Lessening our dependence on fossil fuels (e.g., driving fuel-efficient automobiles and demanding the switch to renewable forms of energy) and trying to find a way to live within the Earth's limited carrying capacity are the keys. Perhaps the most pragmatic advice we can all follow is simply being aware of what each of us does to the environment on a daily basis. That is, be aware of what you purchase and its imminent impact on the environment. Were there alternative purchases that could have been made? Everything we consume ends up somewhere. Will it decompose? How long will that take? As the Maori saying mentioned at the beginning of this book puts it, "Leave nothing but footprints."

We referenced George Carlin's unique view of people trying to save the planet, he argued, "There is nothing wrong with the planet. The planet is fine. The PEOPLE are f*cked! The planet is fine. Compared to people, the planet is doing great. The planet has been here for five and half billion years [and] somehow we're a threat?" Carlin is correct; the planet has been around long enough (more like 4.5 billion years) to play witness to numerous mass extinctions and the rise and fall of many dominant species. And yet, as these species have come and gone, the planet Earth remains. This leads us to realization that while, on the surface, it is a nice sentiment to want to "save the planet," it is humanity and other animal and plant species that we should be fighting to preserve. The planet will last for an indefinite amount time, certainly longer than humans. Civilizations have risen from the ashes and just as quickly, they have disappeared only to be replaced. Consider Machu Picchu which was built at the height of the Inca Empire, which dominated western South America in the 15th and 16 centuries. There were an estimated 150 buildings including bath houses, temples, and sanctuaries and many of these structures were aligned with the sun to correspond with Incan holidays. The city was abandoned (historians debate why, although it often argued that the inhabitants feared the invasion of Spanish explorers and their discovery of this architectural treasure would be destroyed along with the annihilation of the inhabitants) only to be rediscovered in the early 1900s. The buildings had mostly crumbled and been overrun with vegetation. This is a scenario that will play out when humans become extinct, nature will return to dominance and the planet will survive.

It's an old adage, but still true, that the planet Earth is the only planet that we are aware of that can sustain human life. What happens to humanity if the planet Earth can no longer sustain human life? We all know the answer to that question. We now have to choose between doing something about it, or not. The choice is up to us.

Bibliography

AbandonedMines.gov. 2020. "Extend of the Problem." Bureau of Land Management. https://www.abandonedmines.gov/extent_of_the_problem.

Abel, Ernest. 1980. *Marihuana: The First Twelve Thousand Years.* New York: Plenum.

Acton Institute. 2007. *Environmental Stewardship in the Judeo-Christian Tradition: Jewish, Catholic, and Protestant Wisdom on the Environment.* Grand Rapids, MI: Acton Institute.

Agence France-Presse in Ljubljana. 2016. "Slovenia Adds Water to Constitution as Fundamental Right for All." *The Guardian,* November 17. https://www.theguardian.com/environment/2016/nov/18/slovenia-adds-water-to-constitution-as-fundamental-right-for-all.

Ahuja, Anjana. 2001. "Swarms Reach Biblical Proportions." *London Times,* June 19: 11.

The Aldo Leopold Foundation. 2020. "Aldo Leopold." https://www.aldoleopold.org/about/aldo-leopold/.

Alyan, Omer, Ozcan Ozdemir, Omac Tufekcioglu, Bilal Geyik, Dursan Aras, and Deniz Demirkan. 2006. "Myocardial Injury Due to Lightning Strike: A Case Report." *Angiology,* 57: 219–23.

Alzheimer's Association. 2020. "What Is Alzheimer's Disease?" https://www.alz.org/alzheimers-dementia/what-is-alzheimers.

Amadeo, Kimberly. 2019. "Hurricane Irma Facts, Damage, and Costs." *The Balance,* October 21. https://www.thebalance.com/hurricane-irma-facts-timeline-damage-costs-4150395.

American Cancer Society. 2020. "What Causes Cancer?" https://www.cancer.org/cancer/cancer-causes.html.

American Farmland Trust (AFT). 2011. "Ideas for Change in America." http://www.change.org/ideas/view/no_farms_no_food_save_the_land_that_sustains_us.

_____. 2020. "Our Work." https://farmland.org/our-work/

American Heritage Dictionary, 4th edition. 2006. Houghton Mifflin. http://dictionary.reference.com/help/adh4.html.

Andryszewski, Tricia. 2008. *Mass Extinctions: Examining the Current Crisis.* Minneapolis, MN: Twenty-First Century Books.

Animal Network. 2018. "The Effects of Climate Change on Wild Animals." https://animals.net/the-effects-of-climate-change-on-wild-animals/.

Answers.com. 2011. "Oxygen." http://www.answers.com/topic/oxygen.

Anyday Guide. 2020. "Fossil Fools Day." https://anydayguide.com/calendar/2323.

Arcadia. 2017. "The Effects of Climate Change on Plants." https://blog.arcadia.com/effects-climate-change-plants/.

Associated Press. 2013. "Fireball Shows Meteor Risk May Be Bigger." *The Citizen,* November 7: A6.

_____. 2015. "Study: Warming Will Push Extinction." *The Citizen* May 1: A6.

_____. 2017. "Invasive Aquatic Weed Creeps Across US." *The Citizen,* February 5: A6.

_____. 2019a. "More of Amazon Is Lost." *Los Angeles Times,* November 19: A4.

_____. 2019b. "Marine Menace Heads North to Oregon." *Los Angeles Times,* October 25: A6.

_____. 2019c. "New York's Bold Plan on Climate." *Los Angeles Times,* July 21: A15.

_____. 2019d. "Election Swings Lead to a Shift on Climate." *Los Angeles Times,* April 28: A3.

_____. 2020a. "A Plot to Wipe Out 'Murder Hornet.'" *Los Angeles Times,* May 5: A5.

_____. 2020b. "With Water Scarce, Handwashing Is Luxury." *Los Angeles Times,* May 21: A3.

_____. 2020c. "State Removes Climate Change Bond From Ballot—For Now." *The Post-Standard,* August 2: A6.

_____. 2020d. "Book: Virus Downplayed." *The Citizen,* September 10: A7.

The Association for the Advancement of Sustainability in Higher Education. 2020. "Advancing Sustainability in Higher Education." https://www.aashe.org/about-us/who-we-are/.

AsteroidDay.org. 2020. "Asteroid Day Press Release—Events World Wide." https://asteroidday.org/updates/news-updates/asteroid-day-press-release-2020/.

Atlin, Cole. 2014. "Aquatic Invasive Alien Species and the Evolution of Canadian and U.S. Ballast Water Regulations in the Great Lakes—Rowing in Tandem or Muddying the Waters?" *Indiana International & Comparative Law Review,* 24(1): 65–92.

Australian Marine Conservation Society (AMCS). 2020. "Coral Bleaching." https://www.marineconservation.org.au/coral-bleaching/.

Babington, Charles. 2012. "Santorum: Global Warming Is Politics, not Science." *Associated*

Press, February 20. https://www.yahoo.com/news/santorum-global-warming-politics-not-science-164640155.html.

Baetz, Juergen. 2011. "Up in the Sky: It's Not Superman." *The Post-Standard,* October 23: A-15.

Baker, Nathan. 2012. "Cleaner Green Energy." *The Citizen,* August 22, A-1, A-5.

Bakke, Gretchen. 2016. *The Grid: The Fraying Wires Between Americans and Our Energy Future.* London: Bloomsbury.

Baltimore Sun. 2020. "'They're Climbing': Exterminators Say Calls for Rats in Baltimore Homes Have Doubled During Coronavirus Pandemic." April 29. https://douglas.or.networkofcare.org/ph/news-article-detail.aspx?id=112983.

Banerjee, Neela. 2012. "July was Hottest Month on Record." *Los Angeles Times,* August 9: A-7.

Barboza, Tony. 2013. "Long Beach Builds 'Wall of Mulch' to Fight Pollution." *Los Angeles Times,* August 7: AA-3.

_____. 2018. "Climate Report Warns of Bleak Future." *Los Angeles Times,* November 24: A1, A7.

Barnosky, Anthony D., Nicholas Matzke, Susumu Tomiya, Guinevere O.U. Wogan, Brian Swartz, Tiago B. Quental, Charles Marshall, Jenny L. McGuire, Emily L. Lindsey, Kaitlin C. Maguire, Ben Mersey, and Elizabeth A. Ferrer. 2011. "Has the Earth's Sixth Mass Extinction Already Arrived?" *Nature,* 471 (7336): 51–57.

Barrett, Gary W., and Eugene P. Odum. 2000 (April). "The Twenty-First Century: The World at Carrying Capacity." *BioScience,* 50(4): 363–368.

Barry, John Byrne. 1995. "Is the Grass Really Greener?" *Sierra* (November–December), 80 (6): 22–23.

BBC News. 2011. "Japan Tsunami: Aid Workers Tackle Aftermath." March 14. http://www.bbc.co.uk/news/health-12731391.

_____. 2012. "Effects of Volcanic Eruptions." http://www.bbc.co.uk/schools/gcsebitesize/geography/natural_hazards/volcanoes_rev5.shtml.

_____. 2014. "In Pictures: Illegal Logging in Peru." September 4. https://www.bbc.com/news/world-latin-america-28926270.

_____. 2019. "Adani Mine: Australia Approves Controversial Coal Project." June 13. https://www.bbc.com/news/world-australia-48618774.

Becker, Luann, Robert J. Poreda, Andrew G. Hunt, Theodore E. Bunch, and Michael Rampino. 2001. "Impact Event at the Permian-Triassic Boundary: Evidence from Extraterrestrial Nobel Gases in Fullerenes." *Science,* 291: 1530–1533.

Bekoff, Marc, and Sarah Bexell. 2010. "Ignoring Nature: Why We Do It, the Dire Consequences, and the Need for a Paradigm Shift to Save Animals, Habitats, and Ourselves." *Human Ecology Review,* 17(1); 70–76.

Bell, Michael Mayerfeld, and Loka L. Ashwood. 2016. *An Invitation to Environmental Sociology, Fifth Edition.* Los Angeles: Sage.

Bengali, Shashank. 2018. "U.S. Plastic Waste Is Piling Up in Southeast Asia." *Los Angeles Times,* December 30: A3.

_____. 2019. "Asian Giants Vow to Go Green But Tout Coal Abroad." *Los Angeles Times,* May 13: A1.

Benton, Michael J. 2003a. *When Life Nearly Died: The Greatest Mass Extinction of All Time.* New York: Thames and Hudson.

_____. 2003b. "Wipeout." *New Scientist,* 178 (2392): 38–43. http://palaeo.gly.bris.ac.uk/Essays/wipeout/default.html.

_____. 2011. "Mass Extinctions." *New Scientist,* 209 (2802): 43–57. http://www.newscientist.com/data/doc/article/dn19554/instant_expert_9-massextinctions.pdf.

Berlinger, Joshua. 2012. "The New Matt Damon Movie That Blasts Fracking Was Partly Funded by Government of Abu Dhabi." *Business Insider,* October 3. http://www.businessinsider.com/matt-damon-movie-blasts-fracking-backed-by-uae-2012-9.

BestColleges.com. 2020. "Greenest Universities." https://www.bestcolleges.com/features/greenest-universities/.

Bhattarai, Khem Raj, Inger Elizabeth Maren, and Suresh Chandra Subedi. 2014. "Biodiversity and Invisibility: Distribution Patters of Invasive Plant Species in the Himalayas, Nepal." *Journal of Mountain Science,* 11(3): 688–696.

Bissonnette, Zac. 2010. "University Thinks the Right Font Could Help Save the Earth." http://www.dailyfinance.com/2010/03/26/university-thinks-the-right-font-could-help-save-the-earth/.

Bittman, Mark. 2009. *Food Matters: A Guide to Conscious Eating with More Than 75 Recipes.* New York: Simon & Schuster.

_____. 2013. "Let's Not Braise the Planet." *The New York Times,* July 2: A-21.

Blomberg, Lindsey. "Wasted." *The Environmental Magazine,* 22 (5): 15–17.

Bloom, Laura Begley. 2019. "Ranked: 10 Happiest and 10 Saddest Countries in the World." *Forbes,* March 25. https://www.forbes.com/sites/laurabegleybloom/2019/03/25/ranked-10-happiest-and-10-saddest-countries-in-the-world/#6ab30d5e6374.

Bored Panda. 2016. "Russians Throw Away Empty Vodka and Beer Bottles, Ocean Turns Them Into Colorful Glass 'Pebbles.'" https://www.boredpanda.com/vodka-bottle-pebbles-glass-beach-ussuri-bay-russia/?utm_source=search.yahoo&utm_medium=referral&utm_campaign=organic.

Borenstein, Seth. 2015. "World Is a Forest of 3 Trillion Trees." *The Post-Standard,* September 3: A-14.

_____. 2019. "Glacier Melt Accelerates." *The Citizen,* April 9: A6.

_____. 2020a. "Earth Turns Wilder, Cleaner." *The Citizen,* April 14: A7.

_____. 2020b. "As People Stay Home, Earth Turns Cleaner." *The Post-Standard,* April 23: A20.

_____. 2020c. "Study: Carbon Pollution Plunges." *The Citizen,* May 20: A7.

Boyd, David R. 2013. "The Constitutional Right to a Healthy Environment." LawNow.org. https://www.lawnow.org/right-to-healthy-environment/.

Brennan, Andrew, and Y.S. Lo. 2010. *Understanding Environmental Ethics.* Durham, UK: Acumen.

Bright, Chris. 1998. *Life Out of Bounds: Bioinvasion in a Borderless World.* New York: Norton.

Brockmann, Hilke, and Jan Delhey. 2010. "Introduction: The Dynamics of Happiness and the Dynamics of Happiness Research." *Social Indicators Research,* 97: 1–5.

Brones, Anna. 2017. "101 Ways to Use Food Waste." *Paste Magazine,* July 13. https://www.pastemagazine.com/food/zero-waste/101-ways-to-use-food-waste/.

Brown, Ann. 2014. "How Many Breaths Do You Take Each Day?" *The EPA Blog,* April 28. https://blog.epa.gov/2014/04/28/how-many-breaths-do-you-take-each-day/.

Brown, Matthew. 2015. "Experts See Long-term risks from Colorado Mine Spill." *The Post-Standard,* August 13: A13.

_____. 2019. "Old Mines Leak 50M Gallons of Waste Daily." *The Post-Standard,* February 21: A16.

Buell, Lawrence. 1995. *The Environmental Imagination.* Cambridge, MA: Harvard University Press.

Bulos, Nabih. 2020. "In Mideast, Wars Continue Even In a Pandemic." *Los Angeles Times*: A3.

Burkett, Erin M., and David J. Jude. 2015. "Long-term Impacts of Invasive Round Goby Neogobius Melanostomus on Fish Community Diversity and Diets in the St. Clair River, Michigan." *Journal of Great Lakes Research,* 41(3): 862–872.

Butler, Rhett A. 2020. "Calculating Deforestation Figures for the Amazon." https://rainforests.mongabay.com/amazon/deforestation_calculations.html.

Caldeira, Ken, and Michael E. Wickett. 2003. "Anthropogenic Carbon and Ocean pH." *Nature,* 425: 365. https://www.nature.com/articles/425365a#citeas.

California Coastal Commission (CCC). 2019. "The Problem with Marine Debris." https://www.coastal.ca.gov/publiced/marinedebris.html.

California Energy Commission. 2012. "Where Fossil Fuels Come From." http://www.energyquest.ca.gov/story/chapter08.html.

Campbell, Matthew, and Ying Tian. 2019. "The World's Biggest Electric Vehicle Company Looks Nothing Like Tesla." *Bloomberg Businessweek.* https://www.bloomberg.com/news/features/2019-04-16/the-world-s-biggest-electric-vehicle-company-looks-nothing-like-tesla.

Camus, Albert. *Myth of Sisyphus.* 2008. New York: Penguin Books.

Canadian-Universities.net. 2019. "Sustainability in Canadian Universities." http://www.canadian-universities.net/Campus/Sustainability.html.

Canfield, Sabrina. 2017. "Giant Dead Zone in Gulf of Mexico Traced to Meat Industry." *Courthouse News Service,* August 3. https://www.courthousenews.com/giant-dead-zone-gulf-mexico-traced-meat-industry/.

Cappiello, Dina. 2012. "Earth Day: Then and Now." *The Post-Standard,* April 22: A-1, A-8.

CARE. 2020. "End World Hunger and Poverty." https://www.care.org/work/poverty/end-world-hunger-and-poverty.

Carson, Rachel. 1962. *Silent Spring.* Boston: Houghton Mifflin https://earthnworld.com/countries-with-the-largest-forest-area/.

Cary, Zulma. 2019. "Top 15 Countries with the Largest Forest Area." *Earth & World,* June 11.

Castanon, J.I.R. 2007. "History of the Use of Antibiotics as Growth Promoters in European Poultry Feeds." *Poultry Science.* http://ps/fass.org/content/86/11/2466.full.pdf+html.

Catton, W.R. 1980. *Overshoot: The Ecological Basis of Revolutionary Change.* Urbana: University of Illinois Press.

Cayo, Don. 2005. "Gross National Happiness Index Flawed." *Vancouver Sun,* July 1: H-3. http://www.gpiatlantic.org/conference/media/clipping_july1.pdf.

CBS News. 2010. "Following the Trail of Toxic E-Waste," January 8. http://www.cbsnews.com/8301-18560_162-4579229.html.

_____. 2018. How MGM Resorts Is Cutting Down on Food Waste for a Good Cause." February 7. https://www.cbsnews.com/news/mgm-resorts-donating-leftover-food-to-the-hungry/.

CBS This Morning. 2020. "Hurricane Barrels Toward Coast." Original air date, September 15.

Center for Sustainable Systems. 2019. "U.S. Cities Factsheet." http://css.umich.edu/factsheets/us-cities-factsheet.

Centers for Climate and Energy Solutions. 2020. "Wildfires and Climate Change." https://www.c2es.org/content/wildfires-and-climate-change/.

Centers for Disease Control and Prevention (CDC). 1994. "Addressing Emerging Infectious Disease Threats: A Prevention Strategy for the United States Executive Summary." 43 (RR-5): 1–18.

_____. 2017a. "Leading Causes of Death." https://www.cdc.gov/nchs/fastats/leading-causes-of-death.htm.

_____. 2017b. "Heart Disease." https://ephtracking.cdc.gov/showHeartEnv.action.

_____. 2019a. "Plague." https://www.cdc.gov/plague/index.html.

_____. 2019b. "1918 Pandemic." https://www.cdc.gov/flu/pandemic-resources/1918-pandemic-h1n1.html.

_____. 2019c. "About HIV/AIDS." https://www.cdc.gov/hiv/basics/whatishiv.html.

Centre for Alternative Technology. 2012. "Hemp and Lime Usage in WISE—for Insulation, Breathability and Thermal Mass." http://info.cat.org.uk/questions/wise/hemp-and-lime-usage-wise-insulation-breathability-and-thermal-mass.

Chang, Charis. 2020. "How the 2019 Australian Bushfire Season Compares to Other Fire Disasters." *News.com.au.* https://www.news.com.au/technology/environment/how-the-2019-australian-bushfire-season-compares-to-other-fire-disasters/news-story/7924ce9c58b5d2f435d0ed73ffe34174.

Chatham-Stephens, Kevin M., Mana Mann, Andrea Wershof Schwartz, and Philip J. Landrigan. 2012. "First, Do No Harm." *American Educator* (Winter 2011–2012), 35 (4): 22–31.

Chatterjee, Subhankar and Shivika Sharma. 2019. "Microplastics in Our Oceans and Marine Health." *Field Actions Science Reports,* 19: 54–61.

Cho, Renee. 2019. "10 Climate Change Impacts That

Will Affect Us All." *State of the Planet,* December 27. https://blogs.ei.columbia.edu/2019/12/27/climate-change-impacts-everyone/.

Chrisafis, Angelique. 2016. "French Law Forbids Food Waste by Supermarkets." *The Guardian,* February 4. https://www.theguardian.com/world/2016/feb/04/french-law-forbids-food-waste-by-supermarkets.

Christian Ecology. 2012. "A Scriptural Call for Environmental Stewardship." http://www.christianecology.org/Stewardship.html.

The Citizen. 2012a. "Warming Climate Favors Soybeans." February 21: A-5.

_____. 2012b. "Agriculture: Company Growing Mushrooms as Packing Material." March 9: A-3.

_____. 2012c. "Don't Let Invasives Hitch a Ride." March 22: A-4.

_____. 2013. "1,100 Injured in Meteor Explosion." February 16: A-5.

_____. 2015. "A Call to Save Creation." June 21: A-14.

_____. 2020. "Destructive Insect Found in NY." August, 15: A3.

City of Fort Bragg (CA). 2012. "Glass Beach: Frequently Asked Questions." http://www.fortbragg.com/explore/glass-beach-faq/.

Clean Water Action. 2020. "Trump's Attacks on the Clean Water Act." https://www.cleanwater.org/features/trumps-attacks-clean-water-act.

Cleveland Clinic. 2020. "Overview: What Is Coronavirus?" https://my.clevelandclinic.org/health/diseases/21214-coronavirus.

Clubb, Oliver, and Rachel May. 2011. "More Green Ideas: CNY Can Learn from Other Successful Projects." *The Post-Standard,* September 30: A-11.

CNN World. 2010. "TOXIC: Garbage Island." http://articles.cnn.com/2010-02-6/world/vbs.toxic.garbage.island_1_north-pacific-gyre-pacific-ocean-currents?_s=PM:WORLD.

Cocks, Samuel, and Steven Simpson. 2015. "Anthropocentric and Ecocentric: An Application of Environmental Philosophy to Outdoor Recreation and Environmental Education." *Journal of Experiential Education,* 38(3): 216–227.

Cohen, Ariel. 2019. "Will Russia Survive the Coming Energy Transition?" *Forbes,* June 27. https://www.forbes.com/sites/arielcohen/2019/06/27/will-russia-survive-the-coming-energy-transition/#3b6e18cd5577.

Coin, Glenn. 2018. "Most Wanted." *The Post-Standard,* April 8:A8.

_____. 2019. "Heavy Rains Get Stronger, More Frequent." *The Post-Standard,* April 14:A20.

_____. 2020. "Plan Calls for Planting 70,000 Trees in 20 Years." *The Post-Standard,* February 13: A1.

Coldewey, Devin. 2020. "What Is Contact Tracing?" TechCrunch.com. https://techcrunch.com/2020/04/18/what-is-contract-tracing/.

Collomb, Jean-Daniel. 2014. "The Ideology of Climate Change Denial in the United States." *European Journal of American Studies,* 9(1): 1–20. https://journals.openedition.org/ejas/10305.

Conca, James. 2019. "U.S. CO2 Emissions Rise as Nuclear Power Plants Close." *Forbes,* January 16. https://www.forbes.com/sites/jamesconca/2019/01/16/u-s-co2-emissions-rise-as-nuclear-power-plants-close/#61ff7fb70347.

Congress.gov. 2020. "H.R.—Energy Innovation and Carbon Dividend Act of 2019." https://www.congress.gov/bill/116th-congress/house-bill/763.

Conserve Energy Future. 2020. "What Is Land Pollution?" https://www.conserve-energy-future.com/causes-effects-solutions-of-land-pollution.php.

Consumer Reports. 2018. "How to Recycle Old Electronics." https://www.consumerreports.org/recycling/how-to-recycle-electronics/.

Cook, John. 2010. "Earth's Five Mass Extinction Events." *Skeptical Science.* http://www.skepticalscience.com/Earths-five-mass-extinction-events.html.

Cooper, Mary Ann, and Rick Kulkarni. 2012. "Lightning Injuries." *Medscape Reference.* http://emedicine.medscape.com/article/770642-overview.

Corwin, Jeff. 2009. "The Sixth Extinction." *Los Angeles Times,* November 30. http://articles.latimes.com/2009/nov/30/opinion/la-oe-corwin30–2009nov30.

Costanza, Robert, Ralph d'Arge, Rudolf de Groot, Stephen Farber, Monica Grasso, Bruce Hannon, Karin Limburg, Shahid Naeem, Robert V. O'Neill, Jose Paruelo, Robert G. Raskin, Paul Sutton and Marjan van den Belt. 1997. "The Value of the World's Ecosystem Services and Natural Capital." *Nature,* 387: 253–260.

Costanza, Robert, Rudolf de Groot, Paul Sutton, Sander van der Ploeg, Sharolyn J. Anderson, Ida Kubiszewski, Stephen Farber, R. Kerry Turner. 2014 (May). "Changes in the Global Value of Ecosystem Services." *Global Environmental Change,* 26: 152–158.

CO2.earth. 2020. "The World's CO2 Home Page." https://www.co2.earth/.

Crulckshank, Saralyn. 2018. "Climate Change, Superstorms, and Civil Engineering." *Johns Hopkins University Magazine,* September 13. https://hub.jhu.edu/2018/09/13/hurricane-florence-failing-infrastructure/.

Crumley II, Jack S. 2016. *Introducing Philosophy: Knowledge and Reality.* Tonawanda, NY: Broadview.

Csikszentmihalyi, Mihaly. 1997. *Finding Flow.* New York: Basic Books, 1997.

_____. 2008. *Flow: The Psychology of Optimal Experience.* New York: Harper and Row.

Cunningham, Suzanne. 1996. *Philosophy and the Darwinian Legacy.* Rochester, NY: University of Rochester Press.

Cusack, Mary Ellen. 1993. "Comment: Judicial Interpretation of State Constitutional Rights to a Healthful Environment." *Boston College Environmental Affairs Law Review.* https://litigationssentials.lexisnexis.com/webcd/app?action=DocumentDisplay&crawlid=1&doctype=cite&docid=20+B.C.+Envtl.+Aff.+L.+Rev.+173&srctype=smi&srcid=3B15&key=e656ab601fd3670946ef9356e7e54e0a.

Dahler, Don. 2013. "Pesky Beetle Not a Hit with Baseball Bat Maker." *CBSNews.com,* July 4. http://www.cbsnews.com/8301-18563_162-

57592374/pesky-beetle-not-a-hit-with-baseball-bat-maker/.

Davidson, Jordan. 2019. "Giant Floating Solar Farms Could Make Fuel and Help Solve the Climate Crisis, Says Study." *EcoWatch*, June 25. https://www.ecowatch.com/floating-solar-farms-climate-crisis-2638980599.html?rebelltitem=1#rebelltiteml.

Deadliest Space Weather. 2013. "Meteors." The Weather Channel (original air date January 17, 2013).

DeAngelis, Tori. 2008. "Two Faces of Oxytocin." American Psychological Association, *Monitor on Psychology*, 39(2): 30.

De'ath, Glenn, Katharina E. Fabricius, Hugh Sweatman, and Marji Puotinen. 2012. "The 27-Year Decline of Coral Cover on the Great Barrier Reef and Its Causes." *Proceedings of the National Academy of Sciences*, DOI: 10.1073/pnas.1208909109.

De Graff, John. 2009. "Gross National Happiness." *UTNE Reader.* http://www.utne.com/Spirituality/Gross-NationalHappiness-5818.aspx.

DeJohn, Suzanne. 2019. "Invasive Plants: Buy Now, Pay Later." Gardener's Supply Company. https://www.gardeners.com/how-to/invasive-plants/7373.html.

Delaney, Tim. 2001. *Community, Sport and Leisure*, 2nd edition. Auburn, NY: Legend Books.

______. 2004. *Classical Social Theory: Investigation and Application.* Upper Saddle River, NJ: Pearson.

______. 2005. *Contemporary Social Theory: Investigation and Application.* Upper Saddle River, NJ: Pearson.

______. 2008. *Simpsonology: There's a Little Bit of Springfield in All of Us.* Amherst, NY: Prometheus.

______. 2012. *Connection Sociology to Our Lives: An Introduction to Sociology.* Boulder, CO: Paradigm.

______. 2013. *Classical and Contemporary Social Theory: Investigation and Application.* Upper Saddle River, NJ: Pearson.

______. 2014a. *American Street Gangs*, 2nd edition. Upper Saddle River, NJ: Pearson.

______. 2014b. *Classical and Contemporary Social Theory: Investigation and Application.* Boston: Pearson.

______. 2017. *Social Deviance.* Lanham, MD: Rowman & Littlefield.

______. 2019. *Common Sense as a Paradigm of Thought: An Analysis of Social Interaction.* London: Routledge.

______. 2020. *Darkened Enlightenment: The Deterioration of Democracy, Human Rights, and Rational Thought in the Twenty-First Century."* London: Routledge.

Delaney, Tim, and Anastasia Malakhova. 2020. "The Need for Renewable Energy Resources and the Reasons Why the United States and Russia Lag Behind." *Journal of Strategic Innovation and Sustainability*, 15(2): 45–54.

Delaney, Tim, and Ellen Reed. 2013. "A Marxist Look at the Film *Avatar*," pp. 145–163 in *Marxism and the Movies: Critical Essays on Class Struggle in the Cinema*, edited by Mary K. Leigh and Kevin Durand. Jefferson, NC: McFarland.

Delaney, Tim, and Tim Madigan. 2009. *The Sociology of Sports: An Introduction.* Jefferson, NC: McFarland.

______, and ______. 2017. *Friendship and Happiness: And the Connection Between the Two.* Jefferson, NC: McFarland.

______, and ______, editors. 2018. *A Global Perspective on Friendship and Happiness.* Wilmington, DE: Vernon Press.

______, and ______. 2021. *The Sociology of Sports: An Introduction, Third Edition.* Jefferson, NC: McFarland.

Denchak, Melissa. 2018. "Hurricanes and Climate Change: Everything You Need to Know." National Resources Defense Council, December 3. https://www.nrdc.org/stories/hurricanes-and-climate-change-everything-you-need-know.

Dennis, Brady. 2013. "How Flushing Leftover Drugs Threatens Ecosystem." *The Post-Standard*, February 26: D-6.

______. 2016. "Study: Climate Change a Risk to Health." *The Post-Standard*, April 5: A10.

Dennis, Brady, and Andrew Freedman. 2019. "Yup, July Really was the Hottest Month on Record." *The Post-Standard*, August 6: A8.

Department of Agricultural and Consumer Economics (of Illinois). 2020. "Weekly Outlook: Corn Used for Ethanol Update." https://farmdocdaily.illinois.edu/2020/03/corn-used-for-ethanol-update.html.

Des Jardins, Joseph. 2013. *Environmental Ethics: An Introduction to Environmental Philosophy*, 5th Edition. Belmont, CA: Wadsworth.

De Souza, Marcelo. 2019. "More About the Amazon Wildfires." *The Citizen*, August 27: B6.

Devall, Bill, and George Sessions. *Deep Ecology*, 1985. Salt Lake City, UT: Gibbs Smith.

DeWaal, Frans. 1996. *Good Natured: The Origins of Right and Wrong in Humans and Other Animals.* Cambridge, MA: Harvard University Press.

Diabetes.co.uk. 2019. "Diabetes Types." https://www.diabetes.co.uk/diabetes-types.html.

Dickson, Paul. 2001. *Sputnik: The Shock of the Century.* New York: Walker.

Discovery News. 2013. "Obama Win Keeps NASA on Asteroid Odyssey." http://news.discovery.com/space/asteroids-meteors-meteorites/obama-win-nasa-direction-asteroids-121107.htm.

Doran, Elizabeth. 2011. "A Silver Lining to Going Green." *The Post-Standard.* January 13: A-5.

Drake, Frances. 2000. *Global Warming: The Science of Climate Change.* New York: Oxford University Press.

Drug Enforcement Administration (DEA). 2020. "Drug Labs in the United States National Clandestine Laboratory Register Data." https://www.dea.gov/clan-lab.

Dunaway, Finis. 2008. "Gas Masks, Pogo, and the Ecological Indian: Earth Day and the Visual Politics of American Environmentalism." *American Quarterly*, 60 (1): 67–71, 73–82, 84, 86, 88–99, 229.

Earth Day Network. 2020. "About Us." https://www.earthday.org/about-us/.

Earth Observatory. 2020. "Could Satellites Help Head Off a Locust Invasion? https://earthobservatory.

nasa.gov/images/146495/could-satellites-help-head-off-a-locust-invasion.
Earth Talk. 2020. "The Effects of Global Warming on Wildlife." *ThoughtCo,* February 18. https://www.thoughtco.com/how-wildlife-affected-by-global-warming-1203849#citation-2.
EatForTheEarth. 2014. "Food Waste: Causes, Effects, and Solutions." FarmTogetherNow.org. https://farmtogethernow.org/2014/11/08/food-waste-causes-effects-and-solutions/.
Ecofibre Industries Operation. 2010. "Animal Bedding." http://ecofibre.com.au/animal-bedding/.
Ecologist. 2020. "Don't Be a Fossil Fool." March 13. https://theecologist.org/2020/mar/13/dont-be-fossil-fool.
Ecology Fund. 2011. "Homepage." http://www.ecologyfund.com/ecology/_ecology.html.
The Economist. 2011. "Welcome to the Anthropocene." May 28: 11.
_____. 2013. "Welcome to the Plastisphere." July 20: 70.
Edwards, Ferne. 2006. "Dumpster Dining." *Alternatives Journal,* 32 (3): 16–17.
Ehrenfeld, John R. *Sustainability by Design.* 2008. New Haven, CT: Yale University Press.
Eilperin, Juliet. 2012. "The Vanishing Coral of the Great Barrier Reef." *The Post-Standard,* October 15: C-3.
_____. 2020. "Trump to Open Arctic Refuge for Drilling." *The Post-Standard,* August 18: A7.
Eisenstadt, Marnie. 2012. "Wanted: Invasive Plants That Choke Off Native Species in Gardens and Countryside." *The Post-Standard,* September 23: A-1, A-8.
Ellabban, Omar, Haitham Abu-Rub, and Frede Blaabjerg. 2014. "Renewable Energy Resources: Current Status, Future Prospects and Their Enabling Technology." *Renewable and Sustainable Energy Reviews,* 39: 748–764.
Ellis, Eric. 2009. "Biosphere." *The Encyclopedia of Earth.* http://www.earth.org/article/Biosphere.
Ellsmoor, James. 2019. "United States Spend Ten Times More on Fossil Fuel Subsidies Than Education." *Forbes,* June 15. https://www.forbes.com/sites/jamesellsmoor/2019/06/15/united-states-spend-ten-times-more-on-fossil-fuel-subsidies-than-education/#66a07b454473.
Encyclopedia Britannica. 2012. "End-Triassic Extinction." http://www.britannica.com/EBchecked/topic/1523109/end-Triassic-extinction.
_____. 2020. "Exxon Valdez Oil Spill." https://www.britannica.com/event/Exxon-Valdez-oil-spill.
Endangered Species International. 2011. "The Five Worst Mass Extinctions." http://www.endangeredspeciesinternational.org/overview.html.
Enerdata Intelligence. 2019. "Energy Market Reports: Russia Energy Report." https://estore.enerdata.net/energy-market/russia-energy-report-and-data.html.
Energy Efficiency & Renewable Energy (EERE). 2020. "How Do Wind Turbines Work?" https://www.energy.gov/eere/wind/how-do-wind-turbines-work.
Energy Skeptic. 2014. "Cost of Invasive Species in the United States." http://energyskeptic.com/2014/cost-of-invasive-species-in-the-united-states/.
Environmental Building News. 2009. "Porous Paving." Vol. 18, No. 4. http://www.buildinggreen.com/auth/article.cfm/2009/3/26/Pourous-Paving/.
Environmental Defense Fund. 2011. "Global Warming and Extreme Weather." http://www.edf.org/page.cfm?tagid=1405.
Environmental Protection Agency (EPA). 2008. "Why Should You Be Concerned About Air Pollution?" http://www.epa.gov/air/peg/concern.html.
_____. 2010. "History of the Clean Air Act." http://epa.gov/oar/caa/caa_history.html.
_____. 2011a. "Clean Air Act." http://www.epa.gov/air/caa/.
_____. 2011b. "Safe Drinking Water Act (SDWA)." http://water.epa.gov/lawsregs/rulesregs/sdwa/index.cfm.
_____. 2011c. "Our Waters." http://water.epa.gov/type/.
_____. 2011d. "Turning Food Waste into Energy at the East Bay Municipal Utility District." http://www.epa.gov/region9/waste/features/foodtoenergy/food-waste.html.
_____. 2011e. "Volcanic Ash." http://www.epa.gov/aging/resources/other/volcanicash.html.
_____. 2011f. "Study Finds Peat Wildfire Smoke Linked to Heart Failure Risk." http://www.epa.gov/research/htm/wildfire.htm.
_____. 2011g. "Non-Hazardous Waste." http://www.epa.gov/epawaste/basic-solid.htm.
_____. 2012a. "Protecting the Stratospheric Ozone Layer." http://www.epa.gov/air/peg/stratozone.html.
_____. 2012b. "Air and Radiation." http://www.epa.gov/air/basic.html.
_____. 2012c. "Acid Rain in New England: A Brief History." http://www.epa.gov/region1/eco/acidrain/history.html.
_____. 2012d. "Aquatic Biodiversity: Glossary." http://www.epa.gov/bioindicators/aquatic/glossary.html.
_____. 2012e. "Laws and Regulations: Summary of the Clean Water Act." http://www.epa.gov/lawsregs/laws/cwa.html.
_____. 2012f. "Water: Pollution Prevention and Control." http://water.epa.gov/polwaste/.
_____. 2012g. "Contaminated Soil and Sentiments." http://www.epa.gov/nrmrl/lrpcd/contamin_ss.html.
_____. 2012h. "Municipal Solid Waste." http://www.epa.gov/cleanenergy/energy-andyou/affect/municipal-sw.html.
_____. 2012i. "Noise Pollution." http://www.epa.gov/air/noise.html.
_____. 2012lj "Glossary: Hydraulic Fracturing." http://water.epa.gov/type/groundwater/uic/glossary.
_____. 2012k. "Plastics." http://www.epa.gov/osw/conserve/materials/plastics.htm.
_____. 2012l. "Food Recovery Challenge." http://www.epa.gov/foodrecoverychallenge/.
_____. 2012m. "General Information on E-Waste." http://www.epa.gov/osw/conserve/materials/ecycling/faq.html.

______. 2012n. "Green Roofs." http://www.epa.gov/hiri/mitigation/greenroofs.html.

______. 2016a. "Health Effects of Ozone in the General Population." https://www.epa.gov/ozone-pollution-and-your-patients-health/health-effects-ozone-general-population.

______. 2016b. "Climate Change Indicators: Snow and Ice." https://www.epa.gov/climate-indicators/snow-ice.

______. 2017a. "Climate Impacts on Forests." https://19january2017snapshot.epa.gov/climate-impacts/climate-impacts-forests_.html.

______. 2017b. "Medical Waste." https://www.epa.gov/rcra/medical-waste.

______. 2017c. "Brown Marmorated Stink Bug." https://www.epa.gov/safepestcontrol/brown-marmorated-stink-bug.

______. 2018. "Threats to Coral Reefs." https://www.epa.gov/coral-reefs/threats-coral-reefs.

______. 2019a. "Criteria for Definition of Solid Waste and Solid Hazardous Waste Exclusions." https://www.epa.gov/hw/criteria-definition-solid-waste-and-solid-and-hazardous-waste-exclusions.

______. 2019b. "Plastics: Material-Specific Data." https://www.epa.gov/facts-and-figures-about-materials-waste-and-recycling/plastics-material-specific-data.

______. 2020a. "Introduction to Indoor Air Quality." https://www.epa.gov/indoor-air-quality-iaq/introduction-indoor-air-quality.

______. 2020b. "EPCRA Sections 311–312: Emergency Planning and Community Right-to-Know Act (EPCRA) Hazardous Chemical Inventory Reporting Requirements." https://www.epa.gov/epcra/epcra-sections-311-312.

______. 2020c. "National Overview: Facts and Figures on Materials, Wastes and Recycling." https://www.epa.gov/facts-and-figures-about-materials-waste-and-recycling/national-overview-facts-and-figures-materials.

______. 2020d. "Coronavirus and Drinking Water and Wastewater." https://www.epa.gov/coronavirus/coronavirus-and-drinking-water-and-wastewater.

Environmental Recyclers International (ERI). 2015. "Toxins Found in E-Waste." https://www.environmental-expert.com/articles/toxins-found-in-e-waste-462709.

Environmental Working Group. 2012. "2011 EWG Bottled Water Scorecard: How Much Do We Drink?" http://www.ewg.org/bottled-water-2011-how-much-do-we-drink.

Epstein, Marc J. 2008. *Making Sustainability Work*. Sheffield, UK: Greenleaf.

Esch, Mary. 2020. "Curtain Lowers on Nuke Plant Near NYC." *The Citizen*, April 30: A3.

European Commission. 2020a. "Causes of Climate Change." https://ec.europa.eu/clima/change/causes_en.

______. 2020b. "Ozone Regulation." https://ec.europa.eu/clima/policies/ozone/regulation_en.

Fahrenthold, David A. 2010. "Don't Care About Global Warming? That's Just Natural." *The Washington Post* (as it appeared in *The Post-Standard*), December 8: A-11.

Fairlie, Simon. 2010. *Meat: A Benign Extravagance*. White River Junction, VT: Chelsea Green.

Fallen Fruit of Buffalo. 2017. "The Endless Orchard." http://fallenfruit.org/projects/.

Falling Fruit. 2020. "Map the Urban Harvest." http://fallingfruit.org/maps.

Federal Emergency Management Agency. 2012. "Home Fires." http://www.ready.gov/home-fires.

Federal Trade Commission. 2020. "Green Guides." https://www.ftc.gov/news-events/media-resources/truth-advertising/green-guides.

Fieser, Louis F., George C. Harris, E.B. Hershberg, Morley Morgana, Frederick C. Novello, and Stearns T. Putnam. 1946. "Napalm." *Industrial and Engineering Chemistry*, 38 (8): 768–773.

Fimbel, Robert A., Alejandro Grajal, and John G. Robinson, editors. 2001. *The Cutting Edge: Conserving Wildlife in Logged Tropical Forest*. New York: Columbia University Press.

Fisher, Mischa. 2013. "The Republican Party Isn't Really the Anti-Science Party." *The Atlantic*, November 11. https://www.theatlantic.com/politics/archive/2013/11/the-republican-party-isnt-really-the-anti-science-party/281219/.

Fitzgerald, Jim. 2012. "Sandy Uprooted Trees by the Thousands in NY, NJ." *The Post-Standard*, November 18: B-6.

Food and Agriculture Organization of the United Nations (FAO). 2005. "Discard in the World's Marine Fisheries." www.fao.org/tempref/drocrep/fao/008/y5936e/y5936e00.pdf.

______. 2006. "Livestock Impacts on the Environment." http://www.fao.org/ag/magazine/0612sp1.htm.

______. 2015. "Extending the Area Under Sustainable Land Management and Reliable Water Control Systems." *NEPAD—Comprehensive Africa Agriculture Development Programme*. http://www.fao.org/3/Y6831E/y6831e-03.htm.

______. 2019. "The State of Food Security and Nutrition in the World." http://www.fao.org/state-of-food-security-nutrition/en/.

______. 2020a. "World Food Situation." http://www.fao.org/worldfoodsituation/csdb/en/.

______. 2020b. "2018 The State of the World's Forests." http://www.fao.org/state-of-forests/en/.

______. 2020c. "Sustainable Land Management." http://www.fao.org/land-water/land/sustainable-land-management/en/

Food Print. 2020. "The Problem of Food Waste." https://foodprint.org/issues/the-problem-of-food-waste/?cid=5664.

Food Production Daily. 2004. "Half of U.S. Food Goes to Waste." http://www.foodproductiondaily.com/Supply-Chain/Half-of-US-food-goes-to-waste.

Fox-Skelly, Jasmin. 2017. "There Are Diseases Hidden in Ice, and They Are Waking Up." *BBC*, May 4. http://www.bbc.com/earth/story/20170504-there-are-diseases-hidden-in-ice-and-they-are-waking-up.

Freedman, Andrew. 2020. "Greenland's Melting Ice

Raised Sea Levels." *The Post-Standard,* March 19:A17.

Freudenrich, Craig. "How Plastics Work." *How Stuff Works.* https://science.howstuffworks.com/plastic4.htm.

Friedman, Thomas L. 2011. "Our Sustainable Earth." Syndicated column as it appeared in *The Post-Standard,* June 9: A-12.

Fun Times Guide. 2012a. "Who Owns RVs? Some Facts and Figures About RV Sales and the People Who Own." Motorhomes." http://rv-roadtrips.thefuntimesguide.com/2007/08/who_owns_rvs.php.

_____. 2012b. "Green RV Tips: 6 Ways to Go Green When RVing." http://rv-roadtrips.thefuntimesguide.com/2008/05/green_rv_tips_for_going_green.php.

Furqan, Alhilal, Ahmad Puad Mat Som, Rosazman Hussin. 2010. "Promoting Green Tourism for Future Sustainability." *Theoretical and Empirical Researchers in Urban Management,* 5(17):64–74.

Garden Activist. 2020. "Food Gardens Matter!" http://gardenactivist.org/.

Gargulinski, Ryn. 2011. "Nutritional and Environmental Problems Associated with Famine." Livestrong.com, June 8. http://www.livestrong.com/article/466691-nutritional-environmental-problems- associated-with-famine/.

Ghosh, Bibek. 2007. *Crops and Livestock Farming.* Delhi, India: Global Media.

Gillam, Carey. 2020. "Revealed: Monsanto Predicted Crop System Would Damage U.S. Farms" *The Guardian,* March 30. https://www.theguardian.com/us-news/2020/mar/30/monsanto-crop-system-damage-us-farms-documents.

Glanz, James and Campbell Robertson. 2020. "Lockdown Delays at Least 36,000 Lives, Data Show." *The New York Times,* May 20. https://www.nytimes.com/2020/05/20/us/coronavirus-distancing-deaths.html?searchResultPosition=1.

Gleick, Peter. 2009. "Energy Implications of Bottled Water." *Environmental Research Letters,* 4: 664–670.

Glennon, Robert. 2009. *Unquenchable: America's Water Crisis and What to Do About It."* Washington, D.C.: Island Press.

Global Footprint Network. 2020a. "Measure What You Treasure." https://www.footprintnetwork.org/.

_____. 2020b. "Earth Overshoot Day." https://www.footprintnetwork.org/our-work/earth-overshoot-day/.

Global Petrol Prices. 2019. "Russia Electricity Prices." http://www/globalpetrolprices.com/Russia/electricity_proces/.

_____. 2020. "Russia Gasoline Prices, 17-Feb-2020." https://www.globalpetrolprices.com/Russia/gasoline_prices/.

Goerner, Sally J., Robert G. Dyck, and Dorothy Lagerroos. 2008. *The New Science of Sustainability: Building a Foundation for Great Change.* Chapel Hill, NC: Triangle Center for Complex Systems.

Goldenberg, Suzanne. 2011. "Africa Famine: Soaring Food Prices Intensifying Crisis, Report Warns." *The Guardian,* August 16. http://www.guardian.co.uk/environment/2011/aug/16/africa-famine-food-prices-world-bank.

Gonzalez-Garcia, S., A. Hospido, G. Feijoo, and M.T. Moreira. 2010. "Life Cycle Assessment of Raw Materials for Non-wood Pulp Mills: Hemp and Flax." *Resources, Conservation and Recycling,* 54: 923–930.

Goodell, Jeff. 2019. "The World's Most Insane Energy Project Moves Ahead." *Rolling Stone,* June 14. https://www.rollingstone.com/politics/politics-news/adani-mine-australia-climate-change-848315/.

Goodman, Amy and Denis Moynihan. 2018. "Hurricanes and Climate Change." *The Citizen,* October 13: A4.

Gore, Al. 2011. "Climate of Denial: Can Science and the Truth Withstand the Merchants of Poison?" *Rolling Stone,* July 7–21: 76–83, 112–113. Graduation Pledge of Social and Environmental Responsibility. http://www.graduationpledge.org/.

The Graduate. 1967. Original release date, December 22, 1967, by United Artists.

Green, Catherine. 2013. "In L.A. Getting Paid to Go Green." *Los Angeles Times.* June 27: B-1, B-5.

Green and Growing. 2018. "How Does Wastewater Affect the Environment: 4 Facts." https://www.greenandgrowing.org/how-does-wastewater-affect-the-environment/.

Green Building Manual, 2010. "Whole Building Design Guide: Psychosocial Value of Space." http://www.wbdg.org/resources/pschspace_value.phd.

Green Cross International. 2012a. "Our History." http://www.gcint.org/our-history.

_____. 2012b. "Our Mission." http://www.gcint.org/our-mission.

Griggs, Cornelia. 2020. "A New York Doctor's Coronavirus Warning: The Sky Is Falling." An Op-Ed appearing in *The New York Times,* March 19. https://www.nytimes.com/2020/03/19/opinion/coronavirus-doctor-new-york.html.

Grinspoon, David. 2016. *Earth in Human Hands: Shaping Our Planet's Future.* New York: Grand Central Publishing.

Groom, Debra J. 2012a. "Area Schools Get A+ in Energy Savings." *The Post-Standard,* March 15: A-9.

_____. 2012b. "How Now, Sustainable Cow?" *The Post-Standard,* March 20: 1, 6.

Gruber, June, Iris B. Mauss, and Maya Tamir. 2012. "A Dark Side of Happiness? How, When, and Why Happiness Is Not Always Good." *Perspectives on Psychological Science.* May 1, 3: 315–323.

The Guardian. 2018. "Humanity Has Wiped Out 60% of Animal Populations Since 1970, Report Finds." https://www.theguardian.com/environment/2018/oct/30/humanity-wiped-out-animals-since-1970-major-report-finds.

_____. 2020. "'Race Against Time' to Prevent Famines During Coronavirus Crisis." April 16. https://www.theguardian.com/global-development/2020/apr/16/race-against-time-to-prevent-famines-during-coronavirus-crisis.

Guse, Clayton. 2019. "Black Snot and Cancer Risks: NYC Subway Air Is Full of Pollution." *New York Daily News,* February 15. https://www.nydailynews.com/new-york/ny-metro-air-quality-subway-mta-20190215-story.html.

Guy, Jack, and Valentina Di Donato. 2020. "Venice's Canal Water Looks Clearer As Coronavirus Keeps Visitors Away." *CNN Travel,* March 16. https://www.cnn.com/travel/article/venice-canals-clear-water-scli-intl/index.html.

Haddock, Vicki. 2013. "Wendy Northcutt, the Darwin Awards." *Failure Magazine.* http://failuremag.com/feature/article/wendy_northcutt_the_darwin_awards/.

Haines, Andy, and Kristie Ebi. 2019. "The Imperative for Climate Action to Protect Health." *New England Journal of Medicine,* 380: 263–273.

Hall, Michelle, and Amanda DeCamp. 2012. *Kilauea Volcano.* http://www.stfrancis.edu/content/ns/bromer/earthsci/student9/Web%20PageMichAmand/Michelle.htm.

Hall, Trish. 1998. "Seeking a Focus on Joy in Field of Psychology." *The New York Times,* April 28: 1–5.

Hallam, A., and P.B. Wignall. 1997. *Mass Extinctions and Their Aftermath.* New York: Oxford University Press.

Happy Planet Index (HPI). 2020a. "About the HPI: How Is the Happy Planet Index Calculated?" http://happyplanetindex.org/about/.

_____. 2020b. "Happy Planet Index Score." http://happyplanetindex.org/.

Harper, Charles L. 2012. *Environment and Society, 5th edition.* Boston: Pearson.

Harvey-Samuel, Tim, Thomas Ant, and Luke Alphey. 2017. "Towards the Genetic Control of Invasive Species." *Biological Invasions,* 1996): 1683–2703.

Hays, Brooks. 2020. "Climate Change to Blame for Megadrought Emerging Across Western U.S." *UPI.com,* April 17. https://www.upi.com/Science_News/2020/04/17/Climate-change-to-blame-for-megadrought-emerging-across-Western-US/5651587066418/.

Healy, Melissa. 2019. "Doctors Weigh In on Global Warming: It Makes Us Sicker." *Los Angeles Times,* January 24: A2.

Henderson, Bruce. 2011. "Global Warming Fallout: Birds, Trees, Ants…They All Tell Story of Warming." *The Post-Standard,* May 30: C-3.

Henning, Brian G. 2015. *Riders in the Storm: Ethics in an Age of Climate Change.* Winona, MN: Anselm Academic.

Heron, Melonie. 2019. "Deaths: Leading Causes for 2017." *CDC's National Vital Statistics Reports,* 68(4). https://www.cdc.gov/nchs/data/nvsr/nvsr68/nvsr68_06-508.pdf.

Hershfield, Hal E., Cassie Mogilner, and Uri Barnea. 2016. "People Who Choose Time Over Money Are Happier." *Social Psychological & Personality Science,* 7(7): 697–706.

Hickman, Larry A. 2007. *Pragmatism as Post-postmodernism: Lessons from John Dewey.* New York: Fordham University Press.

Hickman, Larry A., and Thomas M. Alexander. 1998. *The Essential Dewey,* Vol. I. Bloomington: Indiana University Press.

History.com 2020. "History of Arbor Day." https://www.history.com/topics/holidays/the-history-of-arbor-day.

Hoffman, Jane, and Michael Hoffman. 2009. "What Is Greenwashing?" *Scientific American* (April). http://www.scientificamerican.com/article.cfm?id=greenwashing-green-energy-Hoffman.

Hogan, Michael C. 2010. "Overgrazing." *Encyclopedia of Earth.* http://www.eoearth.org/article/Overgrazing?topic=49480.

Hohn, Donovan. 2011, *Moby-Duck.* New York: Penguin Books.

Holden, Robert. 2010. "Your Happiness Plan: The Happiness Quiz." http://www.oprah.com/spirit/Take-the-Happiness-Test-Quiz.

Holliday, Michelle. 2014. "What Is Thrivability?" http://thrivableworld.org/what-is-thrivability/.

Horie, Hiromi. 2019. "Japanese Firm Is Working to Clean Up Junk in Space." *The Post Standard,* February 21: A19.

House, Khara E. 2007. "Kermit the Frog and Ford Team Up for Green-Friendly Super Bowl Commercial." *Yahoo! Contributor News.* http://voices.yahoo.com/kermit-frog-ford-team-green-friendly-188407.html?cat=9http://voices.yahoo.com/kermit-frog-ford-team-green-friendly-188407.html?cat=9.

How I Met Your Mother. 2011. "Garbage Island." Season 6, episode 17. First air date: February 21, 2011.

Howarth, Macy E., Christopher D. Thorncroft, and Lance F. Bosart. 2019 (April). "Changes in Extreme Precipitation in the Northeast United States: 1970–2014." *Journal of Hydrometeorology,* 20(4). https://journals.ametsoc.org/doi/full/10.1175/JHM-D-18-0155.1

Howell, Evan A., Steven J. Bograd, Carey Morishige, Michael P. Seki, and Jeffrey J. Polovina. 2012. "On North Pacific Circulation and Associated Marine Debris Concentration." *Marine Pollution Bulletin,* 65: 16–22.

Hower, Jeff. 2013. "Diseases that Canada Geese Carry." Ohio Geese Control, February 4. http://www.ohiogeesecontrol.com/diseases-that-canada-geese-carry/.

Hsu, Tiffany. 2012. "U.S. to Become World's Largest Oil Producer Before 2020, IEA Says." *Los Angeles Times,* November 12. http://www.latimes.com/business/money/la-fi-mo-u.s.-oil-producer-saudi-arabia-iea-20121112,0,6181922.story.

Hubbard, Amy. 2012. "Loss of Greenland Ice Could be Irreversible." *The Post-Standard,* March 26: C-3.

Huebsch, Russell. 2017. "How Are Fossil Fuels Extracted from the Ground?" *Sciencing,* September 29. https://sciencing.com/how-are-fossil-fuels-extracted-from-the-ground-12227026.html.

Humes, Edward. 2013. *Garbology: Our Dirty Love Affair with Trash.* New York: Avery.

Hunt, Kristin. 2019. "How Sports Stadiums Are Becoming More Sustainable." Green Matters. Available: https://www.greenmatters.com/travel/2018/07/31/2f5fvD/sports-stadiums-sustainable-design.

Hyde, Jesse. 2019. "Cattle Ranching Remains Top Threat to the Amazon." *Los Angeles Times*, October 4L A1, A4.

Hydrogenics. 2020. "Fuel Cells." https://www.hydrogenics.com/technology-resources/hydrogen-technology/fuel-cells/.

IBIS World. 2020. "Logging Industry in the US—Market Research Report." https://www.ibisworld.com/united-states/market-research-reports/logging-industry/.

The Independent. 2011. "World's Sixth Mass Extinction May Be Underway—Study." March 7. http://www.independent.co.uk/environment/worlds-sixth-mass-extinction-may-be-underway-study-2234388.html.

Institute for Energy Research (IER). 2019. "Fossil Fuels." https://www.instituteforenergyresearch.org/?encyclopedia=fossil-fuels.

Institute for Health Metrics and Evaluation. 2016. "For the First Time, Environmental Air Pollution Emerges as a Leading Risk Factor for Stroke Worldwide." http://www.healthdata.org/news-release/first-time-environmental-air-pollution-emerges-leading-risk-factor-stroke-worldwide.

International Association for Environmental Philosophy (IAEP). 2020. https://environmentalphilosophy.org/welcome/.

International Atomic Energy Agency. 2012. "Atoms for Peace," speech given by President Dwight D. Eisenhower, December 8, 1953, before the United Nations General Assembly. http://www.iaea.org/About/history_speech.html.

International Renewable Energy Agency (IRENA). 2019. "Renewable Energy Prospects for the Russian Federation (Remap Working Paper). http://www.irena.org/publications/2017/Apri/Renewable-Energy-Prospects-for-the-Russian-Federation-Remap-working paper.

Internet Encyclopedia of Philosophy (*IEP*). 2020. "Cynics." https://www.iep.utm.edu/cynics/.

______. 2020. "Karl Robert Eduard Von Hartmann." https://iep.utm.edu/hartmann/.

In2Greece. 2012. "Information About Santorini" http://www.in2greece.com/english/places/summer/islands/santorini.htm.

Invasive Species Specialist Group. 2004. *100 of the World's Most Invasive Alien Species: A Selection from the Global Invasive Species Database*. Auckland, New Zealand: Hollands. http://www.issg.org/database/species/reference_files/100English.pdf.

Investopedia. 2020. "The World's Top Oil Producers of 2019." https://www.investopedia.com/investing/worlds-top-oil-producers/.

Iran Front Page (IFP). 2020. "Billions of Locusts Swarm Southern Iran." https://ifpnews.com/billions-of-locusts-swarm-southern-iran.

Irish Examiner. 2019. "Climate Change Checkup." May 31: 4–5.

Jackson, David. 2020. "Trump Orders Meat and Poultry Processing Plants to Stay Open During Coronavirus." *USA Today*, April 28. https://www.usatoday.com/story/news/politics/2020/04/28/coronavirus-trump-plans-order-meat-processing-plants-stay-open/3038300001/.

Jackson, Stephen. T. 2020. "Climate Change." *Encyclopedia Britannica*. https://www.britannica.com/science/climate-change/Climate-change-within-a-human-life-span.

Jessa, Tega. "Where Does Space Begin?" *Universe Today*, October 14. http://www.universetoday.com/75710/where-does-space-begin/.

Johnson, Hannah. 2020. "70,000 Sea Turtles Nested on Deserted Beaches Due to COVID-19." STN2.tv. https://stn2.tv/2020/04/18/70000-sea-turtles-nested-on-deserted-beaches-due-to-covid-19/.

Journal of Happiness Studies. 2013. New York: Springer.

Kahn, Amina. 2019. "Missing: Billions of North American Birds." *Los Angeles Times*, September 21: B2.

Kahneman, Daniel, and Angus Deaton. 2010. "High Income Improves Evaluation of Life but Not Emotional Well-being." *Proceedings of the National Academy of Sciences*, 107(38): 16489–16493.

Kaluzniak, Donna. 2012. "Wildfire's Effects on Drinking Water." *Examiner*, July 15. http://www.examiner.com/article/wildfire-s-effects-on-drinking-water.

Kan, Shirley. 2007. "CRS Report for Congress: China's Anti-satellite Weapon Test." Order Code RS22652, April 23, 2007. http://www.fas.org/sgp/crs/row/RS22652.pdf.

Kaplan, Eben. 2006. "Targets for Terrorists: Chemical Facilities." Council on Foreign Relations. http://www.cfr.org/united-states/targets-terrorists-chemical-facilities/p12207.

Kasser, Tim, and Kennon Sheldon. 2009. "Time Affluence as a Path Toward Personal Happiness and Ethical Business Practice: Empirical Evidence from Four Studies." *Journal of Business Ethics*, 84: 243–255.

Keller, R.P., and C. Perrings. 2011. "International Policy Options for Reducing Environmental Impacts of Invasive Species." *BioScience*, 61 (12), 1005–1012.

Keneally, Meghan. 2012. "What a Beautiful Dump! Beach That Used to Be Trash Tip Is Now Covered in Glass After Spectacular Transformation by Waves." *Daily Mail*, January 9. http://www.dailymail.co.uk/news/article-2083862/What-dump-A-beach-used-covered-trash-covered-sea-glass-waves-turned-garbage-natural-art.html.

Keneally, Thomas. 2010. *Three Famines*. Sydney, Australia: Knopf.

Kernohan, Andrew. 2012. *Environmental Ethics: An Interactive Introduction*. Tonawanda, NY: Broadview Press.

Khan, Amina. 2020. "Another Hot Year for the Planet." *Los Angeles Times*, January 16: A2.

Khan, Anwarullah. 2010. "Thousands of Poor Villagers Fear Hunger After Pakistan Suicide Bombing Ends Food Relief." *Canadian Press*, December 26. http://ca.news.yahoo.com/un-food-aid-centres-northwest-pakistan-ordered-temporarily-20101226-005759-385.html.

Kim, Ki-Hyun, Duy Xuan Ho, Richard J.C. Brown, J.M. Oh, Chan Goo Park, and In Cheol Ryu. 2012. "Some Insights into the Relationship Between Urban Air Pollution and Noise Levels." *Science of the Total Environment*, 424 (1): 271–279.

Kim, Soo. 2016. 2019. "Climate Change 'To Increase Turbulence, Flight Times and Fares.'" *The Telegraph,* May 12. https://www.telegraph.co.uk/travel/news/climate-change-to-increase-turbulence-flight-times-and-costs/.

King, Hobart M. 2020. "What Are Oil Sands?" *Geology Today.* https://geology.com/articles/oil-sands/.

Kiprop, Victor. 2018. "What Are Microplastics and Why Are They Bad?" *World Atlas.* https://www.worldatlas.com/articles/what-are-microplastics-and-why-are-they-bad.html.

Knauss, Tim. 2008. "Drive to Innovation." *The Post-Standard,* June 5: A-1.

______. 2016. "NRC Engineers Go Public with Nuke Safety Concern." *The Post-Standard,* March 8: B3.

Knickmeyer, Ellen. 2020. "EPA Drops Regulation Against Contaminant." *The Citizen,* June 19: B8.

______. 2020. "President Ends Some Water Restrictions." *The Citizen,* January 24: A6.

Kopnina, Helen, Haydn Washington, Bron Taylor, and John J. Piccolo. 2018. "Anthropocentrism: More than Just a Misunderstood Problem." *Journal of Agricultural and Environmental Ethics,* 31: 109–127.

Koppel, Ted. 2016. *Lights Out: A Cyberattack, A Nation Unprepared.* New York: Broadway Books.

Kozloff, Nikolas. 2010. *No Rain in the Amazon: How South America's Climate Change Affects the Entire Planet.* New York: Palgrave.

Kristof, Nicholas D. 2012. "Obama's Aid Plan Is Boring, and It Works." Syndicated column in *The Post-Standard,* July 14: A-12.

Krstic, Zee. 2019. "Ireland Will Plant 22 Million Trees Every Year to Fight Climate Change." Marthastewart.com, September 5. https://www.marthastewart.com/2126607/ireland-440-million-trees-planted-fight-climate-change.

Kulkus, Emily. 2011. "Oncenter Green Roof Takes Root." *The Post-Standard,* September 29: A-1, A-8.

Kumar, Sachin, Achyut K. Panda, and R.K. Singh. 2011 (Sept.). "A Review on Tertiary Recycling of High-Density Polyethylene to Fuel." Resources, Conservation and Recycling, 55(11): 893–910.

Lackey, Robert T. 1998. "Seven Pillars of Ecosystem Management." *Landscape and Urban Planning,* 40 (1/3): 21–30.

Lakshmanan, Indira A.R. 2006. "A Living Tree Emits Oxygen. A Dead One Emits Carbon Dioxide." *Boston Globe.* December 6: A-9.

Lallanilla, Marc. 2019a. "The Effects of War on the Environment." *ThoughtCo.* https://www.thoughtco.com/the-effects-of-war-on-environment-1708787.

______. 2019b. "What Are GMOs and GM Foods?" *Live Science,* July 8. https://www.livescience.com/40895-gmo-facts.html.

Lambrou, Peter. 2014. "Is Fear of Happiness Real? *Psychology Today,* May 1. https://www.psychologytoday.com/gb/blog/codes-joy/201405/is-fear-happiness-real

Landers, Jackson. 2012. *Eating Aliens: One Man's Adventure Hunting Invasive Animal Species.* North Adams, MA: Storey.

Langlois, Jill. 2018. "More Give Their Lives for the Land." *Los Angeles Times,* August 7: A4.

Larson, Christina. 2019. "Restoring Forests, One Tree at a Time, to Help Repair Climate." *The Post-Standard,* October 6:E5.

Larson, Christina, and Federica Narancio 2019. "Scientists Witness Final Days of Venezuelan Glacier." *The Post-Standard,* September 29: B1.

Latona, Jessa. 2007. "CAT Media Department Press Release: Climate Friendly Hemp Homes?" April 13. http://www.cat.org.uk/news/news_release.tmpl?command=search&db=news.db&cqSKUdatarq=34080.

Leahy, Stephen. 2019. "Thirsty Future Ahead as Climate Change Explodes Plant Growth." *National Geographic,* November 4. https://www.nationalgeographic.com/science/2019/10/plants-consume-more-water-climate-change-thirsty-future/.

Leakey, Richard. *The Origin of Humankind.* New York: Basic Books.

Lee, Bruce Y. 2017. "CDC Warns: Don't Get Too Close to Your Chickens, Ducks and Geese." *Forbes,* June 3. https://www.forbes.com/sites/brucelee/2017/06/03/cdc-warns-dont-get-too-close-to-your-chickens-ducks-and-geese/#280fc482b938.

Legatum Prosperity Index. 2019. "Executive Summary." Legatum Institute. https://www.prosperity.com/.

Leigh, Andrew. 2006. "Growth Matters." *Aurora Magazine,* Issue 3 (July). http://people.anu.edu.au/andrew.leigh/pdf/GrowthMatters.pdf.

Lemann, Nicholas. 2013. "When the Earth Moved: What Happened to the Environmental Movement?" *The New Yorker.* April 15: 73–76.

Lennard, Natasha. 2019. "Ecocide Should Be Recognized as a Crime Against Humanity, but We Can't Wait for The Hague to Judge." *The Intercept,* September 24. https://theintercept.com/2019/09/24/climate-justice-ecocide-humanity-crime/.

Lenntech. 2011. "Volcanic Eruptions and Environment." http://www.lenntech.com/volcanic-eruptions-environment.htm.

Leopold, Aldo. 1949. *A Sand County Almanac and Sketches Here and There.* New York: Oxford.

Levine, Michael. 1998. "The Trash Mess Won't Be Easily Disposed Of." *Wall Street Journal.* December 15: 14.

Lewis, Martin W. 1992. *Green Delusions: An Environmentalist Critique of Radical Environmentalism.* Durham, NC: Duke University Press.

Lewis, Sophie. 2019. "24,000 Pounds of Garbage Were Just Removed from Mount Everest, Leading to the Discovery of 4 Dead Bodies." *CBS News,* June 6. https://www.cbsnews.com/news/mount-everest-garbage-24000-pounds-removed-leading-to-the-discovery-of-four-dead-bodies-today-2019-06-06/.

Lewis, Tanya. 2013. "Asteroid Killed Off Dinosaurs? New Study Suggests Comet Instead Caused Extinction Event." *Huffington Post,* March 23. http://www.huffingtonpost.com/2013/03/23/asteroid-killed-dinosaurs-comet-extinction_n_2937296.html.

Library of Congress. 2011. "Films Selected to the Library of Congress National Film Registry

1989–2010. http://www.loc.gov/film/nfrchron.html.

Light, Andrew, and Eric Katz, editors. 1996. *Environmental Pragmatism.* London and New York: Routledge. lightningsafety.com/nlsi_info/thunder2.htm. 2013.

Lindell, Catherine. 1996. "Patterns of Nest Usurpation: When Should Species Converge on Nest Niches?" *The Condor,* 98: 464–473.

Lippin, Tobi Mae, Thomas H. McQuiston, Kristin Bradley-Bull, Toshiba Burns-Johnson, Linda Cook, Michael L. Gill, Donna Howard, Thomas A. Seymour, Doug Stephens, and Brian K. Williams. 2006 (April). "Chemical Plants Remain Vulnerable to Terrorists: A Call to Action." *Environmental Health Perspectives,* 114(9): 1307–1311.

Lockheed Martin. 2020. "Space Fence." Lockheed Martin website. https://www.lockheedmartin.com/en-us/products/space-fence.html.

Lomborg, Bjørn. 2001. *The Skeptical Environmentalist: Measuring the Real State of the World.* New York: Cambridge University Press.

_____. 2011. "Does Helping the Planet Hurt the Poor? Yes, If We Listen to Green Extremists." *The Wall Street Journal,* January 22–23: C-1, C-2.

Lorenzen, Janet A. 2012. "Going Green: The Process of Lifestyle Change." *Sociological Forum.* Vol. 27.1: 94–116.

Los Angeles Times (editorial). 2019. "Extracting Too Much Fossil Fuel." November 22: A10.

_____. 2019b. "The Undervalued Trees of LA." January 2: A10.

_____. 2020a. "Clean Cars? Who Needs 'Em?" April 1: A10.

_____. (editorial). 2020b. "Preserve These Species, Not Their Trophies." August 17: A11.

_____. 2020c. "BP to Cut 10,000 Jobs in Revamp." June 9: A9.

Lynas, Mark. 2020. *Our Final Warning: Six Degrees of Climate Emergency.* London: 4th Estate.

Lyubomirsky, Sonja. 2008. *The How of Happiness: A Scientific Approach to Getting the Life You Want."* New York: Penguin.

Macalister, Alexander. 2020. "Pestilence." *Bible Study Tools.* 2020. https://www.biblestudytools.com/dictionary/pestilence/.

MacCannell, Dean. 2011. *The Ethics of Sight-Seeing.* Berkeley, CA: University of California Press.

Macionis, John J. 2010. *Social Problems,* 4th edition. Upper Saddle River, NJ: Pearson.

Madhani, Aamer, and Kevin Freking. 2020. "Trump to Roll Back Environmental Act." *The Citizen,* July 16: A7.

Madigan, Tim, and Peter Stone, Editors. 2016. *Bertrand Russell: Public Intellectual.* Rochester, New York: Tiger Bark Press.

Mahr, Krista. 2019. "Climate Fuels Conflict in Nigeria." *Los Angeles Times,* February 22: A3.

Malloch, Theodore, and Scott T. Massey. 2006. *Renewing American Culture: The Pursuit of Happiness.* Salem, MA: Scrivener.

Malthus, Thomas. 1798. *An Essay on the Principle of Population: As It Affects the Future Improvement of Society.* London: J. Johnson.

Marriott, Alexander. 2011. "The Curse of the Internet: Fake Historical Quotes." http://alexandermarriott.blogspot.com/2011/11/curse-of-internet-fake-historical.html.

Marvin, Rob. 2018. "Tech Addiction By the Numbers: How Much Time We Spend Online." *PC Magazine,* June 11. https://www.pcmag.com/news/tech-addiction-by-the-numbers-how-much-time-we-spend-online.

Marx, Karl, and Friedrich Engels. 1978. *The Marx-Engels Reader,* 2nd edition, edited by Robert C. Tucker. New York: W.W. Norton.

Mason, Greg. 2013. "Borer Horror." *The Citizen,* July 28: A-1.

The Mayo Clinic. 2020a. "Swine Flu (h1N1 flu)." https://www.mayoclinic.org/diseases-conditions/swine-flu/symptoms-causes/syc-20378103.

_____. 2020b. "Heart Disease: Overview." https://www.mayoclinic.org/diseases-conditions/heart-disease/symptoms-causes/syc-20353118.

_____. 2020c. "Stroke." https://www.mayoclinic.org/diseases-conditions/stroke/symptoms-causes/syc-20350113.

_____. 2020d. "Diabetes." https://www.mayoclinic.org/diseases-conditions/diabetes/symptoms-causes/syc-20371444.

McCall, Rosie. 2020. "Melting Glaciers and Thawing Permafrost Could Release Ancient Viruses Locked Away for Thousands of Years." *Newsweek,* February 6. https://www.newsweek.com/melting-glaciers-thawing-permafrost-ancient-viruses-1486037.

McCarthy, Joe. 2016. "Italy Passes Law to Send Unsold Food to Charities Instead of Dumpsters." *Global Citizen,* March 15. https://www.globalcitizen.org/en/content/italy-passes-law-to-send-unsold-food-to-charities/.

McDaniel, Carl. N. 2005. *Wisdom for a Livable Planet.* San Antonio, TX: Trinity University Press.

McDowell, L.R. 2006. "Vitamin Nutrition of Livestock Animals: Overview from Vitamin Discovery to Today." Department of Animal Services, University of Florida. http://www1.agric.gov.ab.ca/$foragebeef/frgebeef.nsf/all/ccf62/$FILE/vitamin-nutrition-mcdowell.pdf.

McFadden, Christopher. 2017. "Seven Cool Biofuel Crops That We Use for Fuel Production." Interesting Engineering, March 19. https://interestingengineering.com/seven-biofuel-crops-use-fuel-production.

McGuirk, Rod. 2013. "Antarctica Concerns Grow as Tourism Numbers Rise." Yahoo News. March 18. http://news.yahoo.com/antarctica-concerns-grow-tourism-numbers-rise-053737703.html.

McKibben, Bill. 2020. "An Earth as Angry and Clever as It Was 50 Years Ago." *Los Angeles Times,* April 21: A11.

McMartin, Barbara. 1994. *The Great Forest of the Adirondacks.* Utica, NY: North Country Books.

McMichael, Anthony, John W. Powles, Colin D. Butler, and Ricardo Uauy. 2007. "Food, Livestock Production, Energy, Climate Change, and Health." *Journal of Environmental Science and Technology,* 370: 1253–63.

McNeely, Jeffrey A. 2000. "The Future of Alien

Invasive Species: Changing Social Views," pp. 171–90, in *Invasive Species in a Changing World,* edited by Harold A. Mooney and Richard J. Hobbs. Washington, D.C.: Island Press.

Medical Pro Disposal. 2018. "What Is Medical Waste? Definitions, Types, Examples & More." https://www.medprodisposal.com/medical-waste-disposal/what-is-medical-waste-medical-waste-definition-types-examples-and-more/.

Meltz, Robert. 1999. "CRS Report for Congress: Right to a Clean Environment Provisions in State Constitutions, and Arguments as to a Federal Counterpart." February 13. http://www.ncseonline.org/nle/crsreports/risk/rsk-15.cfm.

Menon, Rajan, and William Ruger. 2020. "U.S. Security Strategy Has Ignored Microbes." *Los Angeles Times,* April 4: A11.

Michigan Medicine—University of Michigan. 2017. "Second Cause of Hidden Hearing Loss Identified." Disabled World, February 2. https://www.disabled-world.com/disability/types/hearing/hhl.php.

Mies, Maria, and Vandana Shiva. 1993. London: Zed Books.

Milius, Susan. 2010. "Losing Life's Variety." *Science News*: March 13: 20–25.

Miller, Chaz. 2012. "Profiles in Garbage: Food Waste." *Waste Age* (February), 43 (2). http://waste360.com/mag/waste_food_waste_2.

Miller, G. Tyler. 2011. *Living in the Environment,* 17th edition. Belmont, CA: Brooks Cole.

Miller, G. Tyler and Scott Spoolman. 2018. *Living in the Environment, 19th edition.* Boston: Cengage.

Miller, Peter. 2013. "Saving Energy: It Starts at Home," pp.133–143, in *Green.* Boston: Wadsworth.

Mills, C. Wright. 1956. *The Power Elite.* New York: Oxford University Press.

_____. 1958. *The Causes of World War Three.* New York: Simon & Schuster.

_____. 1959. *The Sociological Imagination.* New York: Oxford University Press.

Mollot, G., J.H. Pantel, and T.N. Romanuk. 2017. "The Effects of Invasive Species on the Decline in Species Richness: A Global Meta-Analysis." *Advances in Ecological Research,* 56: 61–83.

Mooney, Chris. 2018. "World in Deep Trouble If Melting Continues." *The Post-Standard,* June 14:A14.

Morcillo, Patricia, Maria Angeles Esteban, and Alberto Cuesta. 2017. "Mercury and Its Toxic Effects on Fish." *AIMS Environmental Science,* 493):386–402.

More, Victoria C. 2011. "Dumpster Dinners: An Ethnographic Study of Freeganism." *Journal for Undergraduate Ethnography,* June, No. 1: 1–13.

Morford, Stacy. 2018. "Climate Change Will Cost U.S. More in Economic Damage Than Any Other Country But One." *Inside Climate News,* September 24. https://insideclimatenews.org/news/24092018/climate-change-economic-damage-america-social-cost-carbon-china-india-russia.

Moriarty, Rick. 2013. "An Eco Slam-Dunk: Flushing with Rainwater." *The Post-Standard.* January 6: A-1, A-8.

Morin, Monte. 2012. "Study Reveals Ancient Greenhouse Gas Emissions." *Los Angeles Times,* October 3. http://articles.latimes.com/2012/oct/03/science/la-sci-humans-climate-change-20121004.

Morris, Viveca. 2020. "How Humans Created a Pandemic 'Highway.'" *Los Angeles Times,* April 2: A11.

Morson, Saul and Morton Schapiro. 2020. "Extremism Imperils Us All." *Los Angeles Times,* March 9: A11.

Mother Nature Network. 2010. "What Is the Great Pacific Ocean Garbage Patch?" February 24. http://www.mnn.com/earth-matters/translating-uncle-sam/stories/what-is-the-great-pacific-ocean-garbage-patch.

Mouawad, Jad, and Clifford Krauss. 2009. "Dark Side of a Natural Gas Boom." *The New York Times,* December 8. http://www.nytimes.com/2009/12/08/business/energy-environment/08fracking.html?pagewanted=all.

Mount St. Helens Institute. 2013. "History of Mount St. Helens." http://mshinstitute.org/index.php/news_and_research/history_of_msh.

Mountain Area Land Trust. 2011. "Homepage." http://www.savetheland.org/.

Mulvaney, Dustin, and Paul Robbins. 2011. *Green Technology: An A-to-Z Guide.* Thousand Oaks, CA: Sage.

Murali, Sangeetha. 2016. "10 Ways to Reuse Waste Paper." *EcoIdeaz.* https://www.ecoideaz.com/showcase/10-ways-re-use-waste-paper.

Myers, David G., and Ed Diener. 1997. "The Science of Happiness," *The Futurist,* September-October: 5–10.

Myers, Kristin. 2020. "Over 4 Million in US will Contract Coronavirus If States Fully Reopen: Wharton Model." *Yahoo! Finance,* May 19. https://finance.yahoo.com/news/over-4-million-will-contract-coronavirus-as-states-reopen-wharton-model-172015472.html.

NASA. 2012. "Comets: Overview." http://solarsystem.nasa.gov/planets/profile.cfm?Object=Comets.

_____. 2020. "The Causes of Climate Change." https://climate.nasa.gov/causes/.

NASA Goddard Space Flight Center. 2004. "Research in the Atmosphere: Ozone and Climate Change." https://www.giss.nasa.gov/research/features/200402_tango/.

_____. 2020. "NASA Ozone Watch." https://ozonewatch.gsfc.nasa.gov/.

National Aeronautics and Space Administration. 2001. "The Ozone Layer." http://www.nas.nasa.gov/About/Education/Ozone/ozonelayer.html.

National Centers for Environmental Information (NCEI). 2020. "Climate Change and Extreme Snow in the U.S." https://www.ncdc.noaa.gov/news/climate-change-and-extreme-snow-us.

National Conference of State Legislatures (NCSL). 2020. "State Renewable Portfolio Standards and Goals." https://www.ncsl.org/research/energy/renewable-portfolio-standards.aspx.

National Geographic. 2012a. "Deforestation." http://environment.nationalgeographic.com/environment/global-warming/deforestation-overview/.

_____. 2012b. "Make Lightning Strike." http://environment.nationalgeographic.com/environment/natural-disasters/lightning-interactive/.

National Oceanic and Atmospheric Administration (NOAA). 2004. "Greenhouse Effect." http://www.oar.noaa.gov/k12/html/greenhouse2.html.

_____. 2007. "Marine Debris." http://marinedebris.noaa.gov/whatis/welcome.html.

_____. 2020a. "Ocean Acidification." https://www.noaa.gov/education/resource-collections/ocean-coasts-education-resources/ocean-acidification.

_____. 2020b. "Coral Bleaching." https://oceanservice.noaa.gov/facts/coral_bleach.html.

National Park Service. 2016. "Climate Change and Wildland Fire." https://www.nps.gov/subjects/climatechange/ccandfire.htm.

_____. 2020c. "What Is an Invasive Species?" https://oceanservice.noaa.gov/facts/invasive.html.

_____. 2020d. "Invasive Species: Great Lakes Region." https://www.regions.noaa.gov/great-lakes/index.php/great_lakes-restoration-initiative/invasive-species/.

_____. 2020e. "What Is a Green Roof?" https://www.nps.gov/tps/sustainability/new-technology/green-roofs/define.htm.

National Park Service (of Channel Islands National Park). 2012. "Restoring Santa Cruz Island." http://www.nps.gov/chis/naturescience/restoring-santa-cruz-island.htm.

National Public Radio (NPR). "Plastics: What's Recyclable, What Becomes Trash—And Why." https://apps.npr.org/plastics-recycling/.

National Ready Mixed Concrete Association. 2011. "Pervious Concrete Pavement: An Overview." http://www.perviouspavement.org/index.html.

National Weather Service (NWS). 2018. "Historic Hurricane Florence, September 12–15, 2018: Overview." https://www.weather.gov/mhx/Florence2018.

_____. 2019. "Climate and Weather." https://www.climateandweather.net/global-warming/climate-change-and-animals.html.

Natural Resources Canada (NRCAN). 2019. "Melting Glaciers, Rising Sea Levels, Thawing Permafrost and Unpredictable Groundwater Levels: The Unsettling Effects of Climate Change." https://www.nrcan.gc.ca/simply-science/21842.

Natural Resources Defense Council (NRDC). 2012a. "LEED Certification Information: What Is LEED Certification?" http://www.nrdc.org/buildinggreen/leed.asp.

_____. 2012b. "Wasted: How America Is Losing Up to 40 Percent of Its Food from Farm to Fork to Landfill. https://www.nrdc.org/sites/default/files/wasted-food-IP.pdf.

The Nature Conservancy. 2020. "Gulf of Mexico Dead Zone." https://www.courthousenews.com/giant-dead-zone-gulf-mexico-traced-meat-industry/.

NBC News. 2012. "Hurricane Sandy News Report." First air date, October 30.

Nelson, Angela. 2016. "15 Things Obama Has Done for the Environment." Mother Nature Network, October 10. https://www.mnn.com/earth-matters/wilderness-resources/stories/things-obama-has-done-environment.

Nelson, Dean. 2011. "Bhutan's 'Gross National Happiness' Index." *The Telegraph,* March 2. http://www.telegraph.co.uk/news/worldnews/asia/bhutan/8355028/Bhutans-Gross-National-Happiness-index.html.

Nersesian, Roy L. 2010. *Energy for the 21st Century: A Comprehensive Guide to Conventional and Alternative Sources,* Second Edition. Armonk, New York: M.E. Sharpe.

Netburn, Deborah. 2014. "Asteroid Impact That Killed the Dinosaurs Also Cooled the Earth." *Los Angeles Times,* May 13. https://www.latimes.com/science/sciencenow/la-sci-sn-asteroid-impact-winter-20140512-story.html.

New York City Department of Environmental Protection. 2012. "Invasive Species." http://www.nyc.gov/html/dep/html/watershed_protection/invasive_species.shtml.

New York State Department of Environmental Conservation (NYSDEC). 2011. Chapter 5: "Natural Gas Development Activities and High-Volume Hydraulic Fracturing." *Supplemental Generic Environmental Impact Statement* (Revised Draft). http://www.dec.ny.gov/docs/materials_minerals_pdf/rdsgeisch50911.pdf/

_____. 2012. "What Is Solid Waste?" http://www.dec.ny.gov/chemical/8732.html.

_____. 2020. "Hemlock Woolly Adelgid." https://www.dec.ny.gov/animals/7250.html.

The New York Times. 1877. "American Acclimatization Society." November 15. http://query.nytimes.com/mem/archive-free/pdf?res=F50912F73B5B137B93C7A8178AD95F438784F9.

New Zealand Travel Office. 2011. "Tourism Rotorua." (pamphlet provided at Rotorua Travel Office and Information Centre).

News.com.au. 2018. "China's Out-of-Control Tiangong 1 Crashed to Earth." https://www.news.com.au/technology/science/space/chinas-outofcontrol-tiangong-1-set-to-crash-into-earth-over-next-24-hours/news-story/960be65c9c36212e311d6a9d1ecb1b3e.

Northcutt, Wendy. 2000. *The Darwin Awards: Evolution in Action.* New York: Dutton.

Oberliesen, Elise. 2013. "AAP Real Possibilities: Keys to Happiness Quiz." http://www.aarp.org/health/healthy-living/info-03-2013/keys-to-happiness-quiz.html.

Ocean Acidification. 2012. "The Other CO2 Challenge." http://oceanacidification.net/.

O'Connor, Colleen. 2012. "Going 'Green' a Fad or for Real?" *The Denver Post,* April 21. http://www.denverpost.com/ci_8996138.

Office of Governor Gavin Newsom. 2019. "Governor Newsom Proclaims State of Emergency of Wildfires to Protect State's Most Vulnerable Communities." Press Release, March 22. https://www.gov.ca.gov/2019/03/22/governor-newsom-proclaims-state-of-emergency-on-wildfires-to-protect-states-most-vulnerable-communities/.

Officer, Charles, and Jake Page. 2009. *When the*

Planet Rages: Natural Disasters, Global Warming, and the Future of the Earth. Oxford: Oxford University Press.

OilPrice.com. 2009. "What Is Crude Oil? A Detailed Explanation on this Essential Fossil Fuel." https://oilprice.com/Energy/Crude-Oil/What-Is-Crude-Oil-A-Detailed-Explanation-On-This-Essential-Fossil-Fuel.html.

Olalde, Mark, and Ryan Menezes. 2020. "Decline of Oil Industry Leaves a Toxic Legacy in California." *Los Angeles Times,* February 9: A1.

O'Laughlin, Jay. 2005. "Conceptual Model for Comparative Ecological Risk Assessment of Wildfire Effects on Fish, with and Without Hazardous Fuel Treatment." *Forest and Ecology and Management,* 211: 59–72.

Olick, Diana. 2019. "Climate Change Will Crush Real Estate Values for Investors Who Don't Prepare, a New Report Says." *CNBC,* April 8. https://www.cnbc.com/2019/04/08/climate-change-will-crush-real-estate-values-for-unprepared-investors-report.html.

Oliver, Rachel. 2008. "All About: Religion and the Environment." *CNN World,* January 28. http://www.cnn.com/2008/WORLD/asiapcf/01/27/eco.about.religion/.

Oncenter. 2011. "Where Syracuse Plays: War Memorial Arena." http://www.oncenter.org/venue/war-memorial-arena.

Organization for Economic Cooperation and Development (OECD). 2020. "Better Life Index." http://www.oecdbetterlifeindex.org/#/11111111111.

Orr, Scott, Sharon Male, and Lamar Graham. 2009. "How Space Junk Threatens Us." *Parade,* May 10: 6.

Osram Sylvania. 2012. "Homepage: New Products." http://www.sylvania.com/en-us/Pages/default.aspx.

Ott, J.C. 2010. "Government and Happiness in 130 Nations: Good Governance Fosters Higher Level and More Equality of Happiness." *Social Indicators Research,* 102: 3–22.

Overbay, James C. 1992. "Ecosystem Management," in *Proceedings of the National Workshop: Taking an Ecological Approach to Management.*" Department of Agriculture, U.S. Forest Service, WO-WSA-3, Washington, D.C., 3–15.

Oxford Dictionary. 2019b. "2017." http://pastglobalchanges.org/calendar/2017.

_____. 2019a. "General Overview." http://pastglobalchanges.org/about/general-overview.

_____. 2020. "Ethics." http://www.oxforddictionaries.com/definition/english/ethics. PAGES (Past Global Changes).

Palmer, Brian. 2012. "Even a Runner Leaves a Carbon Footprint." *The Post-Standard,* November 5: C-3.

Park, Madison. 2016. "Startups See Potential in 'Ugly Food' Rejected by Supermarkets." *USA Today,* March 4. https://www.usatoday.com/story/news/2016/03/04/ugly-food-startup/80815244/.

Park, Michael Y. 2014. "A Brief History of Disposable Coffee Cup." *Bon Appetit,* May 30. https://www.bonappetit.com/entertaining-style/trends-news/article/disposable-coffee-cup-history.

Parris, Kirsten, and Robert McCauley. 2020. "Noise Pollution and the Environment." Australian Academy of Science. https://www.science.org.au/curious/earth-environment/noise-pollution-and-environment.

Patience, Maresa. 2009. "No Ice at North Pole in 10 Years?" *The Post Standard,* October 15: A-15.

Patterson, Bruce D., Frode Mo, Andreas Borgschulte, Mange Hillestad, Fortunat Joos, Trygve Kristiansen, Svein Sunde, and Jeroen A. van Bokhoven. 2019. "Renewable CO2 Recycling and Synthetic Fuel Production in a Marine Environment." *PNAS,* 116(25): 12212–12219.

Paul, Bill. 1986. "Congregation Is Rapidly Coming of Age." *Wall Street Journal.* March 2: 6.

PBS. 2001. "Humans: Origins of Humankind." http://www.pbs.org/wgbh/evolution/humans/humankind/index.html.

_____. 2018. "Rise of the Superstorms." https://www.pbs.org/wgbh/nova/video/rise-of-the-superstorms/.

Pease, Donald E. 2010. *Theodor Seuss Geisel.* Oxford University Press.

Pederson, Martin C. 2011. "The Green Movement: The Tree Huggers Grow Up." *Metropolis Mag,* April 14. http://www.metropolismag.com/story/20110414/the-green-movement-the-tree-huggers-grow-up.

Peel, J.D.Y. 1971. *Herbert Spencer: The Evolution of a Sociologist.* New York: Basic.

Peninsula Living. 2011. "Wisdom of a Natural Treasure." No. 115 (January): 9.

Pennisi, Elizabeth. 2019. "Three Billion North American Birds Have Vanished Since 1970, Surveys Show." *Science,* September 19. https://www.sciencemag.org/news/2019/09/three-billion-north-american-birds-have-vanished-1970-surveys-show#.

Perkiomen Watershed Conservancy. 2011. "Porous Pavement." http://www.greenworks.tv/stormwater/porouspavement.htm.

Peterson, Greg. 2007. "Pharmaceuticals in Our Water Supply Are Causing Bizarre Mutations to Wildlife." AlterNet.org. https://www.alternet.org/2007/08/pharmaceuticals_in_our_water_supply_are_causing_bizarre_mutations_to_wildlife/.

Pew Research Center. 2019. "Public Expresses Favorable Views of a Number of Federal Agencies." https://www.people-press.org/2019/10/01/public-expresses-favorable-views-of-a-number-of-federal-agencies/.

Phillips, Anna. 2019. "States Sue Over Endangered Species Act." *Los Angeles Times,* September 26: B2.

Phillips, Anna M., and Russ Mitchell. 2020. "Trump Rolls Back Obama Fuel Policy." *Los Angeles Times,* April 1: A4.

Pickrell, John. 2006. "Introduction: Human Evolution." *New Scientist,* September 4. http://www.newscientist.com/article/dn9990-instant-expert-human-evolution.html.

Pope Francis. 2015. *Laudato Si: On Care for Our Common Home.* Huntington, IN: Our Sunday Visitor.

Popovich, Nadja, Livia Albeck-Ripka and Kendra Pierre-Louis. 2020. "The Trump Administration Is Reversing Nearly 100 Environmental

Rules. Here's the Full List." *The New York Times*, May 6 (updated). https://www.nytimes.com/interactive/2020/climate/trump-environment-rollbacks.html.

Porter, Sandra J. 2013. "Why Earth's Rising Carbon Dioxide Levels Matter." *The Post-Standard*, June 6: A-21.

The Post-Standard. 2011a. "Ford Focus Uses Recycled Jeans." January 18: A-12.

_____. 2011b. "Green RVing Guide." June 10: 14 (supplement).

_____. 2012. "Quakes Linked to Drilling Spurs Regulations for States." April 21: A-1.

_____. 2013. "Asteroid Flyby Is Just Around the Corner." February 12: D-8.

_____. 2018. "European Officials Agree On Ban of Some Single-Use Plastics." December 20: A17.

Potter, Alison. 2017. "Let's Grab a Coffee—Billions of Them." *Choice*, June 14. https://www.choice.com.au/food-and-drink/drinks/tea-and-coffee/articles/are-takeaway-coffee-cups-recyclable.

Poulter, Sean. 2011. "Will We Learn to Love Ugly Apples and Pears? Change in Law Puts Misshapen Fruit on Supermarket Shelves." *Daily Mail*, November 3. 2056920/Will-learn-love-ugly-apples-pears-Misshapen-fruits-hit-supermarket-shelves-following-change-law.html.

Powell-Smith, Michelle. 2020. "The Seven Deadliest Plagues in History." *History Collection*. https://historycollection.co/seven-deadliest-plagues-history/.

Power-technology.com. 2019. "Top 10 Biggest Wind Farms." https://www.power-technology.com/features/feature-biggest-wind-farms-in-the-world-texas/#:~:text=Top%2010%20biggest%20wind%20farms%201%20Jiuquan%20Wind,Array%20Offshore%20Wind%20Farm%2C%20UK.%20More%20items...%20.

Prade, Thomas, Sven-Erik Svensson, and Jan Erick Mattsson. 2012. "Energy Balances for Biogas and Solid Biofuel Production from Industrial Hemp." *Biomass and Bioenergy*, 40: 36–52.

Priddy, Brenda. 2020. "The Impact of War on Global Warming." *Alternative Daily*. https://www.thealternativedaily.com/impact-of-war-on-global-warming/.

Proctor, Carmel, P. Alex Linley, and John Maltby. 2010. "Very Happy Youths: Benefits of Very High Life Satisfaction Among Adolescents." *Social Indicators Research*, 98 (3): 519–532.

Puskar-Pasewicz, Margaret, editor. 2010. *Cultural Encyclopedia of Vegetarianism*. Santa Barbara, CA: Greenwood.

Rachels, James. 1990. *Created from Animals: The Moral Implications of Darwinism*. New York: Oxford University Press.

Rain Garden Network. 2011. "What Is a Rain Garden?" http://www.raingardennetwork.com/.

Random History. 2010. "Nature's Deadliest Killers: The History of Volcanoes in 10 Great Eruptions." http://www.randomhistory.com/history-of-volcanoes.html.

Read, Richard. 2020. "A Marvel—and a Threat." *Los Angeles Times*, May 18: A1, A10.

Readfearn, Graham. 2018. "WHO Launches Health Review After Microplastics Found in 90% of Bottled Water." *The Guardian*, March 14. https://www.theguardian.com/environment/2018/mar/15/microplastics-found-in-more-than-90-of-bottled-water-study-says.

Reforest Action. 2020. "Reforestation in Morocco." https://www.reforestaction.com/en/reforestation-morocco.

Right Diagnosis. 2015. "Chronic Lower Respiratory Diseases." https://www.rightdiagnosis.com/c/chronic_lower_respiratory_diseases/intro.htm.

Ritchie, Hannah, and Max Roser. 2019. "Meat and Dairy Production." *OurWorldInData.org*. https://ourworldindata.org/meat-production.

_____, and _____. 2020. "Fossil Fuels." *OneWorldInData.org*. https://ourworldindata.org/fossil-fuels.

Robertson, Douglas S., Malcolm C. McKenna, Owen B. Toon, Sylvia Hope, and Jason A. Lillegraven. 2004 (May-June). "Survival in the First Hours of the Cenozoic." *Geological Society of America Bulletin*, 116 (5–6): 760–768.

Rogers, Paul G. 1990. "The Clean Air Act of 1970." http://www.epa.gov/history/topics/caa70/11.htm.

Roos, Dave. 2020. "How 5 of History's Worst Pandemics Finally Ended." History.com, March 27. https://www.history.com/news/pandemics-end-plague-cholera-black-death-smallpox.

Rosen, Julia. 2019a. "A Very Cold Case: Demise of the Dinos." *Los Angeles Times*, March 3: B2.

_____. 2019b. "Why Has Climate Change Become a Taboo Subject." Los Angeles Times, July 12: A2.

Rosen, Julia, and Anna M. Phillips. 2019. "As Land Goes, So the Climate." *Los Angeles Times*, August 9: A2.

Roth, Katherine. 2018. "Going Green: Waste Not." *The Citizen*, October 11: B6.

Roth, Sammy. 2020a. "Clean Energy Seeks Another U.S. Lifeline." *Los Angeles Times*, April 15.

_____. 2020b. "Clean Energy Seeks a Role in Recovery." *Los Angeles Times*, May 14:A8.

_____. 2020c. "Crisis Sends Oil Price Diving Below Zero." *Los Angeles Times*, April 21: A1, A2.

_____. 2020d. "Clean Energy Sector Sheds 100,000 Jobs." *Los Angeles Times*, April 16: A8.

_____. 2020e. "Clean Energy Seeks a Role in Recovery." *Los Angeles Times*, May 14: A8.

Rotman, Michael. 2020. "Cuyahoga River Fire." Cleveland Historical. https://clevelandhistorical.org/items/show/63.

Rottenberg, Josh. 2009 (December 18). "James Cameron Talks *Avatar*: Brave Blue World." *Entertainment Weekly*: 48–51.

Roxan, Julia. 2020. "Mother Nature Transforms the Polluted Ussuri Bay into a Beautiful Glass Beach." *Luxuo*, March 4. https://www.luxuo.com/lifestyle/travel/mother-nature-transforms-the-polluted-ussuri-bay-into-a-beautiful-glass-beach.html.

Rugaber, Christopher S. 2012. "Happiness Gauges Economic Progress." *Daily Breeze*. August 7: A-8.

Ruiz-Grossman, Sarah. 2016. "Starbucks Will Donate Unsold Food to People in Need." *Huffington Post*, March 22. https://www.huffpost.com/

entry/starbucks-donate-unsold-food_n_56f1973be4b03a640a6bfc4c.

Rural Sociological Society. 2010. "Homepage: About Us, Our History." http://ruralsociology.org/index.php?L1=left_home.php&L2=staticcontent/about_us/history.php.

Russell, Bertrand. 1948. *The History of Western Philosophy.* New York: George Allen and Unwin.

_____. 1951. *The Conquest of Happiness.* New York: Signet Books.

Rust, Susanne. 2019. "New Law Orders Nuclear Dump Inquiry." *Los Angeles Times,* December 30: A6.

Rust, Susanne, Louis Sahagun, and Rosanna Xia. 2020. "EPA Suspends Enforcement, Citing Threat From Virus." *Los Angeles Times,* March 28: B1, B4.

Sahagun, Louis. 2014. "The Perilous Plastisphere." *The Post-Standard,* January 7: D8.

Samenow, Jason. 2018. "Red-Hot Planet: All-time Heat Records Have Been Set All Over the World During the Past Week." *The Washington Post,* July 5. https://www.washingtonpost.com/news/capital-weather-gang/wp/2018/07/03/hot-planet-all-time-heat-records-have-been-set-all-over-the-world-in-last-week/.

Sanders, Sam. 2020. "The Surprising Legacy of Occupy Wall Street in 2020." *NPR,* January 23. https://www.npr.org/2020/01/23/799004281/the-surprising-legacy-of-occupy-wall-street-in-2020.

Sapart, C.J., G. Monteil, M. Prokopiou, R.S.W. van de Wal, J.O. Kaplan, P. Sperlich, K.M. Krumhardt, C. van der Veen, S. Houweling, M.C. Krol, T. Blunier, T. Sowers, P. Martinerie, E. Wiltrant, D. Dahl-Jensen and T. Rockmann. 2012. "Natural and Anthropogenic Variations in Methane Sources during the Past Two Millennia." *Nature,* 490: 85–88.

Save Our Towns, Save Our Land. 2011. "About Us." http://saveourlandsaveourtowns.org/aboutus.html.

SavetheRainforest.org. 2011. "Causes of Rainforest Destruction." http://savetherainforest.org/savetherainforest_006.htm.

Schendler, Auden. 2009. *Getting Green Done: Hard Truths from the Front Lines of the Sustainability Revolution.* New York: Public Affairs.

Schopenhauer, Arthur. 1979. *Essays and Aphorisms.* London, England: Penguin Books.

Schuessler, Jennifer. 2012. "Lessons from Ants to Grasp Humanity." *The New York Times.* April 9: C-1, C-5.

Schulze, Peter C. 2002. "I=PBAT." *Ecological Economics,* 40: 149–150.

Science Alert. 2016. "Earth's CO2 Levels Just Crossed a Really Scary Threshold—And It's Permanent." https://www.sciencealert.com/earth-s-co2-levels-just-permanently-crossed-a-really-scary-threshold.

Scott, Paula Spencer. 2016. "Feeling Awe May Be the Secret to Health and Happiness." *Parade,* October 9: 6–8.

Searchinger, Tim. 2011. "Fuel for Thought: How Biofuels Are Making the World's Poor Even Hungrier." *The Post-Standard.* February 20: E-1.

Seattle.gov. 2020. "Seattle Department of Neighborhoods: Beacon Food Forest." http://www.seattle.gov/neighborhoods/programs-and-services/p-patch-community-gardening/p-patch-list/beacon-food-foresthttp://www.seattle.gov/neighborhoods/programs-and-services/p-patch-community-gardening/p-patch-list/beacon-food-forest.

Seligman, Martin E. P. 2011. *Flourish: A Visionary New Understanding of Happiness and Well-Being.* New York: Free Press.

Serna, Joseph. 2020. "How Did Fires Get So Big So Quickly?" *Los Angeles Times,* August 30: A1, A11.

Seuss, Dr. 1991. *The Lorax.* Random House.

Sheehan, Peter M. 2001. "The Late Ordovician Mass Extinction." *Annual Review of Earth and Planetary Sciences,* 29: 331–364.

Shwartz, Mark. 2005. "Selective Logging Causes Widespread Destruction, Study Finds." *Stanford Report,* October 21. http://news.stanford.edu/news/2005/october26/select-102605.html.

Sierra Club of Canada. 2012. "The Environmental Consequence of War." http://www.sierraclub.ca/national/postings/war-and-environment.html.

Sigurdsson, Haraldur. 1999. *Melting the Earth: The History of Volcanic Eruptions.* New York: Oxford University Press.

Simon, Julian. 1995. "More People, Greater Wealth, More Resources, Healthier Environment," in *Taking Sides: Clashing Views on Controversial Issues,* 6th ed., Theodore D. Goldfarb, editor. Guilford, CT: Dushkin.

Simpson-Holley, Martha, and Ian Law. 2007. "Renewable in More Ways Than One: Hemp Returns to UK Farming." *Biologist,* 54 (1): 18–23.

The Simpsons. 1990. "Two Cars in Every Garage and Three Eyes on Every Fish." Episode 14, first aired date, November 1.

Singer, Irving. 2009. *Philosophy of Love: A Partial Summing-up.* Cambridge, MA: The MIT Press.

Singer, Peter. 2002. *One World: The Ethics of Globalization.* New Haven, CT: Yale University Press.

_____. 2011. "Does Helping the Planet Hurt the Poor? No, If the West Makes Sacrifices." *The Wall Street Journal,* January 22–23: C-1, C-2.

Sivinski, Seth, and Joseph Ulatowski. 2019 (Spring). "The Anthropocentrism of the Cosmic Perspective Argument." *Indiana University Press,* 24 (1): 1–18.

Skerritt, Jen. 2020. "How Tyson Helped Create Supply Problem." *Los Angeles Times,* April 30: A9.

Slesinger, Hugh. 2009. "The True Value." *Medical Tourism Magazine,* April 1. http://www.medicaltourismmag.com/article/the-true-value.html.

Sloterdijk, Peter. 1987. *The Critique of Cynical Reason,* translated by Michael Eldred. Minneapolis: University of Minnesota Press.

Smith, Elizabeth. 2012. "Deforestation and the Effects It Has on a Global Scale." *National Geographic.* http://greenliving.nationalgeographic.com/deforestation-effects-global-scale-2214.html.

Smith, Roff. 2011. "Dark Days of the Triassic: Lost World." *Nature,* 479 (7373). https://www.

nature.com/news/dark-days-of-the-triassic-lost-world-1.9375.

Smith, Sebastian and Chris Stein. 2019. "Snarky Trump Tells Greta Thunberg to 'Chill' and See Movies." *Yahoo News*, December 12. https://www.yahoo.com/news/snarky-trump-tells-greta-thunberg-chill-see-movies-204440954.html.

Smithsonian. 2018. "Ocean Acidification." https://ocean.si.edu/ocean-life/invertebrates/ocean-acidification.

Solomon, Caren G. and Regina C. LaRocque. 2019. "Climate Change—A Health Emergency." *The New England Journal of Medicine*, 380:209–211.

Solomon, Gabrielle M., Hiruni Dodangoda, Tylea McCarthy-Walker, Rita Ntim-Gyakari, and Peter D. Newell. 2019. "The Microbiota of Drosophila Suzukii Influences the Larval Development of Drosophila Melanogaster." *PeerJ*, 7. https://www.ncbi.nlm.nih.gov/pmc/articles/PMC6873876/.

Somerville, Richard C.J. 2008. *The Forgiving Air: Understanding Environmental Change*. Boston: American Meteorological Society.

Sounder, William. 2012. "The Undying War Over *Silent Spring*." *The Post Standard*, September 16: E-1.

Space.com. 2013. "Asteroids: Formation, Discovery, and Exploration." http://www.space.com/51-asteroids-formation-discovery-and-exploration.html.

Spencer, Herbert. 1862. *First Principles*. New York: Appleton.

_____. 1908. *Social Statics and the Man Versus the State*. New York: Appleton.

Spero, David. 2015. "Five Environmental Causes of Diabetes." https://www.diabetesselfmanagement.com/blog/five-environmental-causes-of-diabetes/.

Stanley, George D., Jr., Hannah M.E. Shepherd and Autumn J. Robinson. 2018. "Paleoecological Response of Corals to the End-Triassic Mass Extinction: An Integrational Analysis." *Journal of Earth Science*, 29: 879–885.

Statista. 2020. "Electronic Waste Generated Worldwide from 2010 to 2018 (in million metric tons)." https://www.statista.com/statistics/499891/projection-ewaste-generation-worldwide/ .

Steecker, Matt. 2020. "Cornell to Pause Fossil Fuel Investing." *The Citizen*, June 4: A8.

Stephens, S.L., J.K. Agee, P.Z. Fule, M.P. North, W.H. Romme, T.W. Swetnam, and M.G. Turner. 2013. "Managing Forests and Fire in Changing Climates." *Science*, 342(6154): 41–42.

Stokstad, Erik. 2020 (April). "Deep Deficit." *Science*, 368(6488): 230–233.

Strike Alert. 2010. "What Is Lightning?" http://www.strikealert.com/LightningFacts.htm.

Stuart, Tristram. 2009. *Waste: Uncovering the Global Food Scandal*. New York: W.W. Norton.

SuperGreenMe. 2011. "To Preserve the Limited Amount of Natural Resources We Have." http://www.supergreenme.com/go-green-environment-eco:To-preserve-the-limited-amount-of-natural-resources-we-have.

The Sustainability Scale Project. 2011. "Carrying Capacity." http://sustainablescale.org/ConceptualFramework/UnderstandingScale/MeasuringScale/CarryingCapacity.aspx.

Sutton, Paul. 2015. "Can You Put a Dollar Value on Nature?" World Economic Forum. https://www.weforum.org/agenda/2015/03/can-you-put-a-dollar-value-on-nature/.

Svensson, Peter. 2012. "LED Replacements Hit Stores Empty of 100W Bulbs." Syndicated column in *The Post-Standard*, November 13: A-10.

Tallis, Joshua. 2015. "Remediating Space Debris: Legal and Technical Barriers." *Strategic Studies Quarterly*, 9(1): 86–99.

Tansley, Alfred G. 1935. "The Use and Abuse of Vegetational Concepts and Terms," *Ecologist* 16 (3): 284–307.

Tarlach, Gemma. 2018. "The Five Mass Extinctions That Have Swept Our Planet." *Discover*, July 18 https://www.discovermagazine.com/the-sciences/mass-extinctions.

Tasoff, Harrison. 2020. "Pollution and COVID-19." *The Current*, April 29. https://www.news.ucsb.edu/2020/019878/pollution-and-covid-19.

Tassi, Patricia, Odile Rohmer, Sarah Schimchowitsch, Arnaud Eschenlauer, Anne Bonnefond, Florence Margiocchi, Franck Poisson, and Alain Muzet. 2010. "Living Alongside Railway Tracks: Long-Term Effects of Nocturnal Noise on Sleep and Cardiovascular Reactivity as a Function of Age." *Environmental International*, 36: 683–689.

Taylor, Brad W. and Max L. Bothwell. 2014. "The Origin of Invasive Microorganisms Matter for Science, Policy, and Management: The Case of Didymosphenia Geminata." *BioScience*, 64)6): 531–538.

Tedesco, M., T. Moon, J.K. Andersen, J.E. Box, J. Cappelen, R.S. Fausto, X. Fettweis, B. Loomis, K.D. Mankoff, T. Mote, C.J.P.P. Smeets, D, van As, and R.S. W. van de Wal. 2019. "Greenland Ice Sheet." NOAA's Artic Program. https://arctic.noaa.gov/Report-Card/Report-Card-2019/ArtMID/7916/ArticleID/842/Greenland-Ice-Sheet.

Tennesen, Michael. 2008. *The Complete Idiot's Guide to Global Warming*, 2nd ed. Indianapolis, IN: Alpha.

Tesla. 2018. "Electric Cars, Solar Panels and Clean Energy Storage." July 30. htpps://www.tesla.com/.

Texas Commission on Environmental Quality. 2008. "Surface Water Quality Concerns from Wildfires in Texas." http://www.tceq.texas.gov/assets/public/response/drought/waterquality.pdf.

Think Exist. 2012. "Benjamin Franklin Quotes." http://thinkexist.com/quotation/the_constitution_only_gives_people_the_right_to/12881.html.

Thompson, Andrea. 2009. "Edge of Space Found." *Space.com*, April 9. http://www.space.com/6564-edge-space.html.

Thompson, Richard C., Bruce La Belle, Hindrik Bouwman, and Lev Neretin. 2011. "Marine Debris as a Global Environmental Problem: Introducing a Solutions Based Framework Focused on Plastics." *STAP Advisory*. http://www.opc.ca.gov/webmaster/ftp/pdf/public_comment/20110909_MCaldwell_att1.pdf.

Time for Change. 2012. "What Is a Carbon Footprint—

Definition." http://timeforchange.org/what-is-a-carbon-footprint-definition.

Tobin, Dave. 2019. "Carrier Dome Gets the Rainwater Flush." *Syracuse.com,* undated March 22. https://www.syracuse.com/news/2014/12/carrier_dome_gets_the_rainwater_flush.html.

Tribune News Service. 2019. "Study: Earth Warming Is Speeding Up." As it appeared in *The Citizen,* July 26: A5.

Triodos Bank. 2020. "Thinking About Your ISA? Don't Be a Fossil Fool." https://www.triodos.co.uk/articles/2020/thinking-about-your-isa-dont-be-a-fossil-fool%E2%80%A6.

Troitino, Christina. 2018. "Americans Waste About a Pound of Food a Day, USDA Study Finds." *Forbes,* April 23. https://www.forbes.com/sites/christinatroitino/2018/04/23/americans-waste-about-a-pound-of-food-a-day-usda-study-finds/#e3a75ff4ec3b.

Trustworthy Cyber Infrastructure for the Power Grid. 2013. "TCIPG: The Power Grid." http://tcipg.mste.illinois.edu/applet/pg.

UN News. 2019. "10 Million Yemenis 'One Step Away from Famine,' UN Food Relief Agency Calls for 'Unhindered Access' to Frontline Regions." https://news.un.org/en/story/2019/03/1035501.

Union of Concerned Scientists (UCS). 2017. "Is There a Connection Between the Ozone Hole and Global Warming?" https://www.ucsusa.org/resources/ozone-hole-and-global-warming.

_____. 2019. "The State of Science in the Trump Era (2019)." https://ucsusa.org/resources/state-science-trump-era.

United Nations. 2020. "Water." https://www.un.org/en/sections/issues-depth/water/.

United Nations Environment Programme (UNEP). 2019. "Emissions Gap Report: Global Progress Report on Climate Action." https://www.unenvironment.org/interactive/emissions-gap-report/2019/.

_____. 2020a. "Locust Swarms and Climate Change." https://www.unenvironment.org/news-and-stories/story/locust-swarms-and-climate-change.

_____. 2020b. "What Are Your Environmental Rights?" https://www.unenvironment.org/explore-topics/environmental-rights-and-governance/what-we-do/advancing-environmental-rights/what-0.

United Nations Intergovernmental Panel on Climate Change (IPCC). 2019. "Land Is a Critical Resource, IPCC Report Says." https://www.ipcc.ch/2019/08/08/land-is-a-critical-resource_srccl/.

United Nations Water. 2010. "World Water Day 2010: Clean Water for a Healthy World." http://www.scriptphd.com/its-not-easy-being-green/2010/03/21/its-not-easy-being-green-tapped-out-on-bottled-water-world-water-day/.

United States Census Bureau. 2012. "2010 Census Urban Area FAQs." http://www.census.gov/geo/www/ua/uafaq.html.

United States Conference of Catholic Bishops. 2020. *Catholic Teaching on the Environment.* Washington, D.C.: United States Conference of Catholic Bishops.

United States Department of Agriculture (USDA). 2020a. "Emerald Ash Borer Beetle." https://www.aphis.usda.gov/aphis/resources/pests-diseases/hungry-pests/the-threat/emerald-ash-borer/emerald-ash-borer-beetle.

_____. 2020b. "Asian Long Horned Beetle." https://www.invasivespeciesinfo.gov/profile/asian-long-horned-beetle.

United States Department of Health and Human Services. 2003. Cancer and the Environment. https://www.niehs.nih.gov/health/materials/cancer_and_the_environment_508.pdf

_____. 2019. "What Causes Alzheimer's Disease?" National Institute on Aging. https://www.nia.nih.gov/health/what-causes-alzheimers-disease.

United States Department of the Interior. 2015. "9 Animals That Are Feeling the Impacts of Climate Change." https://www.doi.gov/blog/9-animals-are-feeling-impacts-climate-change.

United States Department of Veterans Affairs. 2012. "Public Health: Agent Orange Residue on Post-Vietnam War Airplanes." http://www.publichealth.va.gov/exposures/agentorange/residue-c123-aircraft.asp.

United States Energy Information Administration (EIA). 2017. "Country Analysis Brief: Russia. https://www.eia.gov/international/content/analysis/countries_long/Russia/russia.pdf.

_____. 2019a. "U.S. Energy Facts Explained." https://www.eia.gov/energyexplained/us-energy-facts/.

_____. 2019b. "Renewable Energy Explained." https://www.eia.gov/energyexplained/renewable-sources/.

_____. 2019c. "Coal Explained: Mining and Transportation of Coal." https://www.eia.gov/energyexplained/coal/mining-and-transportation.php.

_____. 2019d. "What Is the Difference Between Crude Oil, Petroleum Products, and Petroleum" https://www.eia.gov/tools/faqs/faq.php?id=40&t=6.

_____. 2020. "Frequently Asked Questions: How Many Nuclear Power Plants Are in the United States, and Where Are They Located" https://www.eia.gov/tools/faqs/faq.php?id=207&t=21.

United States Geological Survey (USGS). 2010. "Volcanic Gases and Their Effects." http://volcanoes.usgs.gov/hazards/gas/index.php.

_____. 2020a. "How Much Water Is There on Earth?" https://www.usgs.gov/special-topic/water-science-school/science/how-much-water-there-earth?qt-science_center_objects=0#qt-science_center_objects.

_____. 2020b. "Acid Rain." https://www.usgs.gov/mission-areas/water-resources/science/acid-rain?qt-science_center_objects=0#qt-science_center_objects.

_____. 2020c. "How Many Active Volcanoes Are There on Earth?" https://www.usgs.gov/faqs/how-many-active-volcanoes-are-there-earth?qt-news_science_products=0#qt-news_science_products.

United States Golf Association (USGA). 2016. "Courses Need." https://www.usga.org/course-care/water-resource-center/how-much-water-golf-courses-need.html.

United States National Park Service. 2003. "Time

It Takes for Garbage to Decompose in the Environment." Mote Marine Lab, Sarasota, Florida. http://des.nh.gov/organization/divisions/water/wmb/coastal/trash/documents/marine_debris.pdf.

Unwin, Jack. 2019. "Renewable Stadiums of the Future." Power Technology, April 4. https://www.power-technology.com/features/best-stadiums-renewable-energy/.

The U.S. National Archives and Records Administration. 2012. "The Declaration of Independence: A Transcription." http://www.archives.gov/exhibits/charters/declaration_transcript.html.

Valleron, Alain-Jacques, Anne Cori, Sophie Valtat, Sofia Meurisse, Fabrice Carrat, and Pierre-Yves Boelle. 2010. "Transmissibility and Geographic Spread of the 1889 Influenza Pandemic." *Proc. Natl. Acad. Sci,* 107(19: 8778–8781.

Vancheri, Barbara. 2012. "John Krasinski and Gus Van Sant Totally Immersed in Film Shot in Pittsburgh." *Pittsburgh Post-Gazette,* June 15. http://www.post-gazette.com/stories/ae/movies/john-krasinski-and-gus-van-sant-totally-immersed-in-film-shot-here-640468/.

Vandenhove, H., and M. Van Hees. 2005. "Fibre Crops as Alternative Land Use for Radioactivity Contaminated Arable Land." *Journal of Environmental Radioactivity,* 81: 131–141.

Van Driesche, Jason, and Roy Van Driesche. 2000. *Nature Out of Place: Biological Invasions in the Global Age.* Washington, D.C.: Island Press.

Vastag, Brian. 2012. "James Cameron Becomes First Solo Explorer to Reach the Deepest Point of Ocean." *The Washington Post,* March 25. http://www.washingtonpost.com/national/health-science/james-cameron-begins-solo-dive-to-the-bottom-of-the-ocean/2012/03/25/gIQAPDwOaS_story.html.

Ventura, Luca. 2019. "The Most Peaceful Countries in the World 2019." *Global Finance,* June 21. https://www.gfmag.com/global-data/non-economic-data/most-peaceful-countries.

Vigliotti, Jonathan. 2017. "Once-Pristine Arctic Choking On Our Plastic Addiction." *CBS This Morning,* October 23. https://www.cbsnews.com/news/microplastics-arctic-circle-ocean-pollution-plastic-waste-around-the-world/.

Vilcinskas, Andreas. 2015. "Pathogens as Biological Weapons of Invasive Species." *PLoS Pathogens,* 11(4). https://www.researchgate.net/publication/274724780_Pathogens_as_Biological_Weapons_of_Invasive_Species.

Vojdani, Aristo. 2019. "Environment and Alzheimer's." *Townsend Letter.* https://www.townsendletter.com/article/435-alzheimers-environment-lifestyle-factors/

Volcano Discovery. 2020. "How Many Volcanic Eruptions Occur Every Year?" https://www.volcanodiscovery.com/es/volcanology/faq/how-many-eruptions-per-year.html.

Von Hartmann, Eduard. 1931. *Philosophy of the Unconscious: Speculative Results According to the Inductive Method of Physical Science.* London: Kegan Paul.

vos Savant, Marilyn. 2011. "Ask Marilyn." *Parade,* May 1: 12.

Wadekar, Neha. 2020. "When Insects Devour Your Livelihood." *Los Angeles Times,* April 1: A3.

Wake, David B., and Vance T. Vredenburg. 2008. "Are We in the Midst of the Sixth Mass Extinction? A View from the World of Amphibians." *Proceedings of the National Academy of Sciences,* 105 (1): 11466-World 11473.

Walker, Scott. 2013. "Last of the Amazon," in *Green,* 13–22. Boston: Wadsworth.

Wallace, Donald H. 2002. "War Crimes," pp. 1699–1706, in *Encyclopedia of Crime and Punishment,* Vol. 4, edited by David Levinson. Thousand Oaks, CA: Sage.

Wallace-Wells, David. 2019. *The Uninhabitable Earth: Life after Warming.* New York: Tim Duggan Books.

Ward, Alvin. 2016. "Jimmy Carter's Short-Lived White House Solar Panels." *Mental Floss,* August 21. https://www.mentalfloss.com/article/84998/jimmy-carters-short-lived-white-house-solar-panels

Warren, R., J. Price, J. VanDerWal, S. Cornelius and H. Sohl. 2018. "The Implications of the United Nations Paris Agreement on Climate Change for Globally Significant Biodiversity Areas." *Climatic Change,* 147: 395–409.

Washington, Haydn. 2013. *Human Dependence on Nature: How to Help Solve the Environmental Crisis.* London: Routledge.

Waste Age. 2004. "Food Waste." http://wasteage.com/mag/waste_food_waste_2.

Watanabe, Teresa. 2020. "UC Divests from Fossil Fuels." *Los Angeles Times,* May 20: B1, B2.

Watling, Eve. 2019. "What Is a Fatberg? The Gross Grease Giants Threatening Cities." *Newsweek,* March 14. https://www.newsweek.com/what-fatberg-1361168.

Watts, Jonathan. 2018. "We Have 12 Years to Limit Climate Change Catastrophe, Warns UN." *The Guardian,* October 8. https://www.theguardian.com/environment/2018/oct/08/global-warming-must-not-exceed-15c-warns-landmark-un-report.

_____. 2019. "Concrete: The Most Destructive Material on Earth." *The Guardian,* February 25. https://www.theguardian.com/cities/2019/feb/25/concrete-the-most-destructive-material-on-earth.

Wax, Emily. 2012. "'Guerrilla Gardeners' Spread Seeds of Social Change." *Washington Post,* April 14. http://www.washingtonpost.com/lifestyle/style/guerrilla-gardeners-spread-seeds-of-social-change/2012/04/14/gIQArAA6HT_story.html.

WBUR. 2018. "Americans Throw Out Millions of Plastic Straws Daily. Here's What's Being Done About It." https://www.wbur.org/hereandnow/2018/05/02/plastics-single-use-straws-oceans.

Weaver, Teri. 2011. "EPA Recognizes Onondaga County, Syracuse as Top 10 Green Communities." *The Post-Standard,* April 21: A-1, A-14.

WebElements.com. 2011. "Oxygen: The Periodic Table on the Web." http://www.webelements.com/oxygen/.

Weir, Bill. 2012. "More Than Mere Magic Mushrooms."

Yahoo News, April 12. http://news.yahoo.com/blogs/this-could-be-big-abc-news/more-mere-magic-mushrooms-154207424.html.

Weise, Elizabeth. 2011. "Study Offers Warning About Next Potential Mass Extinction." *USA Today*, March 2. http://usatoday30.usatoday.com/tech/science/2011-03-02-next-mass-extinction_N.htm#.

Weisman, Alan. 2007. *The World Without Us*. New York: Picador.

Welle, Paul D., Mitchell J. Small, Scott C. Doney, Ines L. Azevedo. 2017. "Estimating the Effect of Multiple Environmental Stressors on Coral Bleaching and Morality." *PLOS One*. https://journals.plos.org/plosone/article?id=10.1371/journal.pone.0175018.

Welna, David. 2020. "The End May Be Nearer: Doomsday Clock Moves Within 100 Seconds of Midnight." *NPR*, January 23. https://www.npr.org/2020/01/23/799047659/the-end-may-be-nearer-doomsday-clock-moves-within-100-seconds-of-midnight.

Weston, Anthony, editor. 1999. *An Invitation to Environmental Philosophy*. New York: Oxford University Press.

Wethe, David. 2019. "What Human-Made Earthquakes Mean for Fracking." *Bloomberg*, November 18. https://www.bloomberg.com/news/articles/2019-11-18/what-human-made-earthquakes-mean-for-fracking-quicktake.

Where Are They Located?" https://www.eia.gov/tools/faqs/faq.php?id=207&t=3.

White, Lynn. 1967. "The Historical Roots of Our Ecological Crisis." *Science*, 155: 1203–1207.

Whoriskey, Peter. 2012. "Coming Soon: Measuring Our Gross National Happiness." April 4. *The Washington Post*, p. A-1, A-18.

Wick, Julia. 2020. "Feeling Climate Despair" There's a Word for That." *Los Angeles Times*, September 13: B3.

Wickelgren, Ingrid. 2011. "Scientific American Quiz: How Happy Are You?" http://www.scientificamerican.com/article.cfm?id=happiness-how-happy-are-you-quiz.

Wignall, Paul. 2004. "Causes of Mass Extinctions," pp. 119–150, in *Extinctions in the History of Life*, edited by Paul Taylor. New York: Cambridge University Press.

The Wilderness Society. 2019. "7 Ways Oil and Gas Drilling Is Bad for the Environment." https://www.wilderness.org/articles/blog/7-ways-oil-and-gas-drilling-bad-environment.

Wildlife Forever. 2012. "News and Events: Top 10 Invasive Species Threats to Hunting and Fishing." http://www.wildlifeforever.org/news-events/namg-supports-wildlife-forevers-fight.

Williams, Matt. 2012. "Republican Congressman Paul Broun Dismisses Evolution and Other Theories." *The Guardian*, October 6. https://www.theguardian.com/world/2012/oct/06/republican-congressman-paul-broun-evolution-video.

Wilson, Craig. 2012. "Happiness Is Overrated. You Can Bank on It." *USA Today*, April 11: D-1.

Wilson, Edward O. 1984. *Biophilia*. Cambridge, MA: Harvard University Press.

_____. 1992. *The Diversity of Life*. Cambridge, MA: Belknap Press.

_____. 1998. *Consilience: The Unity of Knowledge*. New York: Knopf.

_____. 2002. *The Future of Life*. New York: Knopf.

_____. 2014. *A Window on Eternity*. New York: Simon & Schuster.

_____. 2017. *Half-Earth: Our Planet's Fight for Life*. New York: Liveright.

Winterson, Jeanette. 2012. *Why Be Happy When You Could Be Normal?* New York: Grove Press.

Wisconsin Department of Health Services. 2015. "Manure Contamination of Residential Wells." https://www.dhs.wisconsin.gov/publications/p4/p45088.pdf.

Withgott, Jay, and Scott Brennan. 2007. *Environment: The Science Behind the Stories*, 2nd edition. New York: Pearson.

Wood, Nathan and Katy Roelich. 2019. "Tensions, Capabilities, and Justice in Climate Change Mitigation of Fossil Fuels." *Energy Research & Social Science*, 52 (June): 114–122.

Worland, Justin. 2015. "The Weird Effect Climate Change Will Have On Plant Growth." *Time*, June 11. https://time.com/3916200/climate-change-plant-growth/.

World Atlas. 2019. "Countries Least Dependent on Fossil Fuel Sources for Energy Needs." https://www.worldatlast.com/articles/countries-least-dependent-on-fossil-fuel-sources-for-energy-needs.html.

_____. 2020a. "Fossil Fuel Dependency by Country." https://www.worldatlas.com/articles/countries-the-most-dependent-on-fossil-fuels.html.

_____. 2020b. "The World's Most Densely Populated Cities." https://www.worldatlas.com/articles/the-world-s-most-densely-populated-cities.html.

_____. 2020c. "The World's Most War-Torn Countries." https://www.worldatlas.com/articles/the-world-s-most-war-torn-countries.html.

_____. 2020d. "The Deadliest Famines Ever." https://www.worldatlas.com/articles/the-deadliest-famines-ever.html.

_____. 2020e. "How Much of the Surface of the Earth Is Covered By Rainforests?"

The World Bank. 2018. "Global Waste to Grow by 70 Percent by 2050 Unless Urgent Action Is Taken: World Bank Report. https://www.worldbank.org/en/news/press-release/2018/09/20/global-waste-to-grow-by-70-percent-by-2050-unless-urgent-action-is-taken-world-bank-report,

_____. 2019a. "Poverty." https://www.worldbank.org/en/topic/poverty/overview.

_____. 2019b. "Fossil Fuel Energy Consumption (% of total). https://data.worldbank.org/indicator/EG.USE.COMM.FO.ZS.

The World Counts. 2020a. "A Cheap but Dirty Fuel Source…" https://www.theworldcounts.com/stories/Negative-Effects-of-Coal-Mining.

_____. 2020b. "Paper Comes From Trees…" https://www.theworldcounts.com/stories/Paper-Waste-Facts.

World Database of Happiness. 2019. "Archive of

Research Findings on Subjective Enjoyment of Life." https://worlddatabaseofhappiness.eur.nl/hap_nat/nat_fp.php?cntry=55&mode=3&subjects=297&publics=12.

World Environment Day (WED). 2020. "World Environment Day 2020: A Practical Guide." https://p.widencdn.net/e2n0wj/WED_SimpleToolkit.

World Health Organization (WHO). 2016. "An Estimated 12.6 million Deaths Each Year Are Attributable to Unhealthy Environments." https://www.who.int/news-room/detail/15-03-2016-an-estimated-12-6-million-deaths-each-year-are-attributable-to-unhealthy-environments.

_____. 2018. "Health-care Waste." https://www.who.int/news-room/fact-sheets/detail/health-care-waste.

_____. 2019. "World Hunger Is Still Not Going Down After Three Years and Obesity Is Growing—UN Report." https://www.who.int/news-room/detail/15-07-2019-world-hunger-is-still-not-going-down-after-three-years-and-obesity-is-still-growing-un-report.

_____. 2020. "Air Pollution and Child Health: Prescribing Clean Air." https://www.who.int/ceh/publications/air-pollution-child-health/en/.

_____. 2020b. "Electronic Waste." https://www.who.int/ceh/risks/ewaste/en/.

World Hunger Education Services. 2018. "2018 World Hunger and Poverty Facts and Statistics." https://www.worldhunger.org/world-hunger-and-poverty-facts-and-statistics/.

World Nuclear Association. 2020. "Nuclear Power in Russia." http://www.world-nuclear.org/information-library/country-profiles/countries-o-s/russia-nuclear-power.aspx.

World Wildlife Fund (WWF). 2020. "Why Are Glaciers and Sea Ice Melting?" https://www.worldwildlife.org/pages/why-are-glaciers-and-sea-ice-melting.

Worldometers. 2020. "Current World Population." https://www.worldometers.info/world-population/

Wright, Judy. 2020. "Recycling Programs for Electronics, Chemicals Coming Back." *The Citizen*, August 28: A7.

Wulfhorst, Ellen. 2017. "Former U.S. President Carter Goes Solar, Helping Power Rural Hometown." *Reuters*, February 10. https://www.reuters.com/article/usa-energy-solar-idUSL1N1FU1YE.

Xia, Rosanna. 2019a. "As the Seas Rise, Cities Face Climate Change Cost." *Los Angeles Times*, December 5: B1, B4.

_____. 2019b. "Cars Pollute the Sea Too, With Tire Rubber." *Los Angeles Times*, October 3, A1, A12.

Zaman, Jazib. 2019. "Renewable Energy: Benefits, Types, and the Future." *Techengage*, August 4. https://techengage.com/renewable-energy-benefits-future/.

Zanolli, Ashley. 2012. "Sustainable Food Management in Action." *BioCycle* (March), 48–52.

Zaraska, Marta. 2012. "Too Much Happiness Can Make You Unhappy, Studies Show." *The Washington Post*, April 2. http://www.washingtonpost.com/national/health-science/too-much-happiness-can-make-you-unhappy-studies-show/2012/04/02/gIQACELLrS_story.html.

Zolfani, Sarfaraz Hashemkhani, Maedeh Sedaghat, Reza Maknoon, and Edmundas Kazimieras Zavadskas. 2015. "Sustainable Tourism: A Comprehensive Literature Review on Frameworks and Applications." *Economic Reseach-Ekonomska Istrazivanja*, 28(1). https://www.tandfonline.com/doi/full/10.1080/1331677X.2014.995895.

Zolnikov, Tara Rava. 2018 (Jan.). "A Humanitarian Crisis: Lessons Learned from Hurricane Irma." *American Journal of Public Health*, 108(1): 27–28.

Zuskin, E., J. Mustajbegovic, J. Doko Jelinic, J. Pucarin-Cvetkovic, and M. Milosevic. 2007. "Effects of Volcanic Eruptions on Environment and Health." *Arh Hig Rada Toksikol*, 58 (4): 479–86. http://www.ncbi.nlm.nih.gov/pubmed/18063533.

Index